Katerina Kabassi · Phivos Mylonas · Jaime Caro
Editors

Novel & Intelligent Digital Systems: Proceedings of the 3rd International Conference (NiDS 2023)

Volume 1

Springer

Editors
Katerina Kabassi
Department of Environment
Ionian University
Zakynthos, Greece

Jaime Caro
College of Engineering
University of the Philippines
Diliman, Philippines

Phivos Mylonas
Department of Informatics and Computer Engineering
University of West Attica
Egaleo, Greece

ISSN 2367-3370 ISSN 2367-3389 (electronic)
Lecture Notes in Networks and Systems
ISBN 978-3-031-44096-0 ISBN 978-3-031-44097-7 (eBook)
https://doi.org/10.1007/978-3-031-44097-7

This Springer imprint is published by the registered company Springer Nature Switzerland AG
The registered company address is: Gewerbestrasse 11, 6330 Cham, Switzerland

Paper in this product is recyclable.

Preface

The 3rd International Conference on **Novel & Intelligent Digital Systems** (NiDS2023) was held in Athens, Greece, from September 28 to 29, 2023, under the auspices of the Institute of Intelligent Systems (IIS). The conference was implemented hybrid, allowing participants to attend it either online or onsite. The Hosting Institution of NiDS2023 was the University of West Attica (Greece).

NiDS 2023 places significant importance on the innovations within intelligent systems and the collaborative research that empowers and enriches artificial intelligence (AI) in software development. It encourages high-quality research, establishing a forum for investigating the obstacles and cutting-edge breakthroughs in AI. It also stimulates an exchange of ideas, strengthening and expanding the network of researchers, academics, and industry representatives in this domain.

NiDS is designed for experts, researchers, and scholars in artificial and computational intelligence, as well as computer science in general, offering them the opportunity to delve into relevant, interconnected, and mutually complementary fields.

Topics within the scope of NiDS series include, but are not limited to:

Adaptive Systems
Affective Computing
Augmented Reality
Big Data
Bioinformatics
Cloud Computing
Cognitive Systems
Collaborative Learning
Cybersecurity
Data Analytics
Data Mining and Knowledge Extraction
Decision-Making Systems
Deep Learning
Digital Marketing
Digital Technology
Distance Learning
E-Commerce
Educational Data Mining
E-Learning
Environmental Informatics
Expert Systems
Fuzzy Systems
Genetic Algorithm Applications
Human–Machine Interaction
Information Retrieval

Intelligent Information Systems
Intelligent Modeling
Machine Learning
Medical Informatics
Mobile Computing
Multi-Agent Systems
Natural Language Processing
Neural Networks
Pattern Recognition
Personalized Systems and Services
Pervasive Multimedia Systems
Recommender Systems
Reinforcement Learning
Semantic Web Applications
Sentiment Analysis
Serious Gaming
Smart Cities
Smart Grid
Social Media Applications
Social Network Analytics
Text Mining
Ubiquitous Computing
User Modeling
Virtual Reality
Web Intelligence.

The call for scientific papers seeks contributions that present significant and original research findings in the utilization of advanced computer technologies and interdisciplinary approaches to empower, support, and improve intelligent systems.

The international Program Committee consisted of leading members of the intelligent systems community, as well as highly promising younger researchers. The conference (General) chair was Mirjana Ivanovic from University of Novi Sad (Serbia), whereas the Program Committee chairs were Katerina Kabassi from Ionian University (Greece), Phivos Mylonas from University of West Attica (Greece), and Jaime Caro from University of the Philippines Diliman (Philippines). The keynote speakers of NiDS 2023 were: a. Stefano A. Cerri, Emeritus Professor, University of Montpellier (France) with speech title "Towards foundational principles in Interactive AI: From stamp collecting to Physics", and b. Prof. Michael L. Tee, Vice Chancellor for Planning & Development, University of the Philippines Manila with speech title "Healthcare in the time of AI".

The scientific papers underwent a thorough review by two to three reviewers, including Senior Reviewer, using a double-blind process, highlighting our dedication to ensuring NiDS's status as a premier, exclusive, and high-quality conference. We believe that the selected full papers encompass highly significant research, while the short papers introduce intriguing and novel ideas. In the review process, the reviewers' evaluations were generally respected. The management of reviews and proceedings preparation was facilitated through EasyChair.

We would like to thank all those who have contributed to the conference, the authors, the Program Committee members, and the Organization Committee with its chair, Kitty Panourgia, as well as the Institute of Intelligent Systems.

Katerina Kabassi
Phivos Mylonas
Jaime Caro

Committees

Conference Committee

General Conference Chair

Mirjana Ivanovic	University of Novi Sad, Serbia

Honorary Chair

Cleo Sgouropoulou	University of West Attica, Greece

Program Committee Chairs

Katerina Kabassi	Ionian University, Greece
Phivos Mylonas	University of West Attica, Greece
Jaime Caro	University of the Philippines Diliman, Philippines

Program Advising Chairs

Claude Frasson	University of Montreal, Canada
Vassilis Gerogiannis	University of Thessaly, Greece
Alaa Mohasseb	University of Portsmouth, UK

Workshop and Tutorial Chairs

Andreas Kanavos	Ionian University, Greece
Stergios Palamas	Ionian University, Greece

Poster and Demos Chairs

Nikos Antonopoulos	Ionian University, Greece
Gerasimos Vonitsanos	University of Patras, Greece

Doctoral Consortium Chairs

Karima Boussaha	University of Oum El Bouaghi, Algeria
Zakaria Laboudi	University of Oum El Bouaghi, Algeria

Organization Chair

Kitty Panourgia — Neoanalysis Ltd., Greece

Publicity Chair

Sudhanshu Joshi — Doon University, India

The Conference is held under the auspices of the Institute of Intelligent Systems.

iiS
INSTITUTE OF
INTELLIGENT SYSTEMS

Program Committee

Jozelle Addawe	University of the Philippines Baguio, Philippines
Shahzad Ashraf	Hohai University, China
Maumita Bhattacharya	Charles Sturt University, Australia
Siddhartha Bhattacharyya	RCC Institute of Information Technology, India
Karima Boussaha	University of Oum El Bouaghi, Algeria
Ivo Bukovsky	CTU, Czech Republic
George Caridakis	University of the Aegean, Greece
Jaime Caro	University of the Philippines Diliman, Philippines
Adriana Coroiu	Babeş-Bolyai University, Romania
Samia Drissi	University of Souk Ahras, Algeria
Eduard Edelhauser	University of Petrosani, Romania
Ligaya Leah Figueroa	University of the Philippines Diliman, Philippines
Claude Frasson	University of Montreal, Canada
Peter Hajek	University of Pardubice, Czech Republic
Richelle Ann Juayong	University of the Philippines Diliman, Philippines
Katerina Kabassi	Ionian University, Greece
Dimitrios Kalles	Hellenic Open University, Greece
Zoe Kanetaki	University of West Attica, Greece
Georgia Kapitsaki	University of Cyprus, Cyprus
Panagiotis Karkazis	University of West Attica, Greece
Efkleidis Keramopoulos	International Hellenic University, Greece
Petia Koprinkova-Hristova	Bulgarian Academy of Sciences, Bulgaria

Sofia Kouah	University of Larbi Ben M'hidi O.E.B, Algeria
Akrivi Krouska	University of West Attica, Greece
Florin Leon	Technical University of Iasi, Romania
Jasmine Malinao	UPV Tacloban College, Philippines
Andreas Marougkas	University of West Attica, Greece
Phivos Mylonas	University of West Attica, Greece
Stavros Ntalampiras	University of Milan, Italy
Christos Papakostas	University of West Attica, Greece
Kyparisia Papanikolaou	ASPETE, Greece
Nikolaos Polatidis	University of Brighton, UK
Filippo Sciarrone	Universitas Mercatorum, Italy
Cleo Sgouropoulou	University of West Attica, Greece
Geoffrey Solano	University of the Philippines Manila, Philippines
Dimitris Sotiros	WUST, Poland
Oleg Sychev	Volgograd State Technical University, Russia
Christos Troussas	University of West Attica, Greece
Aurelio Vilbar	University of the Philippines Cebu, Philippines
Panagiotis Vlamos	Ionian University, Greece
Athanasios Voulodimos	NTUA, Greece
Ioannis Voyiatzis	University of West Attica, Greece
Laboudi Zakaria	University of Oum El Bouaghi, Algeria

Contents

National Unemployment Recovery Initiative NextGenerationEU: Social Impact of Lifelong Learning in Computer-Aided Design

Zoe Kanetaki[1](✉), Sébastien Jacques[2], Constantinos Stergiou[1], and Panagiotis Panos[1]

[1] University of West Attica, 12241 Egaleo, Greece
zoekanet@uniwa.gr
[2] University of Tours, CEDEX 1, 37020 Tours, France

Abstract. In 2022, the Greek Public Employment Service has launched a call for horizontal upskilling/reskilling programs for the unemployed. The action has been financed by European funds, and more specifically by the Recovery and Resilience Facility. This plan is initiated by the NextGenerationEU instrument and implements lifelong learning in green and digital skills. The study presented in this article, which is fully in line with this context, aims to discuss the initial results in terms of the social impact of the first four groups of trainees in the advanced computer-aided design program. The results show that the high percentage of university graduates reflects the current unemployment situation in Greece, and particularly in the prefecture of Attica. Strong social relationships developed between participants, irrespective of age, gender, economic and social status and level of education. The motivation for this program was not prior training, but rather, for most of the participants, the broadening of their field of knowledge with a view to seeking out new areas and sectors of employment.

Keywords: Computer-aided design · Employment · Engineering education · Meta-COVID-19 · NextGenerationEU · Reskilling · Upskilling

Abbreviations

CAD	Computer Aided Design
RRF	Recovery and Resilience Facility
VET	Vocational Education and Training
LLL	Lifelong Learning
LLC	Lifelong Learning Centers
KEDIVIM	Public Lifelong Learning Centers
KDVM	Private Lifelong Learning Centers
LMS	Learning Management System

K. Kabassi et al. (Eds.): NiDS 2023, LNNS 783, pp. 1–12, 2023.
https://doi.org/10.1007/978-3-031-44097-7_1

1 Introduction

The aim of the Recovery and Resilience Facility is to mitigate the economic and social impact of the COVID-19 pandemic and to make European economies and societies more sustainable, resilient and better prepared for the challenges and opportunities of the green and digital transitions [1]. The Facility is structured around six pillars: green transition; digital transformation; economic cohesion, productivity and competitiveness; social and territorial cohesion; health, economic, social and institutional resilience; and next generation policies [2]. NextGenerationEU (as shown in Fig. 1) is a temporary stimulus instrument of more than €800 billion to repair the immediate economic and social damage caused by the COVID-19 pandemic [3].

For Greece, the Greece2.0 plan aims to strengthen economic and social resilience through reforms and investments that promote labor market activation and skills enhancement [2, 4]. The specific reform is included in Pillar 3: "Employment, Skills and Social Cohesion" under the Component "Strengthening education's digital capabilities and modernization of vocational education and training", called Measure 16913: A New Strategy for Lifelong Skilling: Modernizing and Upgrading Greece's Upskilling and Reskilling System -Sub2: Horizontal upskilling/reskilling programs to targeted populations with a total budget of 74,400€ [5].

In 2022, the Greek Public Employment Service (as shown in Fig. 1) coordinated this measure by issuing a call to national Lifelong Learning Centers (LLCs), both public and private, to submit proposals for the design of educational programs for the unemployed population [6]. The lifelong learning providers are divided into: a) public LLCs established by Greek universities, called KEDIVIM, and b) private LLCs, called KDVM. Most Greek universities responded immediately to the call by offering various skill enhancement programs focusing on green skills and digital skills. The specific call is addressed to 120.000 unemployed personnel, over the age of 18.

Fig. 1. Program funding.

This paper analyzes the social impact of a grant-funded digital skills enhancement program implemented at the University of West Attica's KEDIVIM, School of Engineering, Department of Mechanical Engineering.

The research questions (RQ) posed as the program is still in progress are:

RQ1: Which factors motivated the trainees to participate in this program?
RQ2: What is the profile of the unemployment population?
RQ3: Can we estimate the factors defining the level of motivation of trainees?

This paper is organized as follows. Section 2 presents the background and requirements of the program. The methodological and organizational aspects are described in Sect. 3. Section 4 presents the demographic characteristics of the trainees. The key findings are discussed in Sect. 5, and conclusions and perspectives are presented in Sect. 6.

2 Background and Requirements of the Program

This study focuses on the digital enhancement program called "AutoCAD: Two-dimensional Computer-Aided Design and three-dimensional Modeling". This program is designed to teach two-dimensional computer-aided design methods, as well as three-dimensional modeling and photorealistic rendering. Participants are trained in the development and editing of vector drawings. The program consists of a common "language" used in engineering and addresses the needs and tasks of participants in the use of two-dimensional plans, from mapping spaces, areas, and geographic information, to re-synthesizing a complete study by a computer.

The minimum requirements for participation include a high school education. Knowledge of English is not required. Other prerequisites for participation include the use of a computer and knowledge of the Windows environment.

The length of each program cycle varies from one month to two months (see Table 1). Due to high demand, the first circle began on November 9, 2022 and was completed in two months. The four subsequent circles did not exceed forty days, due to time constraints. It should be noted that the program is still ongoing and will be repeated as long as remaining funds are available.

Table 1. The four cycles of the program

Cycle	Length (weeks)	Number of trainees
A	8	19
B	6	16
C	4	13
D	5	21
E	4	21

Upon successful completion of this training, participants are able to produce two-dimensional Computer-Aided Design (CAD) drawings, create three-dimensional models, apply real materials to solids and backgrounds, create photorealistic views, and simulate the building in terms of sunlight conditions [7]. They will have to pass a knowledge acquisition test provided by a TÜV NORD certification company for certified 3D design professionals. In addition, trainees will receive a training voucher corresponding to 5 €/hour.

In conclusion, the participants will be fully trained in the knowledge needed to produce a complete building permit file, as well as the individual requirements for collaborations with private and public sector engineers, receive a training voucher as well as certification of individuals.

3 Methodological and Organizational Aspects

For the implementation of the educational program, innovative methods [8–10] and technological tools for distance learning [11, 12] were applied in combination with face-to-face teaching. The trainees were provided with asynchronous support for the case studies they will have to implement using screen recording videos and audio (with a full screen recording of the software environment in high resolution). Therefore, the trainees will have the opportunity to asynchronously observe the process of implementing tasks in the software environment with parallel narration and explanation of the steps by an experienced instructor and, at the same time, improve their three-dimensional visualization skills [13, 14].

3.1 Program Design and Implementation

The 200-h program is conducted in mixed learning environments: 35% (70 h) is conducted face-to-face in in the university's CAD lab. Synchronous distance learning is conducted at a rate of 35% (70 h) and is implemented using a distance learning platform (Microsoft Teams). The instructors, who have more than twenty years of experience in teaching CAD systems in many engineering faculties, are able to transmit the new knowledge, remotely, in an interactive way, having the possibility to intervene, in real time, to solve all the trainees' questions at any time. The lecturer's full screen representing the software environment is in screen-sharing mode, allowing learners to be constantly focused on the subject being taught. In this way, the experienced teachers keep the learners engaged throughout the distance learning course. The remaining 30% (60 h) consists of asynchronous learning with the e-Class Learning Management System (LMS).

The trainees can access the asynchronous platform where the learning material is mainly based on videos (85%) and less on reading material (PowerPoints and Software notes), aiming to be more attractive to leaners of different educational levels. Prior to the launch of the program, each program underwent an approval process to certify the quality and quantity of the learning material, as well as its correspondence to 60 h of asynchronous learning.

At the end of each program cycle, the trainees were asked to participate in an online survey that included demographic questions, program and trainer evaluation questions, and a specific field to add their personal experiences. This survey proved to be a valuable tool for extracting information and drawing conclusions about various social aspects of the unemployment sector.

3.2 Program Management and Instructor Profile

The upskilling program is implemented by a group of six instructors, including the program designer and coordinator. Three of them are members of the authors' team. Since the skills offered to the target population are concentrated in various scientific areas aimed at providing new knowledge to several technological disciplines, the team of instructors was selected based on their area of expertise. Therefore, the training material was organized by scientific discipline, covering various tasks related to architecture, interior design, mechanical parts, and civil engineering, as shown in Table 2.

Table 2. Role and area of expertise of instructors.

Instructor's ID	Field of expertise	Program coordinator	Technical Assistance	Teaching Online	Teaching Face-to-face	Asynchronous material
ZK	Architecture	√		√	√	√
KR	Mechanical			√		
PP	Interior design		√	√	√	
EP	Architecture			√	√	
CS	Mechanical					√
GE	Civil				√	

3.3 Objectives of the Requalification Program

The primary goals of the program were initially established by ministerial decree: each development program must provide and enhance the skills necessary for learners to become digitally competent in the contemporary national workforce. Learners must be adequately trained to become competitive and dynamic in the new meta-COVID-19 era. Blended learning environments have been established to familiarize trainees with newly implemented distance learning and to potentially introduce them to new ways of working and collaborating at a distance.

3.4 Objectives of Sustainable Education and Lifelong Learning

In addition to the above, the specific program has set goals for sustainable education to fully realize its transformative potential aiming to achieve the United Nations Sustainable Development Goals (SDGs) [15]:

- *Asynchronous learning must be consistent with lifelong learning goals*: Since the program should last one to two months, trainees should have the opportunity to review the learning material long after the program is over.
- *Course materials must be updatable*: Following specific requests from trainees, the scientific coordinator recorded synchronous online lecture segments focused on specific tasks. These videos were processed in a way that masked the names and profile pictures of the participants and uploaded to the asynchronous learning platform as part of each unit. To this end, and as shown in Fig. 1, a new YouTube channel of educational videos was created. It contains 90 educational videos with a total duration of 54 h.
- *Role of trainers in facilitating relationships*: Trainers also focused on developing learners' social and collaborative skills by encouraging relationships among trainees.

In the context of the social nature of the upskilling program, the asynchronous education material was enhanced with 54 h of asynchronous video training in addition to the original 60 h of certified training (with screen recordings and audio recording methods). This additional material was appropriately processed to match lifelong learning, supporting trainees even after the training program is over.

Fig. 2. Asynchronous support YouTube channel educational videos.

Trainees will have the opportunity, for a period of one year, to enjoy the benefits of the program, such as software license, academic email address, Microsoft (MS) Office 365 license as well as access to the asynchronous learning platform. Therefore, they will be able to access to continuous and personalized training, for an extended time period after the program ends, by performing asynchronous learning on their own time basis.

The above initiatives were implemented by the program's scientific coordinator on a non-profit basis, at the request of the participants, and are considered particularly innovative as they can cover future support, in the context of the program's training sustainability.

4 Trainee Demographics

The total number of trainees engaged in the program to date is 90. The following analysis covers the first four cycles, with the fifth cycle still in progress.

As shown in Fig. 2, of the 65 participants, 70% are women and 30% are men. 29% are graduates of general secondary education, 11% are graduates of vocational secondary education, 13% have vocational training after secondary education, 8% have a theoretical university degree and 40% have a scientific university degree (Fig. 3).

22% of the trainees had previously participated in other development programs and for the remaining 78% it was their first time.

90% reported meeting other classmates during the program and all felt that the instructors contributed positively to meeting others and improving their social skills.

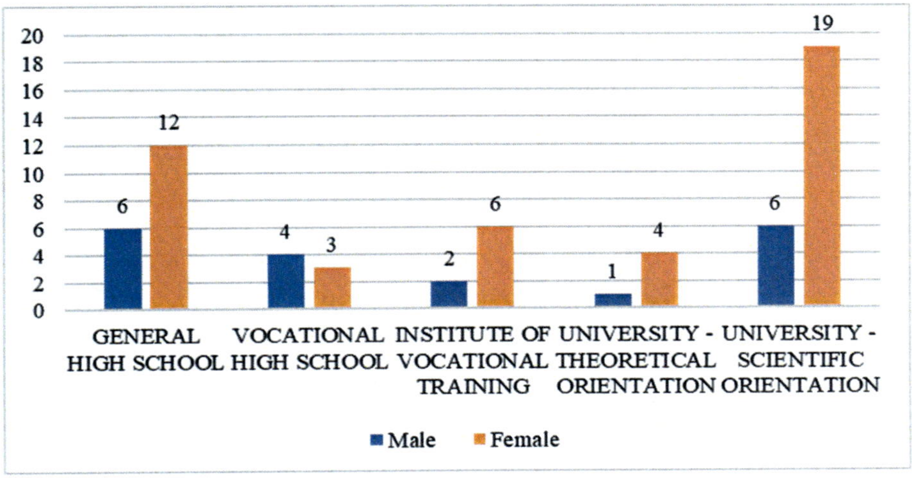

Fig. 3. Gender and educational profile of trainees.

As shown in Fig. 4, the short distance between home and work emerged as an important factor in choosing the program for 70% of participants who reported that it took them between half an hour and an hour to get to the university campus to attend face-to-face classes.

As shown in Fig. 5, 16% of participants live alone.

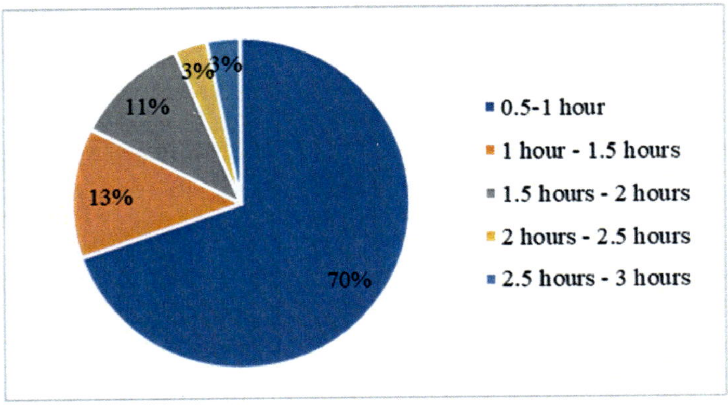

Fig. 4. Commuting distance.

As shown in Table 1, the total number of trainees expected to have completed the program is currently 90. The number of trainees who initially applied to the program was 97, revealing that seven participants dropped out of the program. Of these seven participants, four did not participate in the program because by the time classes began, they had already found employment. Two dropped out of the program after taking the first

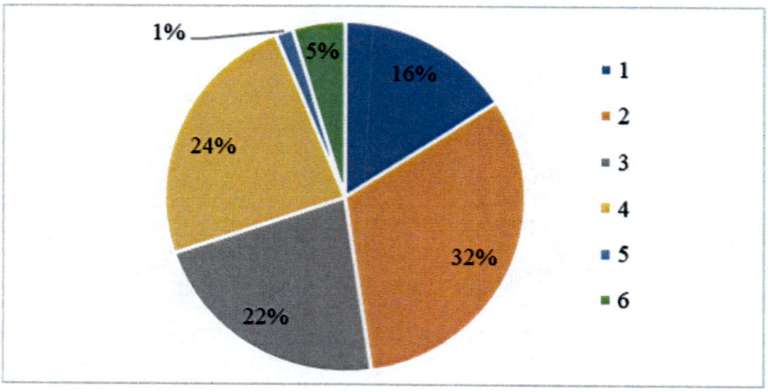

Fig. 5. Number of people sharing the same household.

online course and the last participant left the program after two weeks of participation due to heavy workloads and busy schedules.

5 Results and Discussion

Greece's unemployment rate as a percentage of the total population declines steadily from 14.8 in 2015 to 7.5 in 2022, but remains high compared to the median of the 27 European Union (EU) countries, which is 4.0 in 2022 [16].

The significantly higher percentage of female trainees is due to the fact that many of the women in the program are currently mothers of young children or mothers of older children who stop working to raise their children and who, after several years, seem to have lost touch with the recent advances and techno-logical characteristics of the workforce. Most of them had already worked with CAD software in recent years, especially in the area of two-dimensional design. Their preference for the specific program was to recapture their previous knowledge, update it with new advances, and expand their knowledge to three-dimensional design.

It is worth noting that 48% of the trainees currently have a university degree and 18% have a master's degree or are currently in a postgraduate program. The high percentages of unemployed with a university degree illustrate the current situation in Greece, where qualified personnel have been kept out of the workforce for too long.

The 8% with a university degree in the theoretical field (social sciences, languages and literature, humanities) show that these trainees' experiment with new fields of knowledge in their job search. The same conclusion applies to the 29% of trainees with a highschool education degree, without specific skills: after interviewing them, they all experimented with other specialties, but were attracted to the program because of its digital skills.

In the first two cycles (A and B), the first course was delivered in distance learning environments. Trainees had already received specific instructions on how to connect to the online platform meetings. Slight technical difficulties were encountered, including a lack of computer skills, but these were resolved during the first two online meetings. It was observed that six of the seven resigning trainees were enrolled in Cycles A and B. Therefore, to avoid dropouts due to lack of computer skills, the program was modified for the following cycles (C, D, and E), starting with a six-hour face-to-face lesson in the university computer lab, instead of a 3 h first online meeting.

It is worth noting that the university campus is located in the prefecture of Attica, in the city of Athens, within an ancient olive grove, a green, joyful and conducive environment for student life. It is believed that the university environment creates a sense of belonging that reinforces the goals of lifelong learning. Trainees who were university graduates reported that they were extremely happy to return studying in an academic campus after a long period of work. Other groups, who have never previously experienced university life, have considered the program delivered in mixed learning modes (on campus and online) a great challenge, even in a higher age. As shown in Fig. 6 and Table 3, the asynchronous YouTube channel support garnered 5,278 views over the six months of activity. The analytics data was extracted from YouTube Studio for the period November 20, 2022 to April 29, 2023.

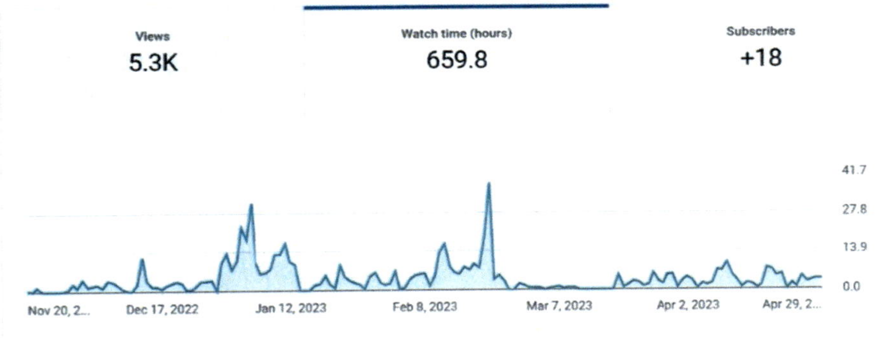

Fig. 6. Asynchronous support through educational videos on the YouTube channel.

Table 3 shows that the most used electronic device for watching the asynchronous videos is the computer. This can be explained by the fact that the program implements the execution of various tasks and it was suggested to the trainees to perform these tasks simultaneously while watching the videos, as an asynchronous medium. Cell phones were mainly used by participants who practiced their tasks on PCs, while watching the videos on their smartphones or tablets. Tablets were no longer the preferred mode for learners, nor were televisions, unlike in the early stages of COVID-19, when there was no time for buying electronic devices and many learners used televisions and even gaming devices (Nintendo Switch), as shown in the 2020–2021 YouTube reports[17].

One week after the end of each cycle, all trainees were asked to take the online certification test. The test took place in the university's computer lab and consisted of 50 multiple choice questions. All trainees passed the test with an average score of 83.6/100, indicating a high level of performance.

Table 3. Electronic devices for watching the asynchronous videos.

	Computer	Mobile phone	Tablet	TV
Views	4,257	853	167	1
Percentage	80.7%	16.2%	3.2%	0.0%
Watch time (hours)	524.0	106.7	28.7	0.3
Percentage	79.4%	16.2%	4.4%	0.1%
Average View Duration	7:23	7:30	10:19	20:50

6 Conclusions and Future work

The Greek Unemployed Skills Upgrading and Reskilling Program began in September 2022 and is currently ongoing, due to available funding. The factors that motivated the trainees to participate in this program were: (a) Retraining, updating and expanding their knowledge to a similar scientific field; (b) Seeking new digital skills to expand opportunities to other areas of employment; (c) Returning to the university environment; (d) Experiencing university life; and (e) Taking their first step into the employment community after several years of absence after raising their children.

The high percentage of university graduates reflects the current unemployment situation in Greece, and most specifically in Attica prefecture. It can be inferred that higher education workforce in the capital city is saturated, causing the unemployed to seek new areas of employment.

The program has been successful not only in creating digital skills and familiarizing trainees with online work environments and group meetings, but also in creating strong social relationships among participants in each cycle, regardless of age, gender, economic and social status, and education level.

Overall, the participants were highly motivated by the program, constantly looking forward to the next day's lesson, practicing their tasks diligently, and even extending the 6-h classes-especially face-to-face by visiting other university labs and contacting their instructors even after the class period ended. High scores were achieved on the certification exam, regardless of the trainees' previous educational level.

This research is limited by the fact that the skill improvement program is still ongoing. Another limitation is that the skill improvement program is taking place in the Attica region, which encompasses the entire Athens metropolitan area, and may not be applicable to other regions.

Future work will include determining the percentage of trainees considered "economically inactive" (individuals who do not intend to seek work) as well as whether they have regain employment in the period following the program [18].

Acknowledgements. The authors would like to Acknowledge the University of West Attica for financially supporting this research, the Center of Longlife learning of the University West Attica, the Greek Public Employment Service, the European Union Recovery and Resilience Facility, NextGenerationEU, TUV Nord and the instructors team.

References

1. Recovery and Resilience Facility. https://commission.europa.eu/business-economy-euro/economic-recovery/recovery-and-resilience-facility_en. Last accessed 30 Apr 2023
2. Greece's recovery and resilience plan, https://commission.europa.eu/business-economy-euro/economic-recovery/recovery-and-resilience-facility/greeces-recovery-and-resilience-plan_en. Last accessed 15 Jan 2023
3. Recovery plan for Europe, https://commission.europa.eu/strategy-and-policy/recovery-plan-europe_en. Last accessed 30 Apr 2023
4. admin_140621: Greece 2.0 – National Recovery and Resilience Plan, https://greece20.gov.gr/en/. Last accessed 15 Jan 2023
5. admin_140621: Projects – Greece 2.0. https://greece20.gov.gr/en/projects/. Last accessed 30 Apr 2023
6. Lifelong learning strategy | Eurydice. https://eurydice.eacea.ec.europa.eu/national-education-systems/greece/lifelong-learning-strategy. Last accessed 30 Apr 2023
7. Sofias, K., Kanetaki, Z., Stergiou, C., Jacques, S.: Combining CAD modeling and simulation of energy performance data for the retrofit of public buildings. Sustainability **15**, 2211 (2023). https://doi.org/10.3390/su15032211
8. Kanetaki, Z., Stergiou, C., Troussas, C., Sgouropoulou, C.: Development of an innovative learning methodology aiming to optimise learners' spatial conception in an online mechanical CAD module during COVID-19 pandemic. In: Frasson, C., Kabassi, K., Voulodimos, A. (eds.) Novelties in Intelligent Digital Systems: Proceedings of the 1st International Conference (NIDS 2021), Athens, Greece, 30 Sep–1 Oct 2021. IOS Press (2021). https://doi.org/10.3233/FAIA210072
9. Geitz, G., Donker, A., Parpala, A.: Studying in an innovative teaching–learning environment: design-based education at a university of applied sciences. Learn. Environ. Res. (2023). https://doi.org/10.1007/s10984-023-09467-9
10. Papakostas, C., Troussas, C., Krouska, A., Sgouropoulou, C.: Exploration of augmented reality in spatial abilities training: a systematic literature review for the last decade. Inform. Educ. **20**(1), 107–130 (2021). https://doi.org/10.15388/infedu.2021.06
11. Jacques, S., Ouahabi, A., Lequeu, T.: Synchronous E-learning in Higher Education during the COVID-19 Pandemic. In: 2021 IEEE Global Engineering Education Conference (EDUCON), pp. 1102–1109 (2021). https://doi.org/10.1109/EDUCON46332.2021.9453887
12. Papakostas, C., Troussas, C., Krouska, A., Sgouropoulou, C.: On the development of a personalized augmented reality spatial ability training mobile application. In: Frasson, C., Kabassi, K., Voulodimos, A. (eds.) Novelties in Intelligent Digital Systems: Proceedings of the 1st International Conference (NIDS 2021), Athens, Greece, 30 Sep–1 Oct 2021, pp. 75–83. IOS Press (2021). https://doi.org/10.3233/FAIA210078
13. Sorby, S.A.: Educational research in developing 3-d spatial skills for engineering students. Int. J. Sci. Educ. **31**, 459–480 (2009). https://doi.org/10.1080/09500690802595839
14. Papakostas, C., Troussas, C., Krouska, A., Sgouropoulou, C.: Personalization of the learning path within an augmented reality spatial ability training application based on fuzzy weights. Sensors **22**, 7059 (2022). https://doi.org/10.3390/s22187059
15. Hanemann, U., Robinson, C.: Rethinking literacy from a lifelong learning perspective in the context of the Sustainable Development Goals and the International Conference on Adult Education. Int. Rev. Educ. **68**(2), 233–258 (2022). https://doi.org/10.1007/s11159-022-09949-7
16. Statistics | Eurostat. https://ec.europa.eu/eurostat/databrowser/view/tps00203/default/table. Last accessed 1 May 2023

17. Kanetaki, Z., et al.: Acquiring, analyzing and interpreting knowledge data for sustainable engineering education: an experimental study using YouTube. Electronics **11**, 2210 (2022). https://doi.org/10.3390/electronics11142210
18. Labour market information: Greece. https://eures.ec.europa.eu/living-and-working/labour-market-information/labour-market-information-greece_en. Last accessed 1 May 2023

Automated Coral Lifeform Classification Using YOLOv5: A Deep Learning Approach

Jannie Fleur V. Oraño[1](✉), Jerome Jack O. Napala[1], Jonah Flor O. Maaghop[2], and Janrey C. Elecito[1]

[1] Southern Leyte State University, Southern Leyte, Philippines
{jfo,jnapala}@southernleytestateu.edu.ph
[2] Visayas State University, Leyte, Philippines
jonahflor.orano@vsu.edu.ph

Abstract. Coral reefs serve as essential coastal environments, offering a diverse range of marine species with habitats, safeguarding shorelines against erosion, and serving as a source of support for both the fishing and tourism industries. However, coral reefs are facing increasing threats due to climate change, overfishing, and pollution. Effective conservation efforts require the critical monitoring of coral reefs' health and diversity. One way to evaluate the diversity and health status of coral reefs is by getting the percent coral cover in the area based on the coral lifeforms encountered. Traditional methods for coral lifeform classification involve manual observation and identification by trained experts, which can be expensive, harmful to coral reefs, and pose risks to surveyors. In this study, the researchers developed an automated coral lifeform classification model using the YOLOv5 deep learning framework. The model was trained on a dataset of coral images, including seven different coral lifeforms, namely Branching, Encrusting, Foliose, Massive, Mushroom, Submassive, and Tabulate. The evaluation of the model's performance involved the use of several metrics, including recall, precision, F1 score, and accuracy. Additionally, a confusion matrix was created to analyze the model's classification results. According to the findings, the coral lifeforms classification model exhibited promising potential for efficient reef monitoring and conservation, having attained an 89.29% accuracy rate overall. This research also highlights the potential of deep learning in advancing coral reef conservation efforts.

Keywords: deep learning · YOLO · coral lifeform · marine conservation

1 Introduction

Coral reefs are made up of numerous individual coral polyps, which are tiny animals that secrete calcium carbonate to form a hard exoskeleton. They are characterized by their hard, calcareous skeletons made of calcium carbonate, which provide the framework for coral reefs. Hard corals are known for their intricate and beautiful structures, which consist of numerous polyps living together in colonies [1]. Species of coral have different shapes, sizes, and colors, and they form complex structures that provide a habitat for a

K. Kabassi et al. (Eds.): NiDS 2023, LNNS 783, pp. 13–22, 2023.
https://doi.org/10.1007/978-3-031-44097-7_2

variety of marine life. Hard corals, also called stony corals or scleractinian corals, are marine invertebrates that belong to the phylum Cnidaria. The formation and structure of reefs rely heavily on the presence of these essential stony corals [2].

The coral reef ecosystem was once an incredibly productive and diverse habitat, abundant with various natural resources that supported human activities. These resources included raw materials for food, medicines, building materials, aesthetics, and a wide array of other necessities [3]. Despite its importance to humans, this habitat is under great pressure from natural and anthropogenic pressures, and the most destructive are caused by humans of which its mortality is close to 100% [4]. Understanding the diversity and abundance of coral lifeforms in coral reef surveys is important for monitoring the health of coral reefs and identifying threats such as climate change, pollution, and overfishing. It is of great importance to monitor changes in coral populations over time to understand the dynamics of coral reefs and identify potential conservation measures in order to protect this very vulnerable marine ecosystem [5].

Coral reef monitoring is conducted by marine biologists to assess the health and diversity of coral reefs by getting the percent coral cover in the area based on coral lifeforms encountered within the quadrat in the transect line. Coral lifeforms refer to the various types of coral species that can be found in these surveys. These lifeforms are important health indicators and diversity of reef habitats as they could influence the number of fish species present in the reef ecosystem [6] and are often used in coral reef monitoring and management efforts.

Although there is a vast amount of coral reef imagery available to marine scientists for annotation, typically less than 5% of these images are analyzed by experts [7]. Moreover, traditional methods for coral lifeform classification involve manual observation and identification by trained experts. This type of coral reef survey has its own disadvantages. It can be expensive due to the equipment, personnel, and logistics involved and can be harmful to the coral reef from physical contact during surveys which can damage the delicate coral structures that can lead to adverse effects on the health of the coral environment and the organisms that reside in them. Lastly, it can pose some risk to the people conducting surveys in an open water environment, particularly in areas with strong currents, rough seas, or the presence of dangerous marine animals. The process of manual design is not only time-consuming, but it also increases the likelihood of human error [8]. Thus, automated technologies are needed for the continuous monitoring of marine ecosystems, complementing the expertise of human professionals and reducing the reliance solely on their feedback [7].

With the latest developments in machine learning and computer vision, the automation of coral lifeform classification has become a promising opportunity. This involves using algorithms to analyze underwater images and identify the various coral types present. Deep learning, a subfield of machine learning, plays a significant role in this process. It involves training neural networks on large datasets to learn patterns by automatically extracting simple features from raw input and transforming lower-level representations into higher ones, making it highly effective for tasks such as image and video recognition [8].

Recent advancements in deep learning and computer vision have paved the way for automated coral classification, allowing algorithms to analyze underwater images and accurately identify different coral life forms. Several studies have successfully implemented deep learning technology to address various challenges in coral classification. For instance, Guntia [9] employed a deep learning classifier to differentiate between types of coral substrates, including soft coral, hard coral boulders, sponge, and hard coral branching. Similarly, [10] implemented the MAFFN_YOLOv5 to detect coral health conditions with an accuracy rate of 90.1%. In addition, the authors in [11] applied image processing and machine learning techniques to classify coral health, while [12] utilized a convolutional neural network to develop a model that can detect unhealthy and dying corals at a primary stage, achieving an accuracy rate of 95%.

For this study, the researchers developed a coral lifeforms classification model utilizing the powerful You Only Look Once (YOLO) deep learning algorithm. The primary intention of this research is to make a valuable contribution to ongoing initiatives that leverage technology for the conservation of coral reefs. The successful implementation of the model generated through this study holds significant potential for advancing the understanding of coral lifeforms and facilitating targeted conservation efforts.

2 Methodology

The workflow of this study is depicted in Fig. 1. The primary step involves acquiring images of coral lifeforms in a certain location using a digital camera. These images undergo preprocessing and are partitioned into three subsets: a training set for model learning, a validation set for adjusting the hyperparameters, and a testing set for assessing the model's performance. The YOLOv5 algorithm was used for the actual model training. The trained model was utilized to make predictions on a new set of images and its performance was assessed based on these results.

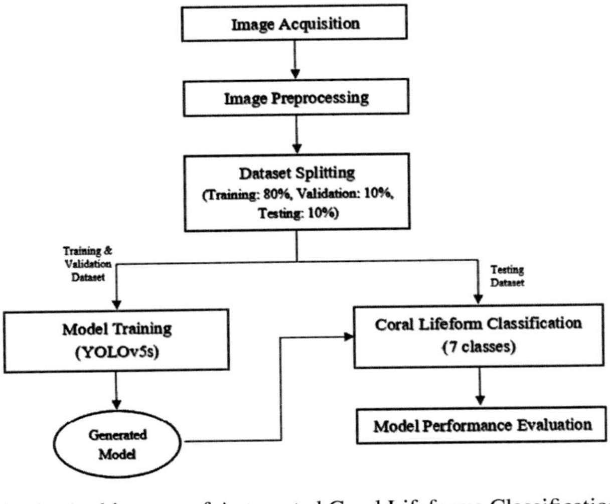

Fig. 1. Architecture of Automated Coral Lifeforms Classification.

2.1 Dataset

The digital images of corals were captured underwater in Sogod Bay, which is situated in Southern Leyte, Philippines. This study encompasses seven (7) distinct types of coral lifeforms: Massive, Mushroom, Foliose, Encrusting, Branching, Submassive, and Tabulate. Prior to analysis or model training, these images underwent preprocessing techniques, including resizing, cropping, and adjustments to brightness and contrast. These steps were implemented to enhance the quality and ensure consistency among the images. It is worth noting that each image was manually classified by a human expert, serving as the foundation for training the model.

In this study, 549 images were used and were partitioned into training, validation, and testing sets using a percentage ratio of 80:10:10. Each set is composed of seven (7) classes representing the types of coral lifeforms. Table 1 displays the distribution of image samples for each class within the different data subsets, while Fig. 2 describes the feature differences of each coral lifeform.

Table 1. The Research Dataset.

Class	Training	Validation	Testing	Class Total
Branching	69	9	9	87
Encrusting	78	10	10	98
Foliose	59	8	8	75
Massive	52	6	6	64
Mushroom	60	8	8	76
Submassive	42	5	5	52
Tabulate	77	10	10	97
Subset Total	437	56	56	549

Fig. 2. Different Types of Coral Lifeforms

2.2 Model Training

The pre-trained YOLOv5 model was used to train the classification model for coral lifeforms. Moreover, the following configurations were defined: 640 pixels image size, 100 epochs, 64 batch size, 0.001 learning rate, and Adam algorithm optimizer.

Throughout the training stage, the model learned about the characteristics or features of the images and their associated class labels. Metrics such as accuracy and loss were tracked and recorded. Also, the model was stored in the designated directory after each epoch.

After the completion of training, the model's performance was evaluated using the validation set and has been tuned by adjusting the hyperparameters and repeating the training and evaluation process until the desired performance is achieved.

2.3 Model Training

After model training, the next crucial step is the evaluation of the model using the test dataset and a set of key performance metrics. These metrics provide a comprehensive understanding of how well the generated model is performing on the unseen dataset and can help identify areas for improvement. The performance measures considered in the evaluation include the accuracy (1), precision (2), recall (3), and F1 score (4).

$$Accuracy = \frac{sum\ of\ correct\ classification}{sum\ of\ all\ classification} \tag{1}$$

$$Precision = \frac{TP}{TP + FP} \tag{2}$$

$$Recall = \frac{TP}{TP + FN} \tag{3}$$

$$F1Score = \frac{2 * Precision * Recall}{Precision + Recall} \tag{4}$$

where:

TP (True Positive): Actual class = True, predicted class = True.
TN (True Negative): Actual class = False, predicted class = False.
FP (False Positive): Actual class = False, predicted class = True.
FN (False Negative): Actual class = True, predicted class = False.

3 Results and Findings

The results of the study show promising outcomes in developing an automated coral lifeform classification model using the YOLOv5 deep learning approach. This section elaborates on the comprehensive findings regarding the model's performance in classifying the seven different coral lifeforms, including Branching, Encrusting, Foliose, Massive, Mushroom, Submassive, and Tabulate. The results include performance measures such as precision, recall, F1 score, and accuracy, along with a confusion matrix to present a full analysis of the model's performance.

Figure 3 illustrates that the highest accuracy attained for Top1 classification is 90.93%, indicating the percentage of images correctly classified by the model with the most probable label prediction. This result reveals that the model's performance was relatively good, as it accurately predicted the label for the majority of the images in the dataset. However, there were still some instances where the model's prediction did not match the actual label, suggesting that the model's accuracy can still be further improved.

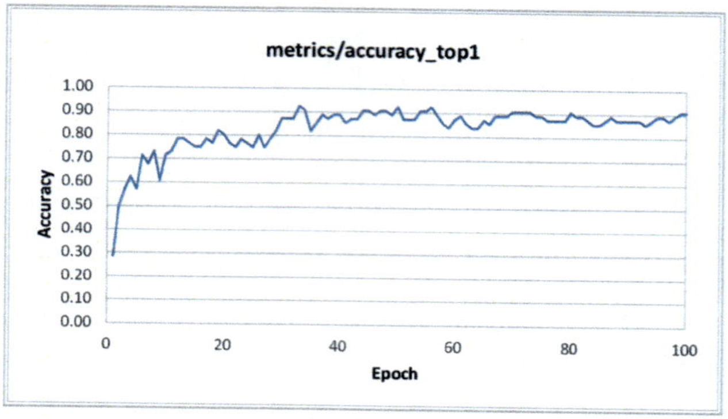

Fig. 3. Top1 Training Accuracy

On the other hand, Fig. 4 indicates that the Top5 achieved the highest accuracy rate of 100% signifying that the model was able to identify the correct label within the top five most probable labels for almost all the images in the dataset. These metrics indicate that the coral lifeform classification model using YOLOv5 is performing well, particularly in terms of the Top5 accuracy even when the most probable label is not the correct one.

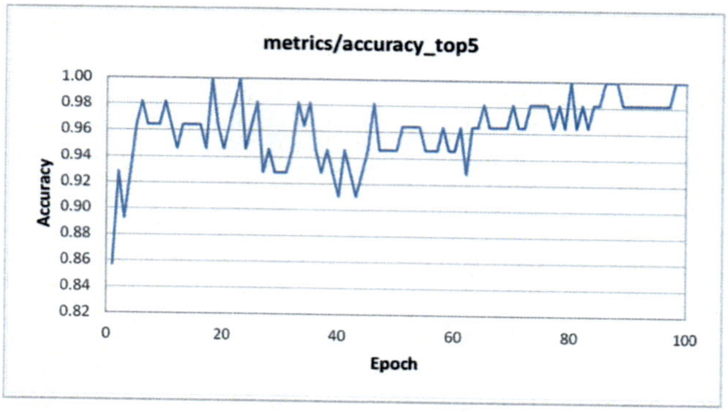

Fig. 4. Top5 Training Accuracy

During the process of creating the coral lifeform classification model, the difference between the actual and predicted labels for both the validation and training sets was measured using loss values. The performance evaluation results show that the training and validation loss values are 0.3 and 0.8, respectively. As illustrated in Fig. 5, the lower training loss value in comparison to the validation loss indicates that the model is successfully classifying the training dataset without overfitting these data. The validation loss value of 0.8 suggests that on average, there is still some level of discrepancy between the model's predictions and true labels in the validation dataset. These findings prove the efficacy of the YOLOv5 deep learning approach in accurately classifying and identifying different coral lifeforms.

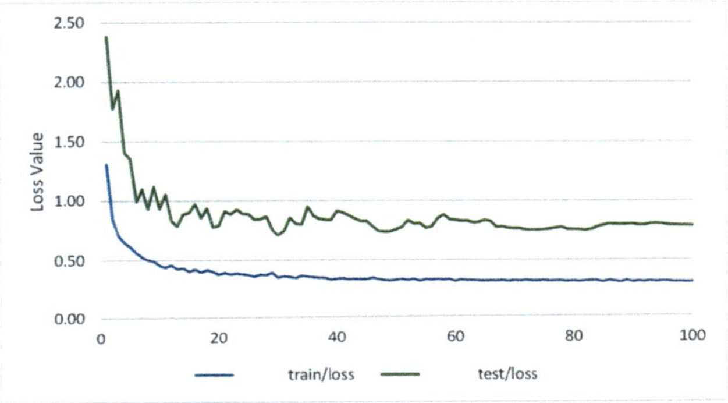

Fig. 5. Training Loss

Shown in Fig. 6 is the confusion matrix as a tool for assessing the performance of the classification model. The confusion matrix displays the actual and predicted labels for each of the seven different coral lifeforms. Every unit in the matrix depicts a fusion of the predicted and actual class labels for a particular category – the count of correct and incorrect predictions for each coral lifeform.

Moreover, the gradient of the color of each cell represents the frequency of correct and incorrect predictions for a given class. The darker the color means higher frequencies, while lighter colors represent lower frequencies. This way, it is easier to visually identify areas where the model is performing well or not.

The diagonal values of the confusion matrix represent the true positives or the instances when the model correctly identified the coral lifeform. Whereas, the off-diagonal values denote the false positive and false negative results, which are cases where the model misclassified the coral lifeform. As reflected in the resulting confusion matrix, higher frequencies were observed along the diagonal area, and lower frequencies were observed off-diagonal, indicating that the majority of the classifications made by the model are correct.

Looking at the results shown in Table 2, the model has a precision value of 100.00% for Massive and Tabulate coral lifeforms, indicating that the model correctly classified all instances of those lifeforms in the test dataset, with no false positives. This is a strong indication that the model is very accurate in identifying these particular coral lifeforms. On the other hand, the precision value of the model for the Submassive coral lifeform is lower, at 60.00%. This indicates that the model misclassified some instances of that

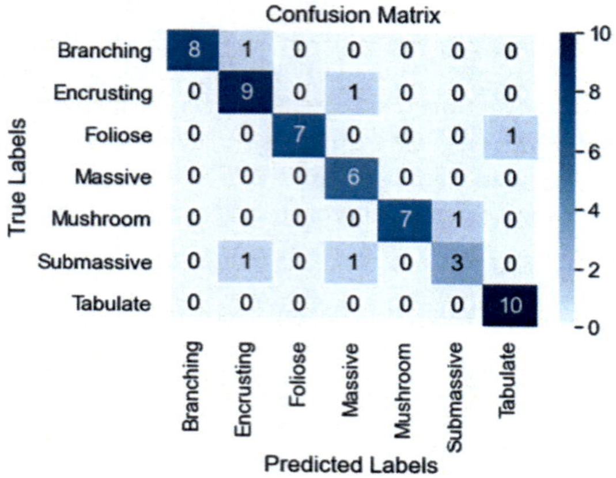

Fig. 6. Confusion Matrix for the Coral Lifeforms Classification Model

lifeform as other types of coral lifeforms and that the model's efficiency in identifying the Submassive coral lifeform may require some improvement.

The model has also achieved a recall value of 100.00% for Branching, Foliose, and Mushroom coral lifeforms, which means that it correctly identified all positive instances of these lifeforms in the test dataset without any false negatives. However, the model has a lower recall value of 75.00% both for Massive and Submassive coral lifeforms, indicating that the model has correctly identified 75.00% of the positive instances of these coral lifeforms, but there are still some instances that were missed or classified incorrectly as negative (false negatives).

Moreover, the highest F1 score is for Tabulate with 95.23%, indicating that the model is accurately identifying positive instances and minimizing false positives and false negatives of Tabulate. The next highest F1 score is Branching with a rate of 94.12%, signifying that the model is fairly accurate in classifying this type of coral lifeform. Foliose and Mushroom classes both obtained an F1 score value of 93.33%, which is also an indicator of a notable classification result. However, the F1 score of Submassive (66.67%) is relatively low, showing that the model struggled more in classifying this lifeform. This could result from different causes, such as a lack of diverse training data for that particular lifeform, or a greater similarity in appearance to other lifeforms, making it harder for the model to differentiate.

Using Eq. 1, the overall performance of the model was evaluated. The model made six (6) instances of misclassifications out of a total of 56 instances. The result shows that the developed model reasonably has a strong performance in accurately classifying the coral lifeforms with an accuracy of 89.29%. This means that out of all the coral images in the test dataset, 89.29% were correctly classified.

The following figures show some of the correct (Fig. 7) and incorrect (Fig. 8) predictions made by the generated model on the testing set.

Table 2. Model Performance Evaluation Results

Class	Accuracy(%)	Precision (%)	Recall (%)	F1 Score(%)
Branching	98.21	88.89	100.00	94.12
Encrusting	94.64	90.00	81.81	85.71
Foliose	98.21	87.50	100.00	93.33
Massive	96.43	100.00	75.00	85.71
Mushroom	98.21	87.50	100.00	93.33
Submassive	94.64	60.00	75.00	66.67
Tabulate	98.21	100.00	90.90	95.23

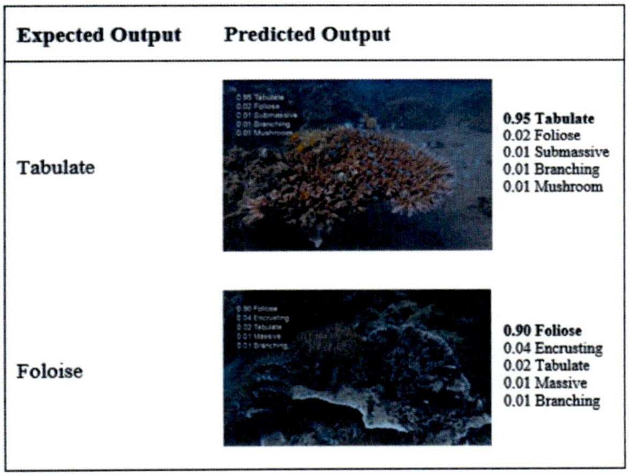

Fig. 7. Correct Model Predictions

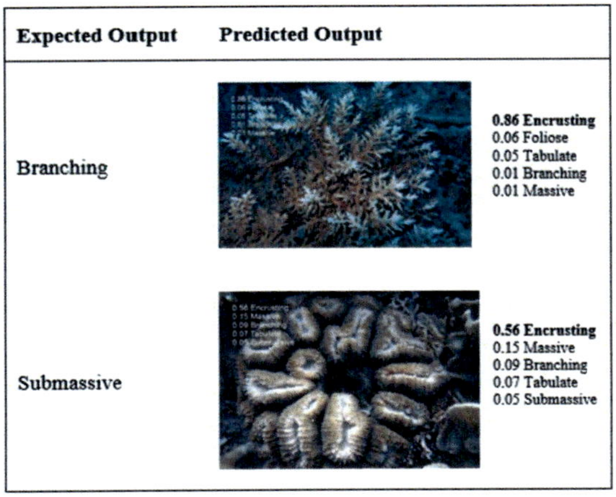

Fig. 8. Incorrect Model Predictions

4 Conclusion and Recommendations

In conclusion, the coral lifeform classification model developed using the YOLOv5 deep learning approach has shown promising results in classifying seven (7) different coral lifeforms, including Branching, Encrusting, Foliose, Massive, Mushroom, Submassive, and Tabulate. The model achieved an overall accuracy of 89.29%, making it a potentially useful tool for efficient reef conservation and monitoring.

Further improvements can be made by expanding the dataset to include more diverse coral lifeforms, as well as exploring the use of other deep-learning approaches and techniques to enhance the model's accuracy and efficiency. Overall, this research showcases the potential of deep learning in advancing coral reef conservation efforts, highlighting the importance of using advanced technologies to protect our oceans.

References

1. Donovan, M.K., et al.: From polyps to pixels: understanding coral reef resilience to local and global change across scales. Landscape Ecol. **38**(3), 737–752 (2023)
2. Siro, G., Pipite, A., Christi, K., Srinivasan, S., Subramani, R.: Marine actinomycetes associated with stony corals: a potential hotspot for specialized metabolites. Microorganisms **10**(7), 1349 (2022)
3. King, M.: Fisheries Biology, Assessment and Management. John Wiley & Sons (2013)
4. Hughes, T.P., et al.: Coral reefs in the Anthropocene. Nature **546**(7656), 82–90 (2017)
5. Adjeroud, M., Kayal, M., Penin, L.: Importance of recruitment processes in the dynamics and resilience of coral reef assemblages. In: Rossi, S., Bramanti, L., Gori, A., Orejas, C. (eds.) Marine animal forests, pp. 549–569. Springer, Cham (2017). https://doi.org/10.1007/978-3-319-21012-4_12
6. Harsindhi, C.J., Bengen, D.G., Zamani, N.P., Kurniawan, F.: Abundance and spatial distribution of reef fish based on coral lifeforms at Tidung Island, Seribu Islands, Jakarta Bay. Aquac. Aquarium, Conserv. Legislation **13**(2), 736–745 (2020)
7. Mahmood, A., et al.: Deep learning for coral classification. In: Handbook of Neural Computation, pp. 383–401. Elsevier (2017). https://doi.org/10.1016/B978-0-12-811318-9.00021-1
8. Sun, M., Yang, X., Xie, Y.: Deep learning in aquaculture: a review. J. Comput. **31**(1), 294–319 (2020)
9. Guntia, R.R., Rorissaa, A.: A dual convolutional neural networks and regression model based coral reef annotation and localization. In: Experimental IR Meets Multilinguality, Multimodality, and Interaction Proceedings of the 13th International Conference of the CLEF Association (CLEF 2022), LNCS Lecture Notes in Computer Science, Springer, Bologna, Italy (2022)
10. Rajan, S.K.S., Damodaran, N.: MAFFN_YOLOv5: multi-scale attention feature fusion network on the YOLOv5 model for the health detection of coral-reefs using a built-in benchmark dataset. Analytics **2**(1), 77–104 (2023)
11. Ani Brown Mary, N., Dharma, D.: A novel framework for real-time diseased coral reef image classification. Multimed. Tools Appl. **78**(9), 11387–11425 (2018)
12. Bahrani, A., Majidi, B., Eshghi, M.: Coral reef management in persian gulf using deep convolutional neural networks. In: 2019 4th International Conference on Pattern Recognition and Image Analysis (IPRIA), pp. 200–204. IEEE (2019)

Mapping Hierarchies and Dependencies from Robustness Diagram with Loop and Time Controls to Class Diagram

Glendel B. Calvo(✉) and Jasmine A. Malinao

Division of Natural Sciences and Mathematics, University of the Philippines Tacloban College, Tacloban City, Philippines
{gbcalvo,jamalinao1}@up.edu.ph

Abstract. Understanding business workflows are crucial in the operation of most, if not all organizations. There are three workflow dimensions that highlight different aspects and information about a system, namely, resource, process, and case. Most modelling frameworks such as the Unified Modelling Language, Business Process Modelling and Notation, Workflow nets, etc. have diagramming tools that use only one or two dimensions. By using a Robustness Diagram with Loop and Time Controls (RDLT), all three workflow dimensions can be represented by one modelling diagram. Currently, there is a lack of literature on representing hierarchical abstractions and dependencies in RDLTs, as well as their mapping to a Class Diagram. This study will propose representations for these hierarchical abstractions and dependencies for RDLTs, mappings to Class Diagrams, and possible improvements to existing RDLT to Class Diagram component mappings by previous literature.

Keywords: Robustness Diagram with Loop and Time Controls · Class Diagram · workflow · mapping · hierarchical abstractions · dependencies

1 Introduction

1.1 Workflows and Workflow Dimensions

A *workflow* is the execution and automation of business processes in which tasks, information, and documents are transferred from one person to another for action in accordance with a set of procedural rules [4]. The three workflow dimensions are resource, process, and case, as per [1].

As defined in [3], the *resource dimension* is a specification of the objects (such as components like class, boundary object, entity object, etc.) with a determinable set of tasks to accomplish, identified as resources. Next, the *process dimension* is a specification of processes, defined as the partial ordering of tasks performed by a system. Finally, the *case dimension* is a specification of cases, where a case represents abstraction of a set of entities, that is processed from some point in the execution workflow until its corresponding output is produced.

K. Kabassi et al. (Eds.): NiDS 2023, LNNS 783, pp. 23–42, 2023.
https://doi.org/10.1007/978-3-031-44097-7_3

1.2 Robustness Diagram with Loop and Time Controls

Definition 1. *RDLT is a modelling diagram that has the capability to represent all three workflow dimensions. Formally, RDLT is defined by* [5] *as a graph representation R of a system (shown in Fig. 1) that is defined as R = (V, E, Σ, C, L, M) where*

- *V is a finite set of vertices, where each vertex has a type* V_{type}*:* $V \rightarrow$ *{'b','e','c'} where'b','e', and'c' means the vertex is either a "boundary", an "entity", or a "controller", respectively as shown in Fig. 2.*
- *A finite set of arcs* $E \subseteq (V \times V)\backslash E'$ *where* $E' = \{(x, y)|x, y \in V, V_{type}(x) \in \{\text{'b','e'}\}, V_{type}(y) \in \{\text{'b','e'}\}\}$ *with the following attributes and user-defined values,*
  - $C : E \rightarrow \Sigma \cup \{\epsilon\}$ *where* Σ *is a finite non-empty set of symbols and* ϵ *is the empty string. Note that for real-world systems, a task* $v \in V$*, i.e.* $V_{type}(v)$ *='c', is executed by a component* $u \in V$*,* $V_{type}(u) \in \{\text{'b','e'}\}$*. This component-task association is represented by the arc* $(u, v) \in E$ *where* $C((u, v)) = \epsilon$*.*

    Furthermore, $C((x, y)) \in \Sigma$ *represents a constraint to be satisfied to reach y from x. This constraint can represent either an input requirement or a parameter C((x, y)) which needs to be satisfied to proceed from using the component/task x to y.* $C((x, y)) = \epsilon$ *represents a constraint-free process flow to reach y from x or a self-loop when x = y.*
  - $L : E \rightarrow Z^+$ *is the maximum number of traversals allowed on the arc.*

- *Let T be a mapping such that* $T((x, y)) = (t1,\ldots, tn)$ *for every* $(x, y) \in E$ *where* $n = L((x, y))$ *and* $t_i \in N$ *is the time a check or traversal is done on (x, y) by some algorithm's walk on R.*
- $M : V \rightarrow \{0, 1\}$ *indicates whether* $u \in V$ *and every* $v \in V$ *where* $(u, v) \in E$ *and* $C((u, v)) = \epsilon$ *induce a sub-graph* G_u *of R known as a* ***reset-bound subsystem*** *(RBS). The RBS* G_u *is induced with the said vertices when M (u) = 1. In this case, u is referred to as the* ***center*** *of the RBS* G_u*.* G_u*'s vertex set* V_G*, contains u and every such v, and its arc set* E_{G_u}*, has* $(x, y) \in E$ *if* $x, y \in V_{G_u}$*.*

  Finally, $(a, b) \in E$ *is called an* ***in-bridge*** *of b if* $a \notin V_{G_u}$ *,* $b \in V_{G_u}$*. Mean-while,* $(b, a) \in E$ *is called an* ***out-bridge*** *of b if* $b \in V_{G_u}$ *and* $a \notin V_{G_u}$ *.*

  Arcs (a, b), (c, d) $\in$ *E are* ***type-alike*** *if:*

  - $\exists y \in V$ where (a, b), (c, d) $\in$ Bridges(y) with Bridges(y) = {(r, s) $\in$ E|(r, s) is either an in-bridge or out-bridge of y}.
  - or $\forall y \in V$, (a, b), (c, d) $\notin$ Bridges(y).

1.3 Class Diagrams

Class diagram is defined by [6] as a graphic representation of a static view of a system. The view is static since it does not support time-dependent behavior. A *class* is a collection of discrete entities (objects) with state and behavior. The main components of the static view are classes and their relationships [3, 6].

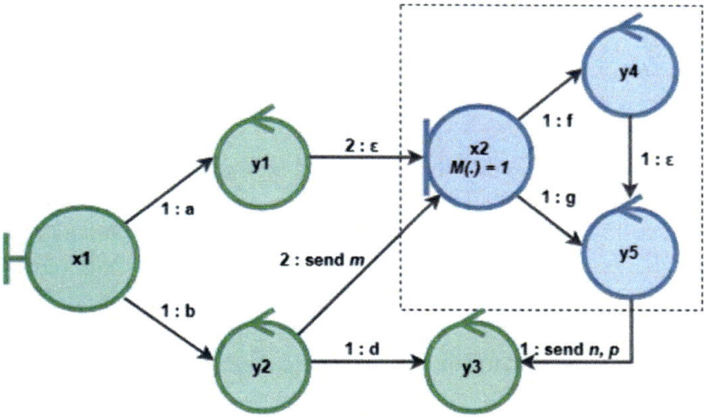

Fig. 1. An example of *RDLT* as adapted from Yiu et al. (2018) [1]

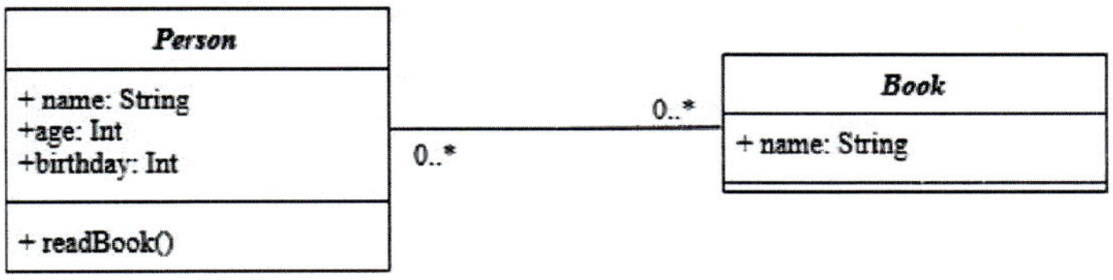

Fig. 2. An example of a *class diagram*

An example of a class diagram shown in Fig. 2 where a person has attributes name, age, and birthday. Additionally, it has an operation *readBook* which requires information from some instance of the class *Book*.

1. *Hierarchical abstractions.*

   As defined by [6], hierarchical abstractions are the type of relationships in class diagrams where several layers of entities are represented. There are several types of hierarchical abstractions: *aggregation*, *composition*, *generalization*, and *specialization*.

   Aggregation represents a part-whole relationship. In this relationship type, the part can exist separately from the whole. For example, even if a band is disbanded, a musician still exists. Meanwhile, *composition* is a strong ownership between two classes in which the part can only belong to one whole. The part cannot exist separately from the whole. An example of this relationship can be depicted with a page that can only exist as part of a book. A page from a student's handbook can only exist in a student handbook. *Generalization* is a directed relationship between two generalizable elements of the same kind, such as classes, or other kinds of elements. This relationship type also allows polymorphism. A bank account is the generalization of a current account and a savings account. *Specialization*, on the other hand, is the relationship type wherein a more specific description of a model element is produced by adding more children [6]. The specialization of a bank account are the current account and the savings account.

2. *Dependencies*

   Based on [6], a *dependency* relationship indicates a semantic relationship between two or more model elements. The following are different types of dependencies as discussed by [3] and [6]:

   The **call dependency** is the statement that a method of one class calls an operation of another class. Next, **derivation dependency** is the statement that one instance can be computed from another instance. Third, **realization dependency** is the mapping between a specification and an implementation of it. Fourth, **refinement dependency** is the statement that a mapping exists between elements at two different semantic levels. Fifth, **substitution dependency** is the statement that the source class supports the interfaces and contracts of the target class and may be substituted for it. Lastly, **trace dependency** is the statement that some connection exists between elements in different models, but less precise than a mapping.

2 Mapping RDLTs to CDs

The proposed RDLT TO CD mapping in this study is divided into two sets: the *static* and *dynamic* aspects of RDLT components.

2.1 Static Aspects of RDLT to CD mapping

The *RDLT* can be mapped into a *package*. According to the UML 2.5 reference manual published by Object Management Group (2017) [3], a package is a grouping of related model elements such as classes, documents, or even other packages. Since RDLT is a system it can be mapped into a package. This is a new mapping. The vertices of a RDLT are boundary objects, entity object, and controller. The *boundary object* is a system component that performs some tasks and communicates with its environment and it can be mapped into a *class with a port* in CD. The presence of a port element indicates that the particular object can interact with the environment (i.e. boundary object). Meanwhile, the *entity object*, which is an internal system component that performs some tasks, can be mapped into a *class without a port* in CD. These two mappings were adopted from [7]. Their proposed mapping only include distinctions between the names of boundary and entity objects when mapped to a class. Next is the *controller* which represents a task done in the system and is mapped to a *method* of a class. This mapping is reused from [7].

These mappings can be seen in Fig. 6 where we see the boundary object x1 in the RDLT from Fig. 11 was converted into a class x1 with a port to indicate it can interact with its environment in the corresponding Class Diagram. Meanwhile, all the controllers owned by x1 is mapped onto operations owned by class x1.

For the m-attribute, the *in-bridges* can be mapped onto a *arcs coming from classes contained in a different package going into the RBS*. In RDLT, the arcs are in-bridges if they are incoming arcs to the center of RBS from the outside. On the other hand, out-bridges are outgoing arcs starting from the RBS and connecting to an outer vertex and can be mapped into an *classes imported into the RBS*. In CD, imports can be used to connect the classes to the RBS package despite the difference in hierarchy level. This is a new mapping. Meanwhile, the center of an RBS is marked by setting a private attribute in its class representation to 1. This mapping is taken from [7]. The RBS of an RDLT can be mapped into a package containing one and only one class. Since an RBS is considered as a system, it can be viewed as a nested RDLT, which we have already mapped to a package in CD.

Connectivities within the RDLT are represented by the arcs connecting the vertices within the system. In this study, the proposed mapping of RDLT's connectivities in CD is a relationship between classes. This mapping is adopted from [7], although their mapping in CD was limited to the association relationship only. For type-alikeness, two arcs in RDLT are considered type-alike if they are both bridges of class x or if neither of them are bridges of class x. This static component of RDLT can be mapped to CD based on where the arcs are coming from. Arcs coming from classes contained in a different package entering the RBS and classes imported into the RBS are considered bridges.

An input constraint in RDLT is the action that needs to be satisfied to proceed from using the task in vertex1 to vertex2. This component can be represented in CD by an attribute belonging to a class. The parameter constraint is a variable that should be passed from one component to another in order to proceed from using the task from vertex 1 to vertex 2. It can be mapped in CD by using an attribute owned by class 1 being referenced as an operation parameter of class 2. Lastly, the L-attribute which is the maximum number of times the arc can be traversed, can be mapped as an attribute owned by a class in CD.

Figure 6 shows how the RBS from the RDLT in Fig. 5 is mapped into a Class Diagram. The mapping of the entity object y1, which is the center of the RBS, is class y1 contained in an RBS package in the Class Diagram. The conversion of in-bridges and out-bridges is signified by the arcs incoming and outgoing(with $<<$ *import* $>>$ labels) on the arrows between the RDLT and RBS package. On the other hand the connectivities of the objects within the sample RDLT in Fig. 5 is mapped with the arrows in Fig. 6 to represent the relationships between classes.

2.2 Dynamic Aspects of RDLT to CD Mapping

The dynamic components of RDLT and it's CD mapping is shown is Table 1. These are the components that involve time-dependent behavior. CD is a static diagram hence instead of mapping the dynamic components as a whole, it can be reduced as a static method.

Table 1. Dynamic RDLT components to CD

RDLT		Class Diagram
2. M-attribute	Reset	A method reset() that resets an RBS' T-attributes to their initial state (upon entry into the RBS) upon exit from the RBS
3. Connectivities	Reset updates on time for type-alike arcs	A method resetT() that sets the T-attribute to a zero array with width equal to the L-attribute
3.1 C-attribute	Check if an arc is unconstrained or constrained	A method checkConstrained() that checks all type-alike connectivities if they satisfy the definition of an unconstrained arc
3.2 L-attribute	Check if an arc can be traversed	A method checkTraversable() that contains a condition that if the L-attribute array contains at least one zero element, the arc can be traversed
3.3 T-attribute	Update the time of checks and traversals on an arc	Setter function setT() for T-attribute

2.3 Mapping Algorithm

Algorithm 1: Build package hierarchy from RDLT to CD

Input: RDLT *R*
Output: Package structure of CD in *R*

```
createPackageTree(RDLT rdlt){
if (rdlt has no nested RDLTs) then
    create new package rdlt
    return
end
else
    for each nest RDLT do
        insertPackage(createPackageTree(nestedRDLT))
    end
end
}
```

Algorithm 2: RDLT to CD

```
Input: RDLT R
Output: CD equivalent of R
createPackageTree(R)
for each vertex V in rdlt do
    create class x add attributes t_attr, c_attr
    add setter methods for t_attr, c_attr
    add method checkConstrained()
    add method checkTraversable()
    if V_type(x) == 'b' then
        add port element to class x
    end
    add class x to package
    if M(x)==1 then
        add private attribute M: Integer = 1 to class x
        add final attributes initial_t = t_attr, initial_c = c_attr
        add private method reset() to class x
    end
    for each controller y owned by V do
        add method y() to class x
        if y has at least one incoming arc not owned by V then
            set y visibility to public
        end
        else
            set y visibility to private
        end
    end
end
for each arc(v1, v2) in E do
    if v2 is not visible to v1 then
        import v2
    end
    if (v1, v2) is not an ownership arc then
        add attribute v1v2_L to class v1(owner)
        if v1(owner) != v2(owner) then
            set v1v2_L visibility to public
        end
        else
            set v1v2_L visibility to private
        end
    end
    if C((v1, v2)) is an input requirement constraint then
        add attribute v1v2_attributename to class v1(owner)
        if v1(owner) != v2(owner) then
            set v1v2_attributename visibility to public
        end
        else
            set v1v2_attributename visibility to private
        end
end
```

Algorithm 3: RDLT to CD Continued

Input: RDLT R
Output: CD equivalent of R
48: **else if** *$C((v_1, v_2))$ is an parameter constraint* **then**
49: **if** $V_{type}\ y$ = 'c' **then**
50: add attribute v_1v_2*_attributename* to class v_1(owner)
51: **end**
52: **else**
53: add attribute v_1v_2*_attributename* to class v_2(owner)
54: **end**
55: **if** *$v_1(owner)$!= $v_2(owner)$* **then**
56: set v_1v_2*_attributename* visibility to public
57: **end**
58: **else**
59: set v_1v_2*_attributename* visibility to private
60: **end**
61: **for** *each arc where owner of v_1 != v_2* **do**
62: **if** *arc (v_1, v_2) has C-attribute extends* **then**
63: make generalization relationship(v_1(owner), v_2(owner))
64: **else if** *arc (v_1, v_2) has C-attribute extends* **then**
65: make specialization relationship(v_1(owner), v_2(owner))
66: **else if** *controller of v_1 calls controller of v_2* **then**
67: make call dependency relationship(v_1(owner), v_2(owner))
68: **else if** *$v_1(owner)$!= $v_2(owner)$* **then**
69: make substitution dependency relationship(v_1(owner), v_2(owner))
70: **else**
71: make unidirectional association(v_1(owner), v_2(owner))
72: multiplicity = 0...1
73: **end**

The Algorithm 1 is the proposed recursive algorithm in building package hierarchy from RDLT to CD. Since RDLT (which is a package) contains nestable hierarchical components, like RBS (which is also mapped as a package), a recursive algorithm is used in order to convert these hierarchies into an empty tree structure representing the package directories. Lines 2–5 defines the base case wherein the structure being is an innermost RDLT, i.e. it does not contain any nested RDLTs. Lines 7–9 define the recursive step which makes use of a function call stack to recursively expand all nested RDLTs.

Algorithm 2 is the proposed mapping for RDLT to CD. Line 1 contains the package hierarchy building step in CD as defined in Algorithm 1. The algorithm from line 2–25 checks the type of vertex, adding their necessary attributes and visibility. Next, line 26–64 checks all of the arcs in the RDLT. Line 27–28 defines the outbridge conversion. Lines 30–38 continues by mapping ownership arcs and setting their visibility. The next section defines mappings for input requirement constraints (lines 39–46), and the parameter constraints (lines 48–60). Finally, lines 61–73 shows the conversions applied to hierarchical abstractions and dependencies.

Algorithm Analysis

The *time complexity* for the proposed algorithm is $O(n^2)$, where n is the number of vertices. On the other hand, the *space complexity* is also $O(n^2)$, where n is the number of vertices.

2.4 Hierarchies Mapping in RDLT to CD

Figure 3 shows the RDLT representation of both generalization and specialization. The vertices x1, y1, and z1 are the children or specializations of vertex w1. All of these children has access to their parent w1 and all its owned controllers hence the flow of the arc is from left to right. The C-condition *extends* indicates that the children (specializations) is based on a generalization (parent). On Fig. 4, the CD mapping of the RDLT representation is shown. The class w1 is the superclass and the classes x1, y2, and z1 are its subclasses.

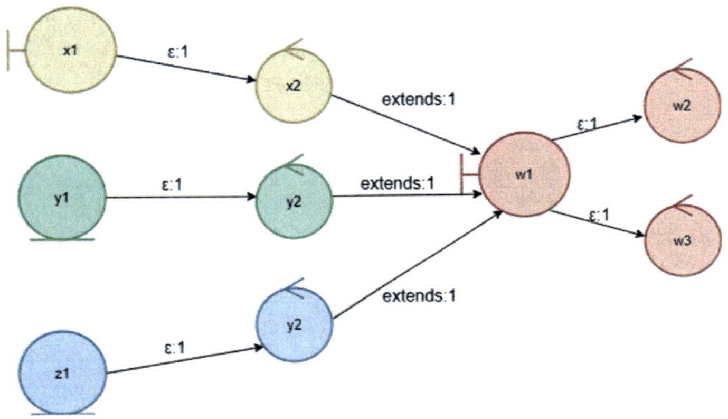

Fig. 3. RDLT representation of Inheritance

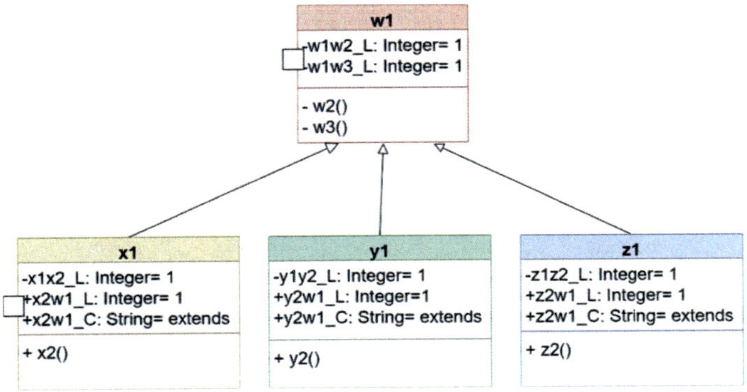

Fig. 4. RDLT to CD mapping of Inheritance

2.5 Dependencies Mapping in RDLT to CD

- **Call dependency.** Figure 5 shows an RDLT with a x4 controller owned by vertex x1, which is mapped into an operation x4 owned by class x1 in CD as shown in Fig. 6, calls another operation, which is controller z2 owned by vertex z1.

- **Trace dependency.** Figure 7 shows a trace dependency relationship between a RDLT and CD. Both of these diagrams represent the same information.
- **Refinement dependency**

  As shown in Fig. 8 as example, the RDLT level 2 is a refinement of the RDLT level 1. Figure 9 shows the CD counterpart of this RDLT representation.
- **Substitution dependency.** Figure 10 is an example of substitution dependency in RDLT. The vertex x1 can be substituted by vertex y1. Figure 11 shows the CD mapping of the represented subsitution dependency.
- **Derivation dependency**.

  From Figure 12, RDLT level 2 can be derived from RDLT level 1. This mapping of RDLT is similar to the example shown in Figure 8 for Refinement dependency, however they have different applications in modeling. With a refinement dependency, the connection reflects a transformation of one model element into a simpler version. On the other hand, a derivation dependency describes a second element whose information can be deduced from the first. Figure 13 shows the derivation of RDLT level 2 from RDLT level 1 in CD.
- **Realization dependency.** The specifications of an Adsorption Chiller in Fig. 14 is realized into an RDLT representation. Figure 15 then shows a package element in CD to which the RDLT representation of the Adsorption Chiller is mapped.

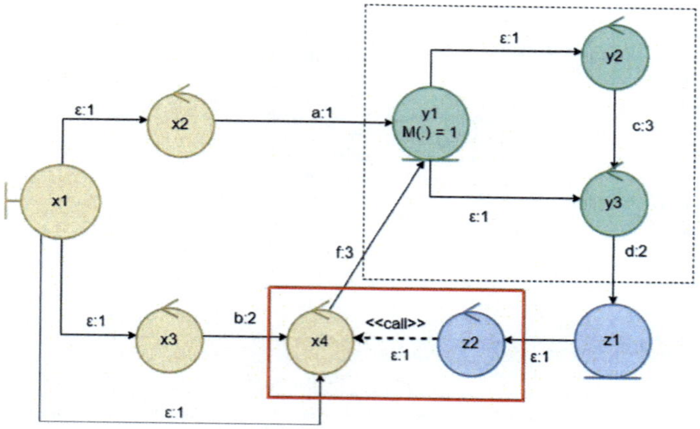

Fig. 5. RDLT representation of *call dependency*

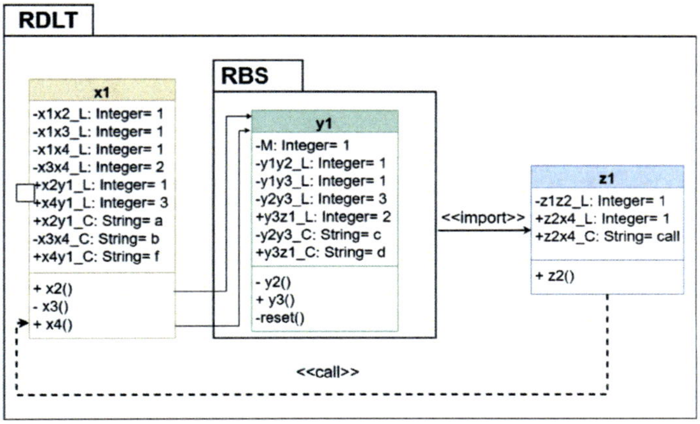

Fig. 6. RDLT to CD mapping of *call dependency*

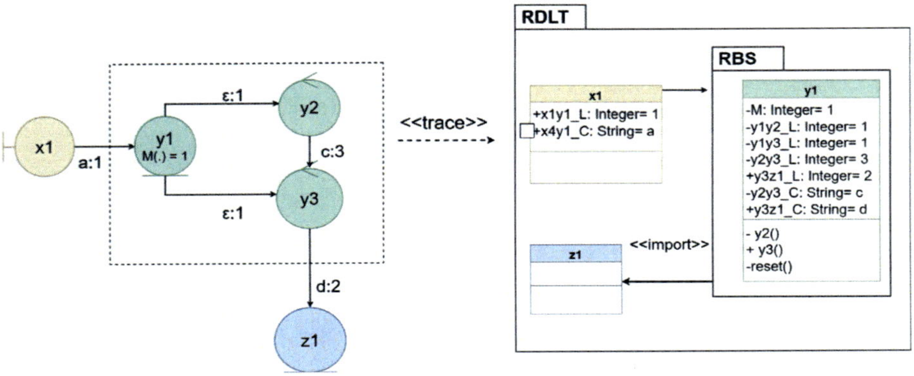

Fig. 7. RDLT to CD mapping of *trace dependency*

2.6 Mapping Verification

The established mappings were validated by generating an extended RDLT from the input RDLT. From this extended RDLT, an activity profile is obtained, then verified. The verification step looks at the extended RDLT if all of its components, i.e., its set of objects and their corresponding set of controllers, are present in the CD where this CD is the output from using the proposed mapping given the extended RDLT as input.

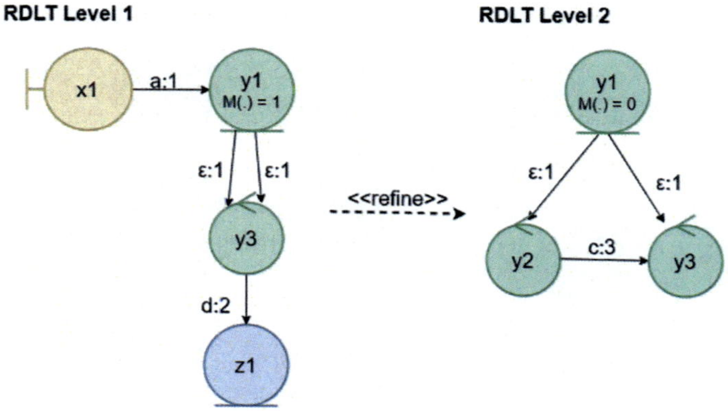

Fig. 8. RDLT representation of *refinement dependency*

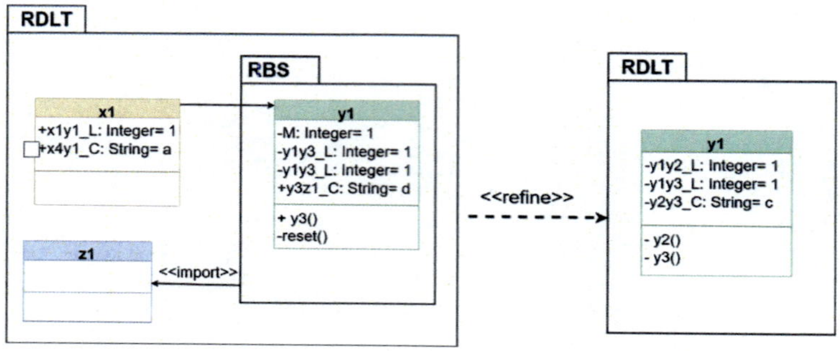

Fig. 9. RDLT to CD mapping of *refinement dependency*

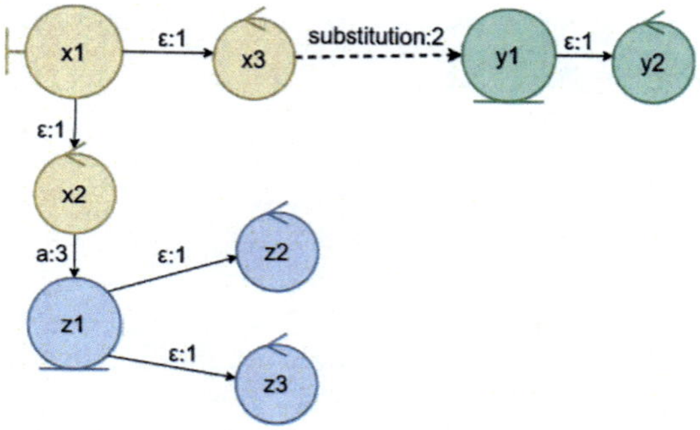

Fig. 10. RDLT representation of *substitution dependency*

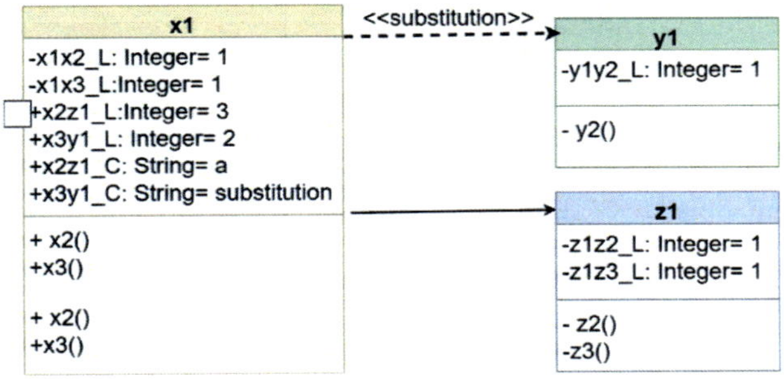

Fig. 11. RDLT to CD mapping of *substitution dependency*

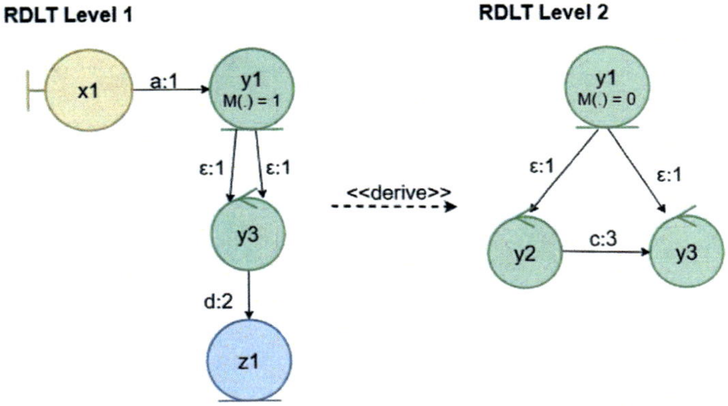

Fig. 12. RDLT representation of *derivation dependency*

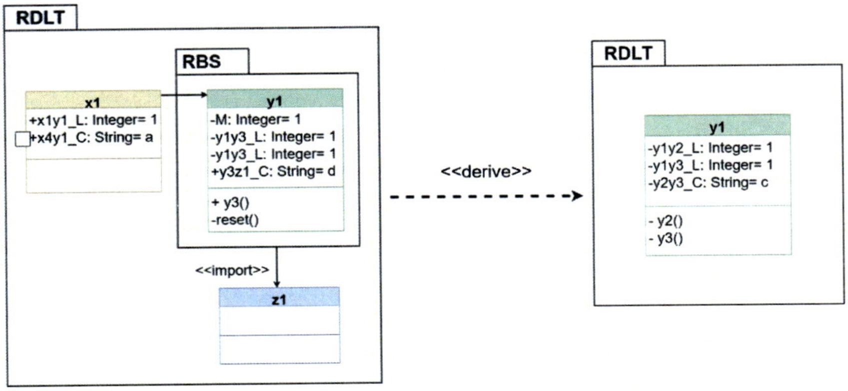

Fig. 13. RDLT to CD mapping of *derivation dependency*

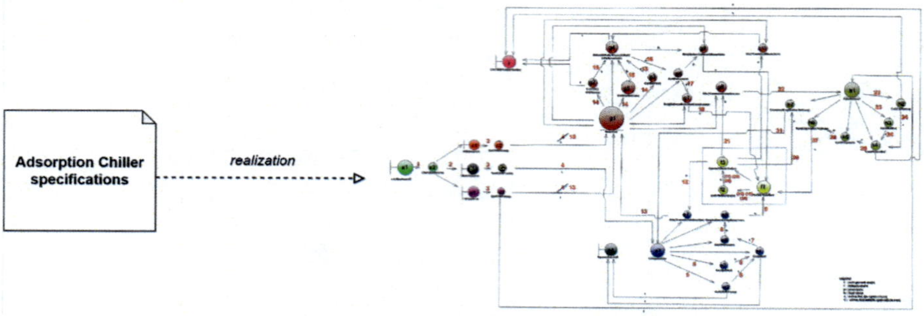

Fig. 14. RDLT representation of *realization dependency*

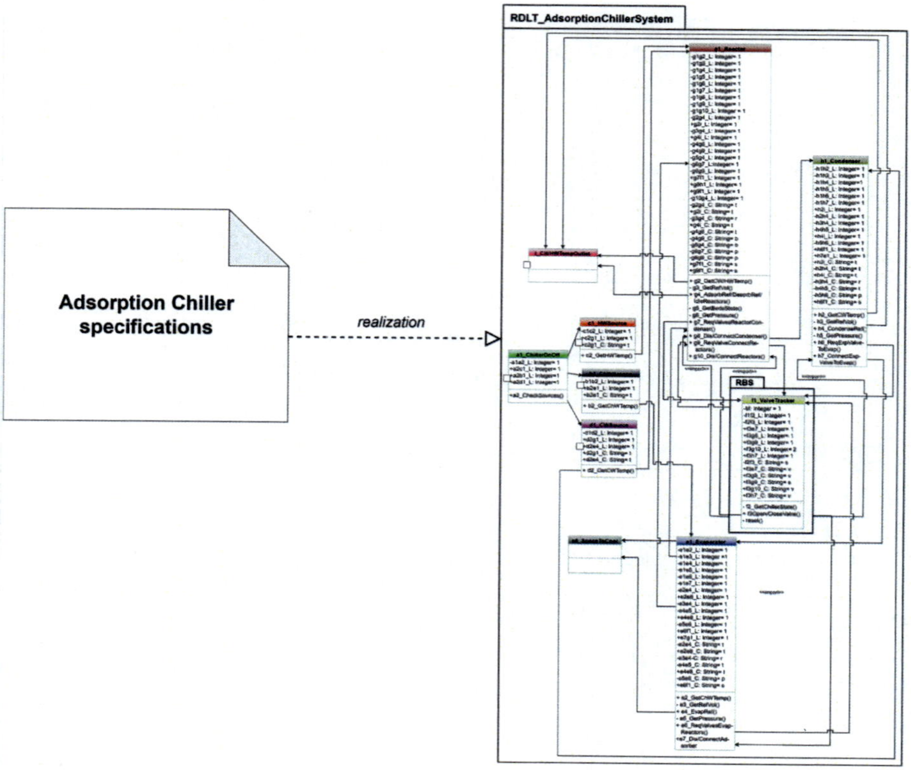

Fig. 15. RDLT to CD mapping of *realization dependency*

3 Results and Discussion

3.1 Illustration of RDLT to CD Mapping with Chiller System

The RDLT was developed in order to model real-life complex systems. However, considering that the RDLT is still relatively new and not as widely used when compared to other visual modeling languages, real-life applications of RDLT are rare. Also, in order to demonstrate the representations being introduced in this paper, it would make sense to use the Chiller System as it has been used as a sort of benchmark by other papers in this topic. This makes comparison between different works much easier to compare, since there is only one model being discussed. The illustration shown in Fig. 16 is the RDLT representation of the Adsorption Chiller System by Malinao (2017) [5].

Figures 17 and 18 shows the comparison of the RDLT to CD mapping by Yiu et al (2018) [2, 7] and the proposed RDLT to CD mapping in this study. The notable difference between the two figures is the use of packages to define visibility boundaries for both RDLT and RBS which is introduced in this study. The use of packages is functional in grouping together the related elements to provide a more organized view on the mapping. In addition, it is immediately noticeable where an RBS is located in an RDLT. Looking at the RBS in Figs. 17 and 18, it is more distinguishable in the former than in the latter. This improvement may not be so evident for smaller systems, but its impact is more pronounced when it comes to modeling complex systems such as the Chiller System. Furthermore, a distinction between boundary and entity objects was added where the boundary objects are mapped into classes with ports to indicate that those classes can interact with the environment. In the previous literature, boundary objects and entity objects were indistinguishable from each other. Another difference is the outbridges. Since the proposed study makes use of packages, it is necessary to use $<<$ *import* $>>$ in order to include the class(where the outbridge is going) in its namespace in order to connect. The use of the reset() method can also be found in the proposed mapping. It indicates that the RBS class can reset its values upon exiting via the outbridges.

3.2 RDLT to CD Mapping of Hierarchies and Dependencies

Inheritance and several dependency representations were successfully formulated for RDLTs. Dependencies are indicated by a dashed edge. To represent inheritance, an intermediary controller is created between the parent and child, which has a C-attribute extends. These representations were then mapped to the pre-existing notations for these relationships in CD. By making use of the hierarchical relationships, related objects in the RDLT can be organized according to similar functionality, and the generated model can be used in a more modular fashion. Because some of the hierarchies have been established, access control also had to be developed in order to control the flow of information in the system between objects at different hierarchical levels. By differentiating between boundary and entity objects in the class diagram, we constrain the volatility in the RDLT and give the model additional descriptiveness in terms of information flow.

These new mappings introduced in this study successfully expanded on previous literature by providing further expressive power to RDLTs. The ability to represent dependencies and hierarchical relationships gives RDLTs the ability to model reality more effectively and accurately.

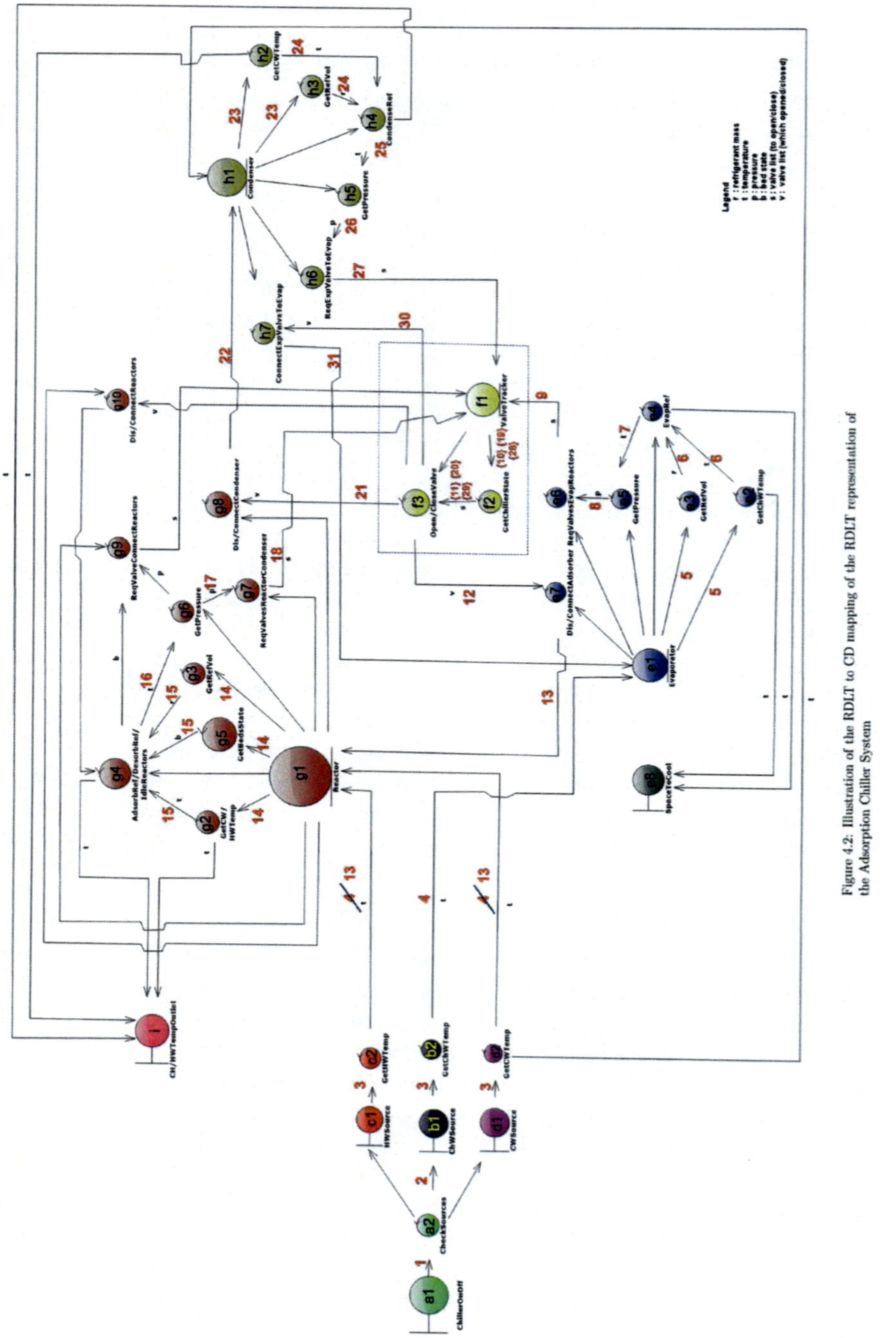

Fig. 16. RDLT representation of the Adsorption Chiller System [5]

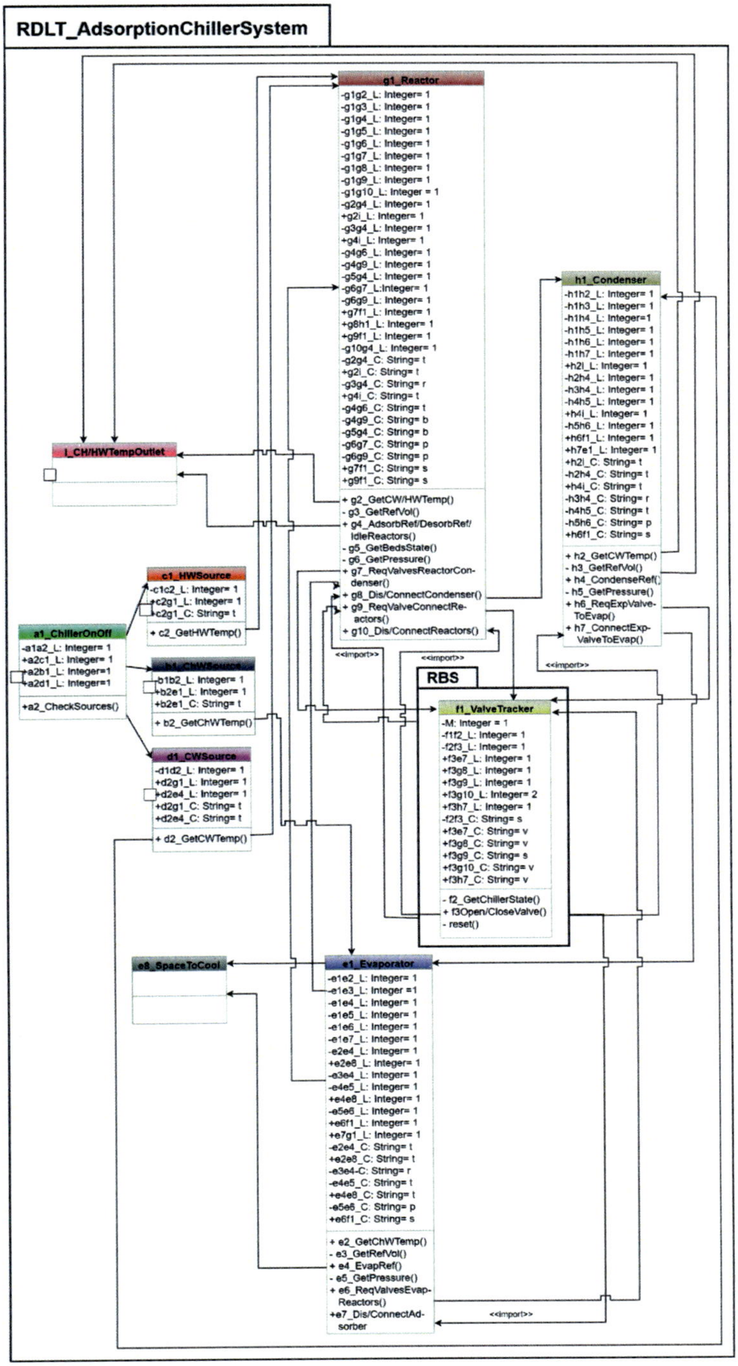

Fig. 17. Illustration of the RDLT to CD mapping of the RDLT representation of the Adsorption Chiller System

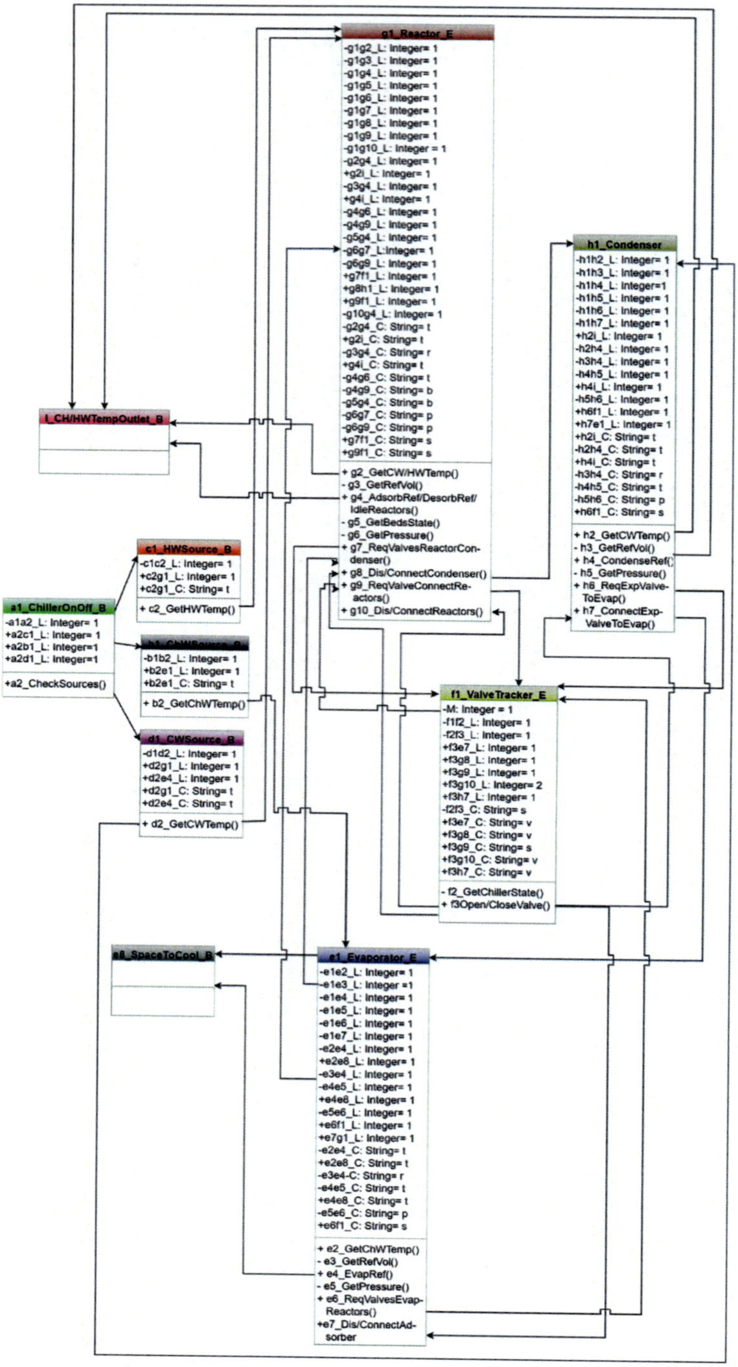

Fig. 18. Illustration of the RDLT to CD mapping of the previous literature with respect to the RDLT representation of the Adsorption Chiller System

4 Conclusions

New RDLT to CD mappings have been successfully established to address the gaps in literature, although further work into developing mappings for dynamic components is still needed. As for the existing mappings in literature, some representations were imprecise in terms of usage within the CD. These were amended as necessary. The call, trace, refinement, substitution, derivation, and realization dependencies have been given a representation in the RDLT, and then mapped to CD to allow verification. The usage for the added mappings was also demonstrated with the use of Chiller System RDLT.

Some of the hierarchical abstractions (i.e., aggregation and composition) and dependencies (i.e., send, creation, instantiation, access, and binding) have not been represented in RDLT during the conduct of this study. For aggregation and composition, the deletion operations necessary in order to illustrate their "has-a" relationship does not have a mapping in RDLT. This component can potentially be represented by making use of the RBS component, where this component can reset its contained values, but this behavior is still challenging to replicate since class deletion would equate to vertex deletion in RDLTs in the current mapping. Alternatively, an auxiliary component in RDLT can be formulated to represent deletion.

Additionally, the access dependency needs more work in better conveying private imports between packages i.e. private connections from one RDLT to another. Another issue that can be addressed in the future is the send dependency due to its lack of representation for asynchronous signal objects in RDLT. A potential remedy for this gap would be to utilize parameter constraints to transmit the signal data. Instantiation relationships were not covered by this study. However, the researcher recommends the use of activity profiles in RDLT as a mapping. It must be noted however that this recommendation does run into some issues during the verification step. In order to pass verification, the number of RDLT components and CD components must be equal. There would be an unequal number of components since the number of objects and the activity profile will not be a 1:1 conversion. As for the binding dependency, the concept of using templates is necessary. There is currently no established mapping for this CD component in RDLT yet.

Lastly, the proposed mappings for the hierarchical abstractions and dependencies were verified to be consistent in terms of their representation. The RDLT to CD mapping of the Chiller System was also verified successfully, indicating that the mappings are indeed correct and applicable to a real world example.

5 Future Research

Further work needs to be done in representing aggregation and composition. The call, send, creation, instantiation, access, and bind dependencies also have to be added into the RDLT. Furthermore, algorithms for vertex deletion that enact the behavior of dependencies should also be developed.

Acknowledgements. My deepest thanks goes to the *Department of Science and Technology-Science Education Institute (DOST-SEI)*. My R.A. 7687 financial scholarship was in-valuable in supporting for my education.

References

1. van der Aalst, W.: Structural characterizations of sound workflow nets. Comput. Sci. Reports **96**(23), 18–22 (1996)
2. Calvo, G., Malinao, J.: Mapping hierarchies and dependencies from robustness diagram with loop and time controls to class diagram. In: 3rd International Conference on Novel and Intelligent Digital Systems (NIDS 2023) (2023)
3. Group, O.M.: Unified modelling language version 2.5 (2017), document Number: formal/2017-12-05
4. Hollingsworth, D.: Workflow management coalition: The workflow reference model, p. 19 (1995). document Number: TC00–1003
5. Malinao, J.: On Building Multidimensional Workflow Models for Complex Systems Modelling. Phd thesis, Vienna University of Technology (2017)
6. Rumbaugh, J., Jacobson, I., Booch, G.: The Unified Modelling Reference Manual 2nd Edition. Pearson Higher Education (2004). ISBN:0-321-24562-8
7. Yiu, A., Garcia, J., Malinao, J., Juayong, R.: On model decomposition of multidimensional workflow diagrams. In: Proceedings of the Workshop on Computation: Theory and Practice (2018)

A Deep Learning Model to Recognise Facial Emotion Expressions

Michalis Feidakis(✉), Gregoris Maros, and Angelos Antikantzidis

University of West Attica, P. Ralli and Thivon 250 Street, GR12241 Egaleo, Greece
m.feidakis@uniwa.gr

Abstract. In the current paper, we present a solution to recognise emotions from facial expressions, based on two publicly available datasets: AffectNet and Fer2013. Our model was trained according to convolutional neural networks (CNN) that are widely used. Towards we reviewed and analysed recent research in the field of deep learning and emotion recognition, and we propose a CNN architecture that includes several layers of convolutions and pooling, followed by fully connected layers. We evaluate the performance of our model on the two datasets and compare it with the state-of-the-art methods. Although our model did not outperform the existing methods in terms of accuracy, our results show that it achieved competitive performance and provides an alternative approach to the problem. The impact of different parameters, such as batch size, learning rate, and dropout, on the model's accuracy was evaluated. This paper contributes to the field of emotion recognition and deep learning by providing a comprehensive analysis of the effectiveness of CNNs for facial emotion recognition and proposing an efficient and accurate model, as well as identifying areas for further improvement in this research field.

Keywords: Emotion recognition · Facial expressions · Deep learning · Convolutional networks

1 Introduction

Affective Computing (AC) is computing that relates to, arises from, or deliberately influences emotions [1]. In literature, there have been various channels to capture emotion data i.e., facial expressions from camera, sound features or voice intonations from a microphone, body posture from sensitive pads, text input analysed and parsed through Natural Language Processing (NLP) complex algorithms, physiological signals that are detected from special sensors i.e., PPG (Photoplethysmography) for Heart Rate, Blood Volume Pulse, et al. Respectively, there is a plethora of AC research studies that try to "sense and respond to" user's affective state or emotion in real-life situations [2–5].

Emotion recognition from facial expressions has been established as a dominant input channel of emotion data, since in our daily life we follow such an emotion de-coding process trying to unveil the correspondent's feelings by their "look in their face" (smile, eye/eye browns, gaze). Towards, Computer Vision and Deep Learning have become

K. Kabassi et al. (Eds.): NiDS 2023, LNNS 783, pp. 43–51, 2023.
https://doi.org/10.1007/978-3-031-44097-7_4

increasingly popular in recent years, with a wide range of applications in areas such as image and video analysis, object detection, and face recognition [3]. In the field of AC, emotion recognition (or affective state detection) from facial expressions constitutes a significant research area, towards effective sense and respond to user's emotions [6].

In the current paper, we present a deep learning model to recognise emotions from facial expressions. Our model was trained and implemented based on two publicly available datasets: AffectNet and Fer2013 [7, 8], deploying convolutional neural networks (CNN) that are widely used. We have achieved accuracy 55% and loss 60% that are quite competitive for the specific dataset [9]. We have also tested a hybrid model combining the two beforementioned datasets improving performance by 5-8%. Future steps involve validating our hybrid model in a humanoid robot (i.e., Soft-bank NAO[6]), in a real human-machine interaction classroom scenario.

1.1 Background

Artificial Intelligence (AI) is a wide scientific area, in which computational models and systems try to mimic or simulate human behaviour and activity. AI has been established for decades and includes many sub-fields like Machine Learning (ML) and Deep Learning (DL). ML involves training a computer system on large datasets, allowing automatic improvements on task performance without being explicitly programmed. ML algorithms recognize patterns in data and make predictions or decisions without human intervention. They are classified into three (3) categories [10]:

i. *Supervised Learning* algorithms that are trained on labelled data and then used to make predictions or reach decisions on new, unseen data.
ii. *Unsupervised Learning* algorithms that are trained on unlabelled data and then used to identify patterns and structure in the data.
iii. *Reinforcement learning* algorithms that are trained according to a reward function and then used to optimize decisions in dynamic environments.

Deep learning (DL) constitutes a new programming paradigm, in which instead of writing coding rules in a programming language, programmers feed data and labels for this data in a model and leave the model to generate those rules. DL today constitutes a dominant AI computing and modelling paradigm in implementing face, emotion, or object recognition.

Facial emotion recognition has been mainly based on the work and findings of Ekman and Friesen [11], who deployed 46 action units (AUs) related to specific facial muscles movement, to recognise six (6) basic emotions (happiness, sadness, anger, fear, surprise, and disgust), plus neutral affective state. These 6+1 basic emotions are recognized and interpreted similarly across cultures and are often referred to as "universal expressions", since they thought to be innate and universally understood by almost all humans [11]. Respectively, Camras and Allison developed the Affect Program, a tool for detecting subtle variations in expressions, recognizing more complex emotions [12]. Their research highlighted the emotion recognition complexity by supporting that emotions are not limited to basic emotions but are also associated with mixed emotions and micro-expressions.

Emotion recognition from facial expressions using computer vision techniques has been widely explored recently due to the latest advancements of DL. Early research involved machine learning techniques (i.e., SVM-Support Vector Machine, DT-Decision Trees, NN-Neural Networks, et al) to solve computer vision problems, which however had limitations in terms of scalability and performance. Model training is based on several datasets like:

- *AffectNet* [7]: Large-scale facial expression dataset containing over 1 million facial expression images collected from the internet. The images have been labelled with one or more facial expression tags, including the 6 basic emotions (joy, sadness, anger, surprise, fear, and disgust), as well as the neutral state.
- *FER2013.* [8]: Comprises images of people expressing different emotions, such as happiness, sadness, anger, surprise, fear, disgust, and neutral. The dataset also includes facial landmarks for each image.
- *CK+* [13]: It contains images of people expressing different emotions, such as happiness, sadness, anger, surprise, and disgust. The dataset includes both poses and spontaneous facial expressions, and facial landmarks for each image.
- *EmoReact* [14]: It contains several videos showing children aged 4-14 years reacting to different stimuli, along with the corresponding annotation of the emotions expressed in the images, in terms of happiness, sadness, anger, surprise and disgust.
- *RAVDESS* [15]: The Ryerson Audio-Visual Database of Emotional Speech and Song is an audio-visual dataset containing over 2,000 audio and video clips of people speaking and singing in different emotional states such as neutral, happy, sad, surprised, anger, disgust and fear. The dataset also includes recordings of speeches and songs, allowing the study of both auditory and visual aspects of emotion recognition.

One of the first DL applications in facial emotion recognition was the deployment of convolutional neural networks (CNNs) to classify facial expressions in images. In 2019, Huang et al., used a CNN to classify the six basic emotions from the FER2013 dataset, achieving an accuracy of 72.8% [16]. Similarly, in 2017, a study by Li et al., used a 3D CNN to classify facial expressions from video sequences, achieving an accuracy of 82.8% on the AffectNet dataset. In 2020, Koujan et al., used a 3D CNN to classify facial expressions from video sequences, achieving an accuracy of 78.5% on the CK+ dataset [17].

In addition to CNNs, there have also been several studies that have explored other deep learning architectures for facial expression recognition. For example, in 2018, a study by Li et al. used a Long-Short Memory (LSTM) network to classify facial expressions from video sequences, achieving an accuracy of 80.9% on the AffectNet dataset. [18]. Mental health disorders such as depression, anxiety, and PTSD (Post-Traumatic Stress Disorder) have been monitored through tracking changes in facial expressions over time. Finally, there are social robots that have been enriched with the capacity to sense and respond to facial emotion expressions [6].

1.2 Problem Statement – Research Questions

Despite the advancements in computer vision and deep learning, emotion recognition from facial expressions remains a challenge due to low accuracy or bias difficulties. For instance, age, gender, or ethnicity variations, as well as differences in lighting and background conditions, prevents generalization of findings and solutions. The lack of large, diverse datasets of facial expressions makes it difficult to train deep learning models, as a result, applications of emotion or affective state recognition remain unexplored.

In our work we tried to address the following issues:

1. What are the performance parameters of deep learning model in real-time emotion recognition from facial expressions?
2. What is the capacity of the deep learning model in recognizing emotions from facial expressions in real-time?
3. What is the impact of different hyper-parameters, such as Learning Rate, Batch Size and Dropout, on model performance in real-world scenarios?
4. What is the model's performance in comparison to other state-of-the-art facial emotion recognition methods when applied to a real-world scenario?

In response to the above questions, we tried to evaluate emotion recognition accuracy, model response time, and real-time recognition.

2 Methodology

In our study, we have exploited AffectNet, a publicly available dataset, since it constitutes the largest dataset for facial expressions recognition -it contains approximately one million facial images collected from the Internet, using 1250 emotion-related keywords in six (6) different languages. About half of the retrieved images (~420KB) have been manually characterized and classified into Ekman's seven (7) distinct classes of facial expressions (Fig. 1) – six (6) basic emotions plus neutral (1) [11] – portrayed into valence and arousal dimensions (2D model). Despite previous reports on the high quality and balanced distribution of AffectNet emotion labels, there are weaknesses that affect the accuracy of related models like the limited variety of faces and images for specific emotion classes, as well as inaccurate labelling.

To address the beforementioned weaknesses, we have also deployed AffectFer, new dataset deriving from cleaning and pre-processing AffectNet - a rather noisy, unbalanced, and hard-to-deploy dataset – and combining to with FER2013. Both datasets have the same structure, simplifying our attempt. Furthermore, both datasets are well known and quite large - AffectNet is one of the largest in the field of facial expression analysis, and Fer2013 is a well-known benchmark dataset in the emotion recognition community. Both contain a diverse set of facial expressions across various ethnicities, ages, and genders, making them ideal for training a robust model.

Also, enough states from AffectNet were discarded because of low their sampling rate. This pre-process resulted in our new AffectFer dataset, reducing the classes to 5 - Surprise was discarded due to limited samples in AffectNet - trying to achieve better performance.

AffectFer advances State-of-the-Art (SotA) in FER. It has been developed in line with the work of [24], following Transfer ML approach [19]. Specifically, we used a pre-trained model in AffectNet and reused it to train a new model in Fer2013. Then, we combined the two models and performed training on the two combined datasets.

Our pre-processing included:

a. Download and unzip the two datasets (AffectNet & FER2013).
b. Use the CSV files provided by Fer2013 and match them with AffectNet.
c. Combine the two datasets with the new AffectFer dataset comprising ~340K images and 11 emotion categories.

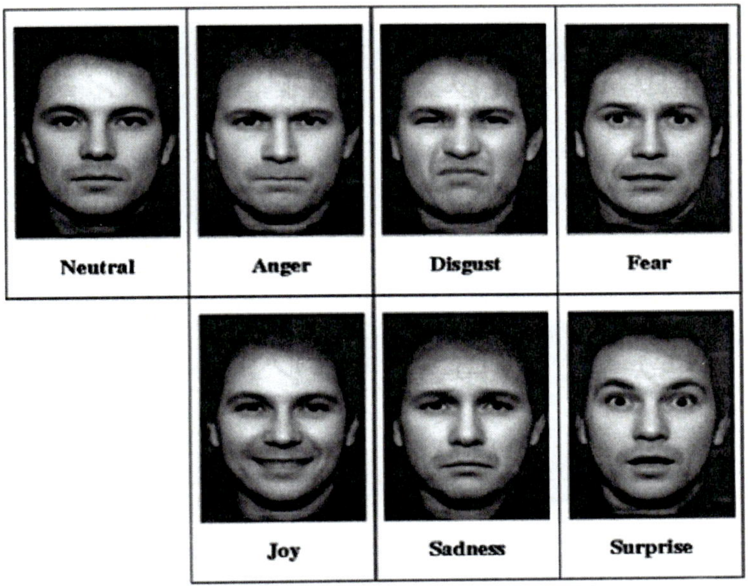

Fig. 1. Ekman's 7 basic emotions and corresponding facial expressions (retrieved 15/02/2023 from [20])

In our experimentation we deployed an nVidia GTX TitanX (3072 CUDA cores) graphics card for computer vision, as well as Keras, Pandas, NumPy, CV2 and TensorFlow python ML libraries. For optimisation, *Adam optimizer* was selected due to its good performance in deep learning tasks [21]. We have deployed hyperparameter tuning and early stopping mechanisms, as well as dropout normalization, to avoid overfit - early stopping technique is used if the model training is automatically interrupted if no improvement is observed in the control data for a pre-determined number of epochs (repetitions). Our model was fine-tuned (i.e., change the model parameters, add extra layers to the neural network, or change the sample size during training) according to a VGG16 pre-trained model [22].

Model evaluation was accomplished by estimating accuracy and loss during both training and validation. Accuracy measures the ratio of the model's correct predictions, while loss measures the deviation between the actual and predicted values, justifying the model performance. Very low training loss and significantly higher validation loss, reveal *overfitting* (or overtraining-the model has learned to predict the training data, but does not generalize to new data). On the other hand, if both training and validation loss are high, then we probably have *underfitting* (the model has not learned enough from the training data and cannot predict accurately either the training data or the test data.

3 Results and Findings

Our model was evaluated by measuring its performance on the test data using various metrics such as accuracy, recall, precision, F1-score and ROC curve Our model achieved an accuracy of 50%-55% on the AffectNet dataset and 60% on the AffectFer dataset, which are quite competitive for the specific dataset [9]. The model showed a steady decline in loss over the epochs, indicating effective learning (Fig. 2).

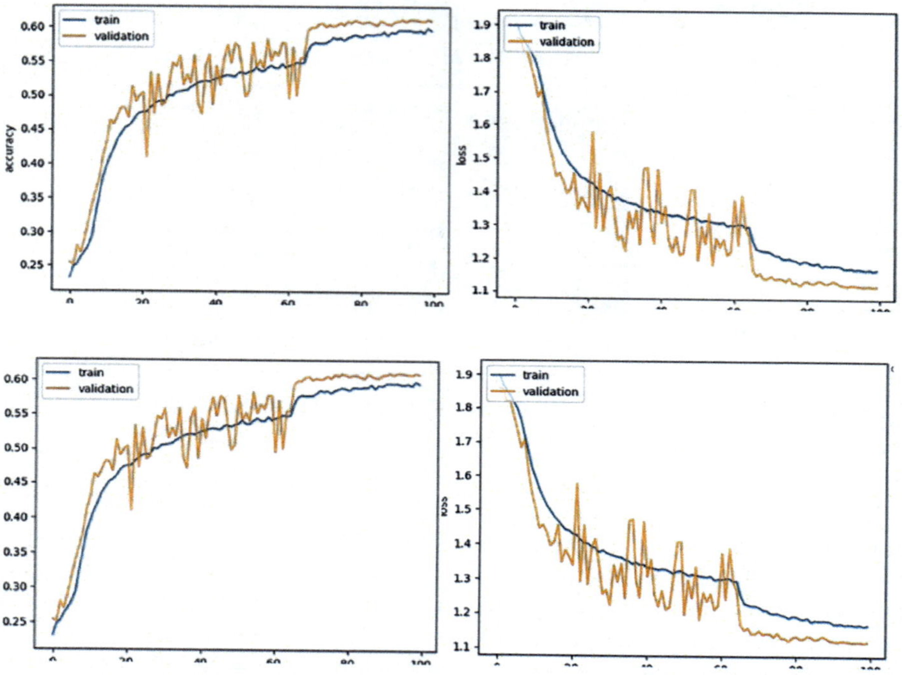

Fig. 2. AffectFer Model Accuracy & Loss curves in both training (top) and test data (bottom)

AffectFer model improved performance by 5–8%. Recognition of joy, sadness and fear achieved better accuracy, in contrast with surprise and anger, mainly due to training data inadequacy, or difficulty in recognizing and classifying certain emotions from images. Deployment of fine-tuning techniques such as data augmentation and hyperparameter tuning improved more the model performance.

Facial emotion recognition is subject to bias difficulties due to the subjectivity of human emotions. Our model interprets facial expressions considering age, gender, or ethnicity variations, as well as lighting, and background conditions, which are difficult to justify during training. Furthermore, DL can contribute significantly to facial emotion recognition since DL algorithms are quite accurate. However, emotion recognition from facial expressions requires from the respondent to stand in front of the camera -sometimes still for 1 sec- to avoid noise from other parts of the body i.e., hands. As a result, much experimentation and fine-tunning might be required, to generalise findings. Our findings also revealed that model's accuracy is highly dependent on the quality of the images used to train the model. Although AffectNet, is quite large dataset, it is difficult to exploit it due to its noise (many low-resolution images, or images with unclear or even wrong data-labels).

4 Conclusions and Next Steps

In this work, we studied the integration of deep learning in real-time emotion recognition from facial expressions. We focused on a CNN model trained on a hybrid dataset of AffectNet and Fer2013. Our findings indicate that emotion recognition from images is a hard task, even for humans, since emotions are complex, often difficult to describe with labels [23]. Even if the model does not recognize exactly the emotion expressed in an image, it can provide useful information about the general feeling that the image evokes in the viewer - often known as opinion mining. Our model can be deployed for analysing people's reaction to various images or improving people's perception of emotions expressed in other understandable media, such as social networks or games. In addition, the analysis of the experiment results can be used to improve the model.

Future steps include investigating other deep learning models, such as recurrent neural networks (RNN) or attention-based models to further improve performance. RNNs can capture dependencies and long-term relations between different frames of face images, recognizing subtle changes in facial expressions over time. Attention models also enrich models with the capacity to focus on specific parts of the face image, improving performance and reducing computational requirements. Finally, another advancement is to integrate sound, speech, or hand gestures close to the face to improve the accuracy and the robustness of emotion recognition.

Acknowledgment. This work is funded by the University of West Attica

References

1. Calvo, R.A., D'Mello, S., Gratch, J., Kappas, A. (eds.) The Oxford handbook of affective computing. Oxford University Press, Oxford, New York (2015). ISBN: 978-0-19-994223-7
2. Cai, Y., Li, X., Li, J.: Emotion recognition using different sensors emotion models, methods and datasets: a comprehensive review. Sensors **23**(5), 2455 (2023). https://doi.org/10.3390/s23052455. Feb.
3. Li, S., Deng, W.: Deep facial expression recognition: a survey. IEEE Trans. Affective Comput. **13**(3), 1195–1215 (2022). https://doi.org/10.1109/TAFFC.2020.2981446. Jul.
4. Pal, S., Mukhopadhyay, S., Suryadevara, N.: Development and progress in sensors and technologies for human emotion recognition. Sensors **21**(16), 5554 (2021). https://doi.org/10.3390/s21165554. Aug.

5. Heo, S., Kwon, S., Lee, J.: Stress detection with single PPG sensor by orchestrating multiple denoising and peak-detecting methods. IEEE Access **9**, 47777–47785 (2021). https://doi.org/10.1109/ACCESS.2021.3060441
6. Spezialetti, M., Placidi, G., Rossi, S.: Emotion recognition for human-robot interaction: recent advances and future perspectives. Front. Robot. AI **7**, 532279 (2020). https://doi.org/10.3389/frobt.2020.532279. Dec.
7. Mollahosseini, A., Hasani, B., Mahoor, M.H.: AffectNet: a facial expression database for valence and arousal recognition. IEEE Transactions on Affective Computing **10**(1), 18–31 (2016)
8. FER2013 Dataset: available (June 23, 2023) in https://datasets.activeloop.ai/docs/ml/datasets/fer2013-dataset
9. AffectNet Benchmark: (Facial Expression Recognition (FER))'. https://paperswithcode.com/sota/facial-expression-recognition-on-affectnet, accessed 20 Apr. 2023
10. Goodfellow, I., Bengio, Y., Courville, A.: Deep learning. In: Adaptive computation and machine learning. The MIT Press, Cambridge, Massachusetts (2016)
11. Ekman, P., Friesen, W.V.: Facial Action Coding System (Jan. 14, 2019). https://doi.org/10.1037/t27734-000
12. Camras, L.A., Allison, K.: Children's understanding of emotional facial expressions and verbal labels. J Nonverbal Behav **9**(2), 84–94 (1985). https://doi.org/10.1007/BF00987140
13. Kanade, T., Cohn, J.F., Tian, Y.: Comprehensive database for facial expression analysis. In: Proceedings Fourth IEEE International Conference on Automatic Face and Gesture Recognition (Cat. No. PR00580), pp. 46–53. Grenoble, France: IEEE Comput. Soc (2000). https://doi.org/10.1109/AFGR.2000.840611
14. Nojavanasghari, B., Baltrušaitis, T., Hughes, C.E., Morency, L.-P.: EmoReact: a multimodal approach and dataset for recognizing emotional responses in children. In: Proceedings of the 18th ACM International Conference on Multimodal Interaction, pp. 137–144. ACM, Tokyo Japan (Oct. 2016). https://doi.org/10.1145/2993148.2993168
15. Livingstone, S.R., Russo, F.A.: The ryerson audio-visual database of emotional speech and song (RAVDESS): a dynamic, multimodal set of facial and vocal expressions in North American English. PLoS ONE **13**(5), e0196391 (2018). https://doi.org/10.1371/journal.pone.0196391. May
16. Huang, Y., Chen, F., Lv, S., Wang, X.: Facial Expression Recognition: A Survey. Symmetry **11**(10), 1189 (2019). https://doi.org/10.3390/sym11101189. Sep.
17. Koujan, M.R., Alharbawee, L., Giannakakis, G., Pugeault, N., Roussos, A.: Real-time facial expression recognition. "In The Wild" by Disentangling 3D Expression from Identity (2020). https://doi.org/10.48550/ARXIV.2005.05509
18. Li, C., Yang, M., Zhang, Y., Lai, K.W.: An intelligent mental health identification method for college students: a mixed-method study. IJERPH **19**(22), 14976 (2022). https://doi.org/10.3390/ijerph192214976. Nov.
19. Bozinovski, S.: Reminder of the first paper on transfer learning in neural networks, 1976. IJCAI **44**(3) (Sep. 2020). https://doi.org/10.31449/inf.v44i3.2828
20. Mizgajski, J., Morzy, M.: Affective recommender systems in online news industry: how emotions influence reading choices. User Model User-Adap Inter **29**(2), 345–379 (2019). https://doi.org/10.1007/s11257-018-9213-x. Apr.
21. Zhang, Z.: Improved adam optimizer for deep neural networks. In: 2018 IEEE/ACM 26th International Symposium on Quality of Service (IWQoS), pp. 1–2. IEEE, Banff, AB, Canada (Jun. 2018). https://doi.org/10.1109/IWQoS.2018.8624183
22. Simonyan, K., Zisserman, A.: Very Deep Convolutional Networks for Large-Scale Image Recognition. arXiv, (Apr. 10, 2015). Accessed: 14 May 2023. [Online]. Available: http://arxiv.org/abs/1409.1556

23. Feidakis, M.: A review of emotion-aware systems for e-learning in virtual environments. In: Formative Assessment, Learning Data Analytics and Gamification, pp. 217–242. Elsevier (2016). https://doi.org/10.1016/B978-0-12-803637-2.00011-7
24. Kalsum, T., Anwar, S.M., Majid, M., Khan, B., Ali, S.M.: Emotion recognition from facial expressions using hybrid feature descriptors. IET Image Processing **12**(6), 1004–1012 (2018). https://doi.org/10.1049/iet-ipr.2017.0499. Jun.

Technical University of Crete February 2023 Readers' Satisfaction from Online News Websites

Klouvidaki Maria(✉), Tsafarakis Stelios, and Grigoroudis Evangelos

Technical University of Crete, Chania, Greece
{mklouvidaki,tsafarakis,egrigoroudis}@tuc.gr

Abstract. As online journalism is expanding, the requirements of the readers are increasing. The need for engaging and sustaining online users/readers is more important than ever. This study aims to investigate the level of readers' satisfaction towards the services of news websites in Greece. Through a market survey 357 responses to a specifically designed questionnaire were collected and analyzed with the use of Multicriteria Satisfaction Analysis, for interpreting data and assessing the quality level of the components, the services, and the information of online news providers. The results indicated that the readers' satisfaction fluctuated at moderate levels. Users visit websites daily to be informed about current affairs even though the quality of the news needs to be improved. Therefore, information must be addressed in a more effective way.

Keywords: News · online journalism · satisfaction · readers · MUSA · Greece

1 Introduction

The success and popularity of a blog or website depends on the satisfaction of its readers. Lin and Sun (2009) showed that the higher the level of customer's satisfaction, the higher his/her e-loyalty. For Lu and Lee (2010) the value of a blog is influenced by repeat readers and content sharing. Loyal blog readers consume time and effort on their preferred blogs. As a matter of fact, loyal readers tend to suggest their favorite blog to friends and relatives, to convince people to join or subscribe to the blog. According to Hsu et al. (2014), blogs can win readers' loyalty by meeting their demands and preferences. Considering the above, this study aimed to investigate if the readers are satisfied with the online news websites in Greece. The findings of the study are expected to contribute to the establishment of online reader satisfaction criteria, aiming to improve the overall quality of online information in Greece. Furthermore, results will assist online news providers to reconsider their marketing strategies to attract more visitors. Considering the key role that information plays in democratic societies the results gain growing importance.

K. Kabassi et al. (Eds.): NiDS 2023, LNNS 783, pp. 52–61, 2023.
https://doi.org/10.1007/978-3-031-44097-7_5

2 Literature Review

Consumer loyalty is about repeated purchases, so readers' loyalty is about repeated visits to a blog. The study that was conducted by Hsu et al. (2014) identifies three parameters for blog readers' satisfaction: recreation, information, and social exchange. These three dimensions are the keys for the increase of blog loyalty. The main goal of a blog is to convince its readers that it can satisfy the information purposes better than other blogs and at the same time it can attract their interest. Additionally, when a blog can achieve readers' satisfaction, it has also the ability to develop attitudinal and behavioral loyalty. Hence, Nambisan and Nambisan (2008) talked about social exchange, which is the social and relational positive consequences of the interaction between bloggers and readers. Moreover, a plus to the aforementioned information is that many readers through accessing the blogs try to satisfy some recreational needs. Recreational needs are the pleasure, the experience, and the fun someone can receive through participation in a blog. This psychological joy provides the feeling of delight that can raise engagement and revisits for a blog. According to Li (2011), online readers do not repeatedly visit websites because they find them enjoyable. It is more important (Li, 2011) when a reader can fulfill their recreational needs to visit again and again the same social network. When blog readers are satisfied with the content they acquire, they tend to admire the blog, and bloggers (if they receive positive feedback) believe that they receive valuable information. In addition, the quality and the content of a blog may increase the readers' belief that the blog is expertized in the field of their interest. Thus, a blog reader could be pleased with a blog or a website if it meets his/her needs for social approval. That situation can lead to the enhancement of their preference and attitude toward a blog. When there are few opportunities for readers to interact with others and share ideas, they prefer to be more active to stay current. Therefore, favorable attitudes are enhanced by subjective norms and behavioral loyalties by the blogs that many people suggest. Apart from this, when the content is accessible, easy to share, and provides solutions to the readers, it can achieve the users' loyalty. Readers can also fulfill their goals, and gain the information they need, by sharing useful links. As a result, when the readers gain everything they need, they will visit the website again and again. Once the value of a blog/website is recognized by the users, a relationship is about to be built up with other users and bloggers. It should be highlighted that Hassan et al. (2015) conducted a survey to find out at which level the online versions of newspapers can satisfy the informational needs of the readers. The power of the internet has increased the accessibility of users to online information (Patel, 2010). According to the Internet world stats, Nigeria has the biggest group of internet users in the African continent. Nigeria has 67.3 million online users; Egypt has 43 million and South Africa has 23.7 million. The study of Hassan et al. (2015), is based on the uses and gratification theory, which was first used by Elihu Katz in 1959. According to this theory, individuals are free to make their choice and read the media they prefer, and individuals are well-informed about the incentives around entertainment and information. These incentives are used like guides because they can satisfy the needs of the users. Importantly, the media preferences are combined with the special needs and characteristics of the users, the content of the blog, and finally the interaction of readers with the content. Hassan et al. (2015), claim that readers choose their preferred newspaper according to their satisfaction with the shared content.

Apart from this, readers have the privilege to choose the media messages that gratify them based on their information and news needs. They can choose between print and online newspapers. Following Okonofua's (2012) survey method, 50% of readers read online newspapers very often and 7.5% read online newspapers very rarely. This survey claims that online papers do not satisfy the readers' needs as much as print newspapers. Additional information comes from the survey conducted by Mathew et al. (2013). The survey indicated that print newspapers started to decline when online newspapers appealed. At the same time, Ekareafo et al. (2013), suggested the management of the newspapers. They argued that the most important thing for the quality of print newspapers is the combination of technical and editorial decisions. The existing literature review supports the hypothesis that satisfaction is dependent on many factors. The aim of this survey is to determine the readers' satisfaction with online websites.

3 Method and Results

The MUSA system is a customer-oriented tool for evaluating service/product quality. The survey-based data are used to analyze customer satisfaction, and the data are then disaggregated using an original method of preference disaggregation. It primarily relies on ordinal regression techniques to implement the multicriteria analysis (Grigoroudis and Siskos, 2010). The presented satisfaction survey concerns the readers' satisfaction survey of online news websites. The survey took place in Greece and was conducted within the period January-February 2023. The final input data consist of 357 questionnaires from online readers all over Greece. Most readers are highly educated, which means that their level of knowledge is above average, so they are people who have an opinion on issues that concern society and current affairs. They have strong criteria for satisfaction since most of them are quite young, between 26–49 years old.

4 Overall and Partial Satisfaction

As shown in Table 1, overall, readers are somewhat dissatisfied at 17.8%, 27.8% are neutral and 37.8% are somewhat satisfied.

Table 1. Value Functions

	Global	Crit1	Crit2	Crit3	Crit4	Crit5	Crit6	Crit7
Value Functi	0	0	0	0	0	0	0	0
	17,814	18,914	6,815	20,856	15,344	14,700	21,196	20,720
	27,886	37,828	13,630	41,713	30,689	29,400	42,392	41,441
	37,886	56,742	20,445	62,569	46,033	44,100	63,588	62,161
	100	100	100	100	100	100	100	100

Regarding criterion 1 related to the information provided by websites, 18.9% of readers are somewhat dissatisfied, 37.8% are neutral and 56.7% are somewhat satisfied. In criterion 2 as to whether the content is interesting, 6.8% replied that 13.6% were somewhat dissatisfied that they were neutral and 20.4% were somewhat satisfied. Furthermore, criterion 3 concerns the content and whether it is up to date with the current affair. 20.8% are somewhat dissatisfied, 41.7% are neutral and 62.5% are somewhat satisfied. About criterion 4 concerning the interaction of users and journalists, 15.3% are somewhat dissatisfied 30.6% are neutral, and 46% are somewhat satisfied. Criterion 5 concerns the level of knowledge of journalists and shows that 14.7% of users are somewhat dissatisfied, 29.7% neutral and 44% somewhat satisfied. Next, criterion 6 relates to the content of the websites, whether it is entertaining. 21.1% are somewhat dissatisfied, 42.3% are neutral and 63.5% are somewhat satisfied. Criterion 7 deals again with content and whether readers can find solutions by reading a website on issues they face. 20.7% are somewhat dissatisfied, 41.4% are neutral and 62.1% are somewhat satisfied.

As shown in Tables 2 and 3 there is strong dissatisfaction among respondents.

Table 2. Frequencies

	Global	Crit1	Crit2	Crit3	Crit4	Crit5	Crit6	Crit7
FREQUENCIE	0,107	0,177	0,191	0,104	0,140	0,267	0,110	0,121
	0,233	0,247	0,292	0,253	0,272	0,256	0,188	0,174
	0,410	0,329	0,340	0,365	0,374	0,315	0,284	0,340
	0,225	0,188	0,157	0,205	0,174	0,132	0,301	0,295
	0,025	0,059	0,020	0,073	0,039	0,031	0,118	0,070

According to the diagrams, 10.7% of readers are very dissatisfied with the websites, 23.3% are somewhat dissatisfied 41% are neutral, 22.5% are somewhat satisfied and only 2.5% are very satisfied with the overall image of the websites. The results show that there is a serious problem with partial satisfaction. Readers only meet their needs for information, interest, and up-to-date content to a minimum. At the same time, there seems to be a lack of interaction. All these elements raise concerns about the content provided by the news sites. According to the results of the primary survey, the websites do not even meet the basic criteria for the satisfaction of readers.

Table 3. Criteria satisfaction results

	Sat_Indices	Dem_Indices	Imp_Indices	Weights
Global	0.266	0.443		
Crit1	0.337	0.243	0.074	11.103
Crit2	0.118	0.727	0.272	30.814
Crit3	0.406	0.166	0.060	10.069
Crit4	0.276	0.386	0.099	13.686
Crit5	0.219	0.412	0.112	14.286
Crit6	0.469	0.152	0.053	9.907
Crit7	0.431	0.171	0.058	10.135

Based on the surveys studied in the literature review, online newspapers are unable to meet the needs of readers as printed newspapers do. Moreover, the journalistic websites -based on the replies to this questionnaire- have low user loyalty rates and high bounce rates. Since readers are not happy, it makes sense that they are not loyal followers of any website. Furthermore, as it was shown in the literature review when a website can best serve users' needs whilst concentrating on retaining them there, it gets to differentiate itself from the competition. Criterion 1 is generally related to the information received by readers. 17.7% are very dissatisfied 24.7% are somewhat dissatisfied 32.9% are neutral 18.8% are somewhat satisfied and only 5.9% are very satisfied. Next, criterion 2 reflects the content and whether this is interesting. 19.1% are very dissatisfied, 29.2% are somewhat dissatisfied 34%, 15.7% are neutral and 2% are very satisfied. At the same time, criterion 3 examines whether the content of the websites is up to date. 10.4% are very dissatisfied, 25.3% are some- what dissatisfied, 36.5 are neutral, 20.5% are somewhat very satisfied and 7.3% are satisfied. Criterion 4 concerns the interaction of users and journalists. 14% are very dissatisfied 27.2% are somewhat dissatisfied, 37.4% are neutral, 17.4% are somewhat satisfied and 3.9% are very satisfied. In criterion 5 for the knowledge level of journalists, 26.7% are very dissatisfied, 25.6% are somewhat dissatisfied, 31.5% are neutral, 13.2% are somewhat dissatisfied and 3% are very satisfied. This finding indicates that readers do not believe journalists possess a high level of knowledge, which also displays their level of skepticism regarding the accuracy of the material. The news and the pieces that journalists write will not be trustworthy if they are not well informed. Moreover, in criterion 6 for entertainment provided by websites, 11% are very dissatisfied, 18.8% are somewhat dissatisfied, 28.4% are neutral, 30.1% are somewhat satisfied and 11.8% are very satisfied. Most of the readers are quite satisfied with the content they read (somewhat satisfied 30,1%) as they fulfill their demand for recreation. This statis- tic is significant since it shows that readers don't trust journalists' reporting while also showing that they are generally content with entertainment-related content. In other words, news websites don't seem to be fulfilling their essential duty of providing accurate and current news updates. Finally, in criterion 7 on website solutions for everyday issues, 12.1% are very dissatisfied, 17.4% are somewhat satisfied, 34% neutral, 29.5% are somewhat satisfied and 7% are very satisfied. In accordance with criterion 7, readers seem impartial, and as a result, they are neither unsatisfied nor satisfied with any potential answers offered to any problems they may encounter daily. Readers become devoted fans of a medium and even go so far as to spread the word about it to acquaintances and friends, as it has been shown in the research of (Li, 2011) when they find solutions to issues that interest them and that satisfy their demands for entertainment.

5 Overall Satisfaction Analysis

Furthermore, the action diagram in Fig. 1 about performance and importance shows that criteria 4 and 5 are very low in performance and high in importance. If websites wish to improve their services, the criteria with the lowest satisfaction index need to be improved. These efforts should focus on the feedback from bloggers and other readers and on the education level of the journalists.

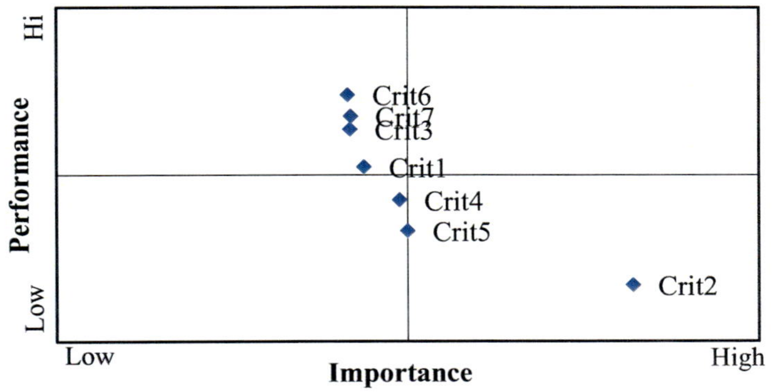

Fig. 1. Action diagram for main satisfaction criteria

Nevertheless, readers are very satisfied with the solutions they find reading online websites and they strongly believe that they meet their recreational needs. Additionally, criterion 2 is high which means that the shared content is quite satisfying for the readers. Criterion 1 is also high in performance and significance, which means that readers are generally happy with the content uploaded by websites as well as with criteria 3, 7, and 6 which are even higher in performance and significance. The content of the websites is well updated, readers get the answers they need on issues that are important, and believe they have such a nice relationship with journalists and other readers. It seems that all readers are happy because (a) they get proper information (b) they meet their needs for really important issues and find solutions and (c) there is good communication and a strong relationship with journalists. This result shows that readers build relationships of trust. As mentioned above, readers want to find on a website everything they care about to become loyal followers. This is the most difficult part; websites must focus on and stand out against competitors. To win a user they will have to provide a set of services that meet most of its needs.

6 Criteria Satisfaction Analysis

Analyzing the partial satisfaction dimensions allows for identifying the criteria characteristics that affect the strong and weak sides of online journalistic websites. Figure 1 is an action diagram that shows that there is not a significant gap between what readers want and what readers get:

- Readers are quite satisfied with the information they get from online websites. It is very important, that Greek online websites provide the appropriate information to the readers through articles and reporting.
- The articles are reliable and up to date regarding readers' ratings and these are the strongest pros of the online websites.
- One improvement margin appears in the feedback from bloggers and other readers. This generic result is combined with the next question about the journalists' education level. These criteria are quite low in satisfaction, which is particularly worrying. It

seems that the websites do not provide such high-level feedback and the bloggers are not well educated to meet the requirements of readers.
- The customers are satisfied because the websites cover their recreational needs. That means that readers can find informative and at the same time fun content.
- It is worthwhile to mention that readers feel satisfied with the solutions they find while reading online websites. It seems that websites can offer solutions for problems readers may face.

Taking into consideration the results of the criteria satisfaction, websites focus their improvement priorities on the content, on the fast and informative feedback, and finally on the education level of the journalists. Journalists should improve their level of knowledge. Generally, there is confusion as on the one hand readers are satisfied with the information, they receive but at the same time are not happy with the level of knowledge of journalists. Consequently, the information they receive could be questioned on many levels by people who are judged to be low-knowledge journalists. Moreover, readers seem to need interaction with bloggers/journalists and other readers. According to the responses websites are trying to approach people with attractive con- tent but they do not pay attention to comments and to interact with each other. Therefore, an action that is an important criterion for readers as shown by the bibliographic review is not sufficiently covered. Furthermore, readers are pleased with the information they read and the up-to-date content, but criteria 6 and 7 are higher. Meeting their leisure needs and providing solutions to problems they are currently facing is more important for them. In fact, it is obvious that websites to a large extent do not meet most readers' needs for it and cannot easily gain user loyalty. In addition, when satisfaction and engagement are not covered, it is not easy to make accurate measurements and predictions on readers' preferences. The improvement diagram in Fig. 2 shows that there is a significant difference between performance and effectiveness. It should be noted that the criteria for the feedback and the education level of bloggers are low in performance level and at the same time quite high at the effectiveness level.

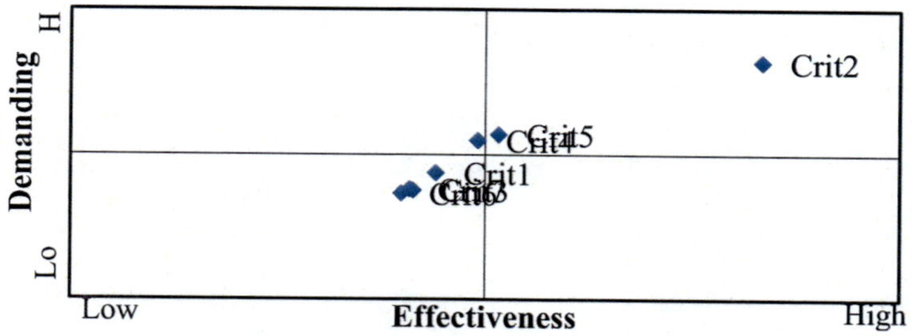

Fig. 2. Improvement diagram for main satisfaction criteria

Additionally, the content and the information provided are very satisfying for readers on the performance side and very low on the effectiveness side. Regarding the satisfaction criteria, the results of the MUSA method reveal the following:

- A leverage opportunity appears for the reliability of the articles. The articles seem to be accurate enough and the information cross-referenced.
- The education level of the journalists is very satisfying, and readers consider it a very important criterion.
- Regarding the general content and the information, readers are dissatisfied. Improvement efforts should focus on these, especially due to their low level.
- Similarly, according to the readers' criteria, the articles are not up to date. The satisfaction index is low so the websites should check and upgrade the articles.
- Moreover, the criterion about the recreational needs of the readers has a low satisfaction index. Readers are not satisfied because the websites do not meet their needs. This is such a huge con for a website as it could bring high levels of bounce rate and low user loyalty.
- Another interesting finding related to the solutions provided needs to be mentioned. It seems that readers show low satisfaction with the solutions they find on websites for different issues they may face.

Thus, websites should invest more in content and information. They should pay more attention to the readers' satisfaction criteria if they really want to increase engagement and acquisition.

It is important to mention that according to the improvement chart, readers are not satisfied with the content as it is not updated as often as it should be and therefore not accurate. This shows that users feel and probably think that the information they read is incorrect and current, implying they really aren't receiving the right information. In contrast, criterion 1 focuses on the broad information a website provides. As low to high, this criterion has indeed been ranked. Information cannot be regarded as accurate or reliable if it's not updated. At the same time, the improvement chart shows that websites should invest more in entertainment content. Readers are not satisfied enough, and the content is not so interesting as to satisfy additional needs beyond mere information. The websites may be overloaded with economic, political, and other social content, so they lose in alternative and special content. Furthermore, websites based on readers' responses do not provide a variety of information and solutions to potential problems they may encounter. This shows that readers don't feel that the media is aware of their real needs and that they're not even trying to meet them. Additionally, interactivity with journalists and other readers is low-to-high and linked to the fact that readers do not feel that the media meet their needs to the maximum. A website doesn't have a close relationship with readers if it is unable to comprehend users' demands. If there is interaction, however, websites frequently get reader criticism and input, and when they pay attention to them, they develop and fulfill a few of the aforementioned needs. We can tell that readers are only modestly satisfied with the websites (26.6%) by looking at the two charts in their entirety and the average satisfaction index. The sixth criterion (recreational needs) as well as the second solution they learn about by reading websites and the third (updated content) are the ones that people are most satisfied with. Moreover, the criteria with the lowest satisfaction rates are 2nd, meaning that readers do not find interesting content, followed by 5th and 4th for the low - as readers see - level of knowledge of journalists and lack of interaction with journalists. All these results are interesting, as on the one hand readers choose the websites for their entertainment and to find different kinds of solutions, and on the other, they do not consider the content reliable, or interesting, nor do they consider the level of journalists high. As a result, the conclusion that comes if you consider the average efficiency index of 4.43%, together with the effectiveness rate of each criterion, clearly shows that readers choose the news sites not because they really want to be informed but want to have fun and get other things out of the process. They are more interested in reading about possible solutions to various issues they face and in obtaining various information. They do not trust what journalists obviously write on important issues since they consider that they have a low level of education while finding the general content of the websites boring. At the same time, the average demand rate reflects the satisfaction of readers at 4.43%, so readers' expectations are only met at 4.43%. Websites as highlighted in the Okonofua's survey (2012), cannot meet the needs of readers as printed newspapers do.

7 Discussion

To sum up, the survey has shown that many of the research cases that were essentially the criteria for satisfaction do not meet readers' needs to the same extent. Readers are in total 26.6% satisfied with news websites, a percentage not too high. As pointed out by the Researchers Lin and Sun (2009), a high level of satisfied e-customer leads to better e-loyalty. Therefore, readers, since they are not satisfied to a large extent, do not become loyal users. Moreover, as Lu and Lee (2010) state, the value of a blog is influenced by repeat readers and content sharing. When readers are dissatisfied, they don't share the articles they read and cannot become repeat readers. At the same time, our research showed that readers doubt the credibility of the articles they read, so they have no trust in the media and journalists. According to Huang et al. (2008), many readers ignore outdated information, so bloggers should take great care to provide accurate information if they want to keep receiving repeat visitors. This criterion is also linked to criterion 5 which is low in the average satisfaction index. Readers are not satisfied with the level of knowledge of journalists, and since what they write is not accurate, there is a piece of controversial information. Additionally, less than half of the 33.7% of respondents are satisfied with the information they receive from the websites. Although 40.6% of respondents think the articles are current and up to date, the accuracy and depth of knowledge seem to be in question. The study that was conducted by Hsu et al. (2014), argues that blogs can achieve readers' loyalty through the satisfaction of their needs and through the target of their preferences. In our own survey, it was found that46.9% of readers are satisfied and think they are entertained through browsing and visiting the websites. Nambisan and Nambisan (2008) talked about social exchange and recreational needs and Oliver (1997) argued that satisfaction provided a pleasure of consumption related to fulfillment. Criterion 4, which relates to interaction and feedback, collected 27.6% of positive responses. Readers do not feel that there is communication and a tied relationship between journalists, so they consider that there is no connection with the website. This means they can't trust a website that doesn't try to make bonds with its readers, to understand them and try to satisfy their needs.

References

Ekhareafo, D.O., Asemah, E.S., Edegoh, L.O.: The challenges of newspaper management in information and communication technology age: The Nigerian situation. British Journal of Arts and Social Sciences **13**(1), 1–14 (2013)

Grigoroudis, E., Siskos, Y.: Customer satisfaction evaluation: Methods for measuring and implementing service quality, Vol. 139. Springer Science & Business Media (2010)

Hassan, I., Latiff Azmi, M.N., Engku Atek, E.S.: Measuring Readers' Satisfaction with Online Newspaper Content: A Study of Daily Trust. American Journal of Innovative Research and Applied Sciences (2015)

Hsu, C.P., Huang, H.C., Ko, C.H., Wang, S.J.: Basing bloggers' power on readers' satisfaction and loyalty. Online information review (2014)

Huang, L.S., Chou, Y.J., Lin, C.H.: The influence of reading motives on the responses after reading blogs. Cyberpsychol. Behav. **11**(3), 351–355 (2008)

Li, D.C.: Online social network acceptance: a social perspective. Internet Res. **21**(5), 562–580 (2011)

Lin, G.T.R., Sun, C.C.: Factors influencing satisfaction and loyalty in online shopping: an integrated model. Online Inf. Rev. **33**(3), 458–475 (2009)

Lu, H.P., Lee, M.R.: Demographic differences and the antecedents of blog stickiness. Online Inf. Rev. **34**(1), 21–38 (2010)

Mathew, J., Ogedebe, P.M., Adeniji, S.B.: Online Newspaper Readership in North Eastern Nigeria. Asian J. Soc. Sci. Huma. **12**(2) (2013)

Nambisan, S., Nambisan, P.: How to profit from a better 'virtual customer environment.' MIT Sloan Manag. Rev. **49**(3), 53–61 (2008)

Okonofua, A.G.: Readership of Online Newspapers by Users of Select Cyber Cafés in Uyo Urban. Paper presented at the Pre-Conference of International Federation of Library Association, Mikkeli, Finland (2012)

Oliver, R.L.: Satisfaction: A behavioral perspective on the consumer. Routledge, McGraw-Hill International editions, Marketing & Advertising Series, Singapore, pp. 13–20, 135–137, 391–392 (1997)

Patel, A.: The survival of newspaper in the digital age of communication. Stern, New York University, New York, Thesis, Leonard N (2010)

Artificial Intelligence-Based Adaptive E-learning Environments

Fateh Benkhalfallah(✉) and Mohamed Ridda Laouar

Laboratory of Mathematics, Informatics and Systems (LAMIS),
Faculty of Exact Sciences and Natural and Life Sciences,
Echahid Cheikh Larbi Tebessi University, Tebessa, Algeria
{fateh.benkhalfallah,ridda.laouar}@univ-tebessa.dz

Abstract. Intelligent adaptive e-learning has become an inescapable necessity, a major issue in the last few years, especially with the pandemic that is hitting the world besides the digital and technical development that we are going through today. The shift to adaptive e-learning is crucial and provides an alternative way to traditional learning to tailor learning and uniquely respond to the needs and characteristics of each learner, offering them a personalised learning approach and a better learning experience. Therefore, the focus has shifted to a system based on artificial intelligence to help learners to manage the teaching and learning process, which leads to improving their efficiency, better developing their skills and abilities, saving effort and time, as well as improving their level of motivation. The main objective of this paper is to study and give a comprehensive overview of the latest adaptive e-learning systems and to highlight the role and importance of artificial intelligence in promoting the adaptation of this system while pointing out some future ideas and recommendations whose implications will be of benefit to both students and researchers.

Keywords: Artificial intelligence · Adaptive e-learning · Intelligent e-learning · Innovative learning methods · Educational system

1 Introduction

People born in this century will live their lives in smart cities. For cities to thrive for their citizens, education must be at the heart of their concerns. These cities need educational programs that produce graduates with research and access to knowledge skills, modern and up-to-date knowledge, practical abilities and cooperative tendencies [1, 2].

The development of contemporary and sophisticated communication systems has propelled humanity to a new stage in the field of information and communication technologies. As a result, communication methods have moved from the physical structure that humans used in the form of signals and sounds to the electronic structure based on automated information systems, which has affected the teaching and learning process from a traditional process to an adaptive and intelligent electronic process whose teaching and learning format depends on the interaction between the learner and the

K. Kabassi et al. (Eds.): NiDS 2023, LNNS 783, pp. 62–66, 2023.
https://doi.org/10.1007/978-3-031-44097-7_6

intelligent electronic educational media, making it more effective and efficient and having the characteristic of being sustainable, thus providing better opportunities for all types of communication and reciprocal educational interactions [3, 4].

Intelligent applications and systems have emerged in which artificial intelligence has been used in the structure of those educational systems represented in learning management systems, as their development involves the integration of artificial intelligence technologies. Among the applications of artificial intelligence in learning is adaptive learning, which seeks to adapt the learning system to the abilities, interests and levels of knowledge and skills of the learner, in addition to their speed of learning and completion of tasks based on an integrated methodology to use advanced technology to bring about positive changes in traditional educational methodologies and create a stimulating environment for developing creativity and innovation competences, interactive participation and the development of digital literacy [5, 6].

This paper provides a broad overview of adaptive and intelligent e-learning systems. First, the authors introduce the general context of the study. Next, they present the research problem statement, followed by an outline of the objectives, and then they discuss the state of the art and related work, after which they present the ongoing work, besides the future research directions. Finally, they finish with a synthesis of the major outcomes.

2 Research Problem Statement

In recent years, particularly during the epidemic period witnessed by the world, life has changed completely as the problem of face-to-face education has arisen, which was a major obstacle in light of the effects of the Corona pandemic, including modifying unexpected restrictions. The need to provide technologies to overcome the barrier of time and space and to provide material to the learner at the time allowed by their circumstances, abilities, and interests, prompted us to adopt this problem whose needs have attracted the interest of many researchers mainly in the educational and computer science sectors.

Previous studies have emphasised the importance of integrating future skills into curricula and activating them in the educational process, as well as the need to pay attention to their development in modern technologies and to deal with them using advanced scientific methods similar to artificial intelligence techniques, due to their effectiveness in developing and improving learners' performance, awareness and improving their study methods [7, 8].

Therefore, the main problem of the study is to know how to combine information and communication technologies, modern techniques and models of artificial intelligence to provide an efficient, useful and adaptive e-learning system available anywhere and anytime.

3 Outline of Objectives

The overall vision of the objective is to formulate a strategy for an e-learning system capable of automatically selecting and adapting electronic content, activities, and assessments and presenting them to the learner based on the information the system gathers about the learner before and during the learning process and based on their previous experience, thus providing opportunities for availability and learning at the most appropriate time and place for the learner. Its importance also lies in carrying out the process of generating a unique educational experience for each learner based on the learner's personality, interests and performance to facilitate learning and achieve objectives such as learner's academic improvement, learner's satisfaction, effective learning process. It also aims to develop the cognitive and performative aspects of digital application skills and the development of future information awareness.

Moreover, developing digital skills, training and educational programmes, as well as reviewing and improving school curricula through the use and application of algorithmic techniques based on artificial intelligence, such as machine and deep learning algorithms, to achieve the best experience in education.

Furthermore, focusing on the benefits and forms of applications of artificial intelligence in education, presenting a set of models of learning methods that show the need to provide an adaptive learning environment, in addition to addressing the most important innovative teaching methods that enhance adaptive learning.

4 State of the Art

Despite offering flexible access and learning opportunities, e-learning does not deliver education that is adapted to the requirements, talents, qualities and learning styles of learners. Consequently, adaptive e-learning environments have been developed to offer appropriate education that meets the needs, abilities and characteristics of each learner, thus providing a personalised education for each learner. Many researchers have been interested in the various demands of adaptive e-learning in recent years, most notably in the period of a pandemic that the world is experiencing. In light of this, some work has been done on adaptive e-learning systems [6]. In this section, the authors will provide literature on adaptive and intelligent e-learning systems. The review will cover various related works that have been suggested in this area.

According to [9], the term 'artificial intelligence' (AI) refers to a wide range of complex technologies and algorithms (e.g., data mining, machine learning and automatic natural language processing). Artificial intelligence in education first refers to "intelligent tutoring systems", which look for solutions to problems such as automatically boosting operators' performance. Currently, in [10], the author presented AI as a strategy for exploiting massive data to perform complex processes.

The integration of artificial intelligence into instructional practices is still in its infancy, even though it has been predicted to advance in the evolutionary phase of human applications. The article [11] discusses some categories of educational applications of AI that have drawn interest from a range of scholars.

According to [12], to achieve the target of a "one person, one-course programme", the intelligent teaching method may offer personalised lessons based on students' language, knowledge structure, emotional state, learning style and other features, and generate a customised learning environment through feedback to respond to the different needs of students. Some automatic assessment systems are based on natural language processing technologies like "Project Essay Grade", "IntelliMetric" and "Criterion", which were developed fast and allow for speedy assignment grading and correction. The use of educational robots in classrooms is spreading quickly as an effective teaching tool.

According to [13], many academics in China are also conducting research and analysis on the current situation and potential development of AI education and teaching applications. For instance, some scholars have discussed and expressed optimism for the applications of the consolidation and teaching model, specific learning areas, as well as the future of AI and education in the learner model area.

The difficulties in implementing AI in education are linked to the educational experience, the claiming partnerships between government agencies and educational institutions, safety ethics of cutting-edge technology and technological management of the balanced development of man-machines [14]. Conversely, the field of special education has successfully used AI. By applying some technical approaches, it may be possible to improve the organs of particular people. Their physical or intellectual shortcomings can be addressed. Furthermore, it is beneficial to meet the needs of different special individuals to the greatest extent feasible and to support their individualised learning. Educational objectives, learning strategies, subject matter, instructional models, learning environments and educational resources have all been studied by some scholars concerning future patterns of incorporating artificial intelligence and education. In the not-too-distant future, the further development of AI will also have an impact on instructors' functions [15].

5 Ongoing Works and Future Directions of the Research

E-learning has become a reality nowadays. However, most existing systems do not put enough effort into making the e-learning system as effective as possible for all learners with heterogeneous interests, educational levels and learning speeds.

In this topic, we are in the process of developing our approach to providing an adaptive, convivial, efficient and intelligent online learning environment. We mention some of our contributions which include the development of an adaptive e-learning environment based on artificial intelligence, a generic approach to help learners succeed in their courses and to offer them the appropriate support unit adapted to their preferences, as well as the restructuring of courses to improve their performances.

Online learning is a very broad area of research. However, we intend to focus our efforts and future research orientations in some main directions such as the continuity to work for the research of new adaptation techniques in distance learning systems, the use of different modern technologies, as well as integration in our proposed approach a recommendation system in different directions, besides that, incorporating deep learning techniques to refine the quality of the offered services and improve some tasks of the system.

6 Conclusion

It is difficult to pay attention to the learning progress of each learner in the traditional teaching system based on a unified learning system for all, so it is necessary to create an appropriate environment that is more efficient, usable, widely accessible and permanent and allows learners to learn autonomously outside the limits of time and place and to meet their growing needs, as their educational attitudes can be assessed and analysed for better learning effectiveness and improved performance by providing them with appropriate guidance, which is represented in an e-learning system based on artificial intelligence, which is considered a fundamental objective and imperative for shaping the educational present and moulding its future characteristics, to build and establish an advanced society in line with the implications of the knowledge age. Artificial intelligence also assists in the evaluation to obtain e-learning objectives to improve the impact of efficiency and to manage the teaching and learning process smoothly. The future analysis and recommendations presented in the paper will be useful and can be used to help learners and guide researchers to reflect on modern learning mechanisms and future development in the field of learning to aim for continuous growth.

References

1. Liu, D., Huang, R., Wosinski, M.: Smart Learning in Smart Cities. Springer (2017)
2. Molnar, A.: Smart cities education: an insight into existing drawbacks. Telemat. Inform. **57**, 101509 (2021)
3. Khan, M.A., Khojah, M.: Artificial intelligence and big data: the advent of new pedagogy in the adaptive e-learning system in the higher educational institutions of Saudi Arabia. Educ. Res. Int. **2022**, 1–10 (2022)
4. Truong, H.M.: Integrating learning styles and adaptive e-learning system: current developments, problems and opportunities. Comput. Human. Behav. **55**, 1185–1193 (2016)
5. Dolenc, K., Aberšek, B.: TECH8 intelligent and adaptive e-learning system: integration into technology and science classrooms in lower secondary schools. Comput. Educ. **82**, 354–365 (2015)
6. Benkhalfallah, F., Bourougaa, S.: Adaptive system to improve the acquisition of educational knowledge for learners. Int. J. Org. Collect. Intell. **12**(3), 1–13 (2022)
7. Yang, W.: Artificial Intelligence education for young children: why, what, and how in curriculum design and implementation. Comput. Educ. **3**, 100061 (2022)
8. Pedro, F., Subosa, M., Rivas, A., Valverde, P.: Artificial intelligence in education: challenges and opportunities for sustainable development (2019)
9. Koper, R., Tattersall, C.: New directions for lifelong learning using network technologies. Br. J. Edu. Technol. **35**(6), 689–700 (2004)
10. Khan, M.A.: Netizens' perspective towards electronic money and its essence in the virtual economy: an empirical analysis with special reference to Delhi-NCR, India. Complexity **2021**, 1–18 (2021)
11. Koparan, T., Güven, B.: The effect of project based learning on the statistical literacy levels of student 8th grade. Eur. J. Educ. Res. **3**(3), 145–157 (2014)
12. Moreno-Marcos, P.M., Alario-Hoyos, C., Muñoz-Merino, P.J., Kloos, C.D.: Prediction in MOOCs: a review and future research directions. IEEE Trans. Learn. Technol. **12**(3), 384–401 (2018)
13. Renz, A., Hilbig, R.: Prerequisites for artificial intelligence in further education: identification of drivers, barriers, and business models of educational technology companies. Int. J. Educ. Technol. High. Educ. **17**(1), 1–21 (2020)
14. Khan, M.A., Vivek, V., Khojah, M., Nabi, M.K., Paul, M., Minhaj, S.M.: Learners' perspective towards e-exams during COVID-19 outbreak: evidence from higher educational institutions of India and Saudi Arabia. Int. J. Environ. Res. Public Health **18**(12), 6534 (2021)
15. Guan, C., Mou, J., Jiang, Z.: Artificial intelligence innovation in education: a twenty-year data-driven historical analysis. Int. J. Innovat. Stud. **4**(4), 134–147 (2020)

CoMoPAR: A Comprehensive Conceptual Model for Designing Personalized Augmented Reality Systems in Education

Christos Papakostas[1(✉)], Christos Troussas[1], Panagiotis Douros[2], Maria Poli[3], and Cleo Sgouropoulou[1]

[1] Department of Informatics and Computer Engineering, University of West Attica, Egaleo, Greece
{cpapakostas,ctrouss,csgouro}@uniwa.gr

[2] Department of Social Work, University of West Attica, Egaleo, Greece
pdouros@uniwa.gr

[3] Department of Interior Architecture, University of West Attica, Egaleo, Greece
mpoli@uniwa.gr

Abstract. A challenging undertaking that involves an in-depth comprehension of the educational objectives, the target population, and the available technology is the design and development of personalized augmented reality systems for utilization in education. This paper introduces CoMoPAR, a high-level conceptual model that can be used as a starting point for designing and developing such systems, as a solution to this critical requirement. The User Interface Layer, Augmented Reality Layer, Personalization Layer, Content Layer, Analytics Layer, and Backend Layer are the five layers that constitute CoMoPAR. The combination of these layers enables an efficient and successful method for developing educational systems that address the problems of traditional educational systems. The indicated conceptual model presents a solid framework for developing and developing individualized augmented reality educational environments, that hold the potential to revolutionize the field of education while offering students a more engaging and efficient learning environment. Teachers, programmers, and researchers interested in creating customized augmented reality learning environments may utilize this conceptual framework.

Keywords: Augmented Reality · Educational System · Personalized Training · Conceptual Framework

1 Introduction

An emerging technology named augmented reality (AR) has demonstrated tremendous promise in a variety of fields, including training [1, 2], entertainment, and advertising. It makes it possible to superimpose digital content over the real world, providing users an immersive and participatory experience. Another area of study that has been gaining ‘popularity recently is personalization, which enables systems to adapt to each user's unique needs and preferences, enhancing the overall user experience and better learning outcomes.

K. Kabassi et al. (Eds.): NiDS 2023, LNNS 783, pp. 67–79, 2023.
https://doi.org/10.1007/978-3-031-44097-7_7

By enabling learners to interact with educational content in a more individualized and interactive way, augmented reality, and customization in the context of education [3–8] and training [6, 9–11] have the potential to completely transform conventional learning methods.

However, creating customized augmented reality systems is a challenging endeavor that necessitates a thorough comprehension of the objectives of education, the target audience, and the available technology. A thorough framework is required to direct the design and creation of such systems. Such a framework can be used to construct and create customized AR systems that take into account the particular requirements and preferences of learners.

In view of the above, this study provides a conceptual model for customized augmented reality systems that combines augmented reality with personalization to improve the learning experience. The user interface, augmented reality, personalization, content, analytics, and backend layers are just a few of the various layers that make up the model. Each layer oversees a distinct task, and by combining these layers, an individualized AR system that offers each learner a customized and immersive learning experience is created.

The User Interface Layer delivers a user-friendly interface that makes it easier for users to interact with the augmented reality elements of the system. The Augmented Reality Layer utilizes augmented reality technology's capabilities to offer a comprehensive and engaging learning environment. The Personalization Layer employs data-driven personalization approaches to offer a customized learning environment that takes into consideration each learner's interests and needs. The educational material offered by the Content Layer is specialized to the learner's ability level and learning goals. The Analytics Layer makes use of data analytics methods to offer notions into the educational process and guide customization choices. The Backend Layer is in charge of monitoring system operations and assuring the security and dependability of the system.

The presented conceptual model can be a helpful reference for researchers, educators, and developers who are interested in creating personalized AR systems. The study goes into detail about each of the model levels, describing their various elements and purposes. The study also explores the possible advantages and drawbacks of tailored augmented reality (AR) systems in education and training.

In conclusion, the proposed conceptual model offers a solid framework for creating customized augmented reality learning systems. The model presents a comprehensive strategy that takes into account all of the system's parts and how they work together to create an enjoyable learning experience. The combination of augmented reality (AR) and personalization offers an opportunity to fundamentally change the educational landscape by offering students the opportunity to utilize a more effective and engaging learning environment.

2 Literature Review

The utilization of augmented reality (AR) as a tool for enhancing teaching and learning in educational settings has demonstrated potential. Frameworks for the design process of effective augmented reality (AR) systems for education have been proposed. We study the literature on conceptual frameworks for augmented reality (AR) systems in education.

The ARCS (Attention, Relevance, Confidence, Satisfaction) model by [12] is one of the first frameworks presented for creating AR systems in education. The ARCS model was utilized in the creation of AR learning environments to enhance learner motivation and engagement. It bases its structure on four essential factors that affect learner motivation [13].

The SAMR (Substitution, Augmentation, Modification, and Redefinition) model by Puentedura (2010) is another framework presented to build AR systems in education. The SAMR model offers a framework for evaluating how much technology, such augmented reality, is incorporated into the teaching and learning process. To make sure that AR is used in a way that improves the learning experience, the SAMR model has been employed in the design of educational AR systems [14].

The Planetary System GO system, developed by [15], has a three-part architecture: Platform server, Web application, and Mobile app. Database on the system's server is in the process of managing all the data in the system.

In conclusion, as the potential of augmented reality (AR) in education continues to be further explored, the incorporation of frameworks in the creation of AR systems in education becomes increasingly crucial. The frameworks covered in this review offer suggestions for creating efficient AR systems that improve teaching and learning. The usefulness of these frameworks in the design of AR systems in education requires more investigation.

CoMoPAR incorporates layers including user interface, content, analytics, and back-end layers to provide a thorough and holistic approach to building tailored AR systems for education. In contrast, several of the studies that have been published have only focused on particular facets of AR design, namely user interface design or content production.

Additionally, CoMoPAR emphasizes the significance of individualization in AR systems for education by including a personalization layer which takes into account particular learner preferences and characteristics. While some of the references acknowledge the significance of personalization in AR systems, CoMoPAR offers a more thorough and in-depth method of putting adaptation into practice.

Finally, CoMoPAR emphasizes the use of analytics in augmented reality (AR) systems for education in order to analyze student progress and modify the system as necessary. This differs from some research, which pay less emphasis on the application of analytics in AR design.

3 Presentation of CoMoPAR

An advanced and layered architecture known as the Conceptual Model of Personalized Augmented Reality (AR) Systems (CoMoPAR) integrates a number of technologies in order to create an immersive and interactive learning environment. A personalized AR system's general architecture (Fig. 1) is as follows:

1. User Interface Layer: This layer offers the user with the means to communicate with the system. It covers products having an AR-capable camera, such as smartphones, tablets, smart glasses, and other wearable devices. In order to track the user's movements and orientation, the user interface layer additionally has sensors including GPS, accelerometers, and gyroscopes.
2. Augmented Reality Layer: The second layer, known as augmented reality, is in charge of superimposing digital data over the real world. It projects digital content onto real-world items after using computer vision algorithms to find and track them. Software libraries like ARKit, ARCore, Vuforia, and Wikitude are part of this layer.
3. Personalization Layer: The personalization layer is in charge of adapting the content and user experience for each individual user. User profiles, preferences, and learning goals are all included. This layer analyzes user activity utilizing machine learning techniques and offers customized suggestions.
4. Content Layer: The information presented on the augmented reality layer is delivered by this layer. It comprises text, audio, photos, and other types of multimedia material. Quizzes, games, and simulations are also a part of this layer of instruction.
5. Analytics Layer: This layer captures information on how users engage with the system and how effectively the educational content performs. This information is used to enhance the personalisation layer and enhance system performance as a whole.
6. The user data, instructional content, and system performance metrics are all stored and managed on servers and databases in the backend layer. Additionally, it has APIs for integrating with exterior systems and services.

Overall, the hardware, software, and data analytics that make up the personalized AR system architecture are intricately combined. Its fundamental objective is to deliver a tailored learning experience that blends the best elements of the real world and the digital world.

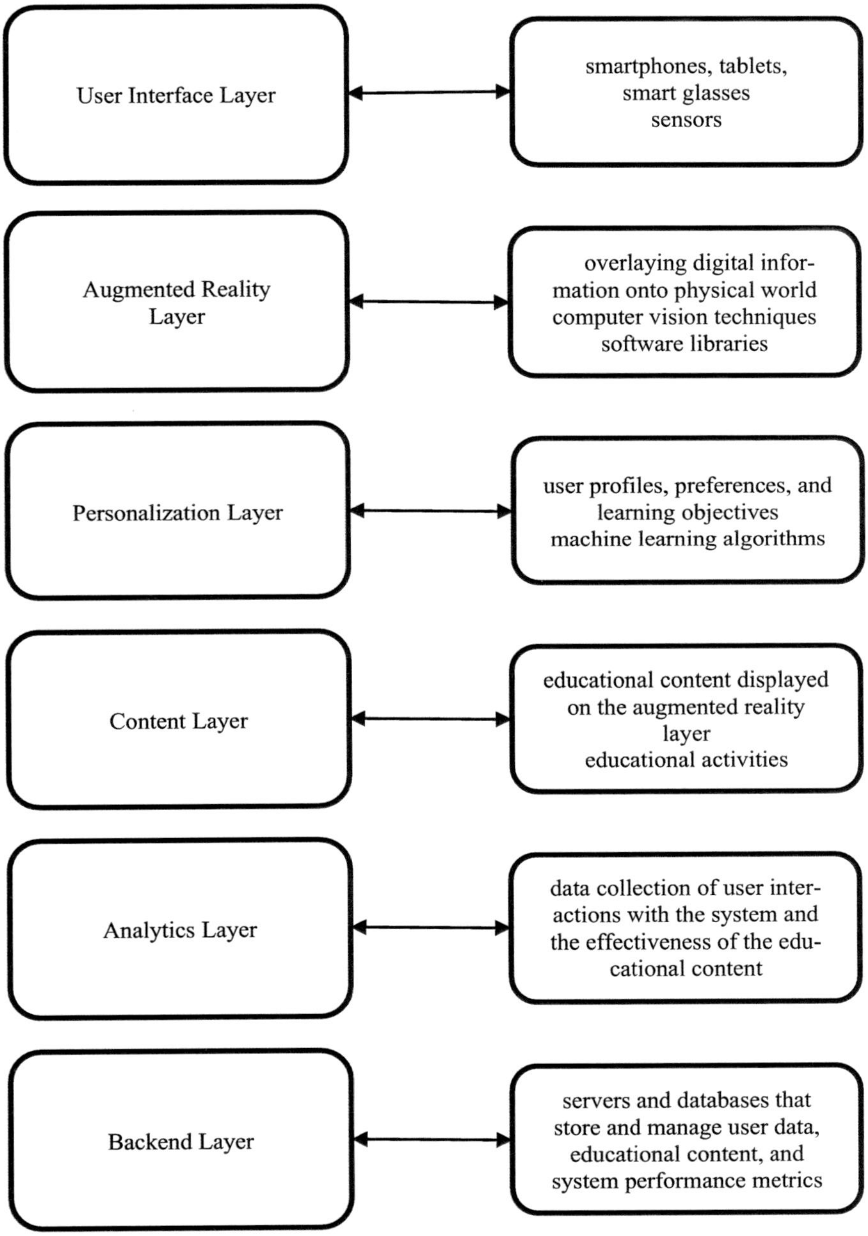

Fig. 1. Layered System Architecture

4 Discussion of CoMoPAR

4.1 User Interface

The first layer in the conceptual framework for customized augmented reality systems is the User Interface (UI) layer. It offers the user interface and has the devices and sensors needed to make this engagement with the system possible. For learners to interact with the educational information in an immersive and interactive way, an intuitive and simple user interface (UI) is essential.

The User Interface layer's components are as follows:

1. Devices: Any wearable device with an AR-capable camera is included in the UI layer, as are smartphones, tablets, smart glasses, and other mobile devices. Users interact with the augmented reality system primarily through these devices. Augmented Reality Layer. The device choice is determined by the particular use case and the intended audience. For instance, a smart glass-based AR system would be better suited for industrial training whereas a smartphone-based AR system might be better suited for younger learners.
2. Sensors: The UI layer also incorporates sensors like GPS, accelerometers, and gyroscopes to track the user's movements and orientation. For the system to be able to provide the user with accurate and current data, these sensors are required. For instance, the GPS sensor can be used to offer location-based information, while the accelerometer and gyroscope can be used to detect the user's movements and enable gesture-based interactions.
3. User Interface Design: An intuitive and interesting user experience depends on the user interface's design. The UI layer includes both the user experience (UX) design and the graphical user interface (GUI). The GUI contains the design, color scheme, and visual elements like icons, buttons, and menus that are utilized to present the user with educational content. The complete user experience is considered during UX design, including the system's flow, usability, and the user's ability to accomplish their learning goals.
4. Interaction Design: Users' interactions with the system are the main priority of interaction design. It entails the creation of gestures, voice instructions, and other forms of interaction that let students interact with the lesson material. For learners to concentrate on the instructional content rather than the technology, a fluid and intuitive user experience that is based on interaction design is essential.
5. Accessibility: To enable students with particular requirements or impairments to access the instructional content, the UI layer should also take accessibility elements like font size, color contrast, and text-to-speech capabilities into consideration.

In conclusion, the User Interface layer is essential for producing a seamless and simple user experience that enables students to interact and immerse themselves in the instructional content. Devices, sensors, interaction design, user interface design, and accessibility features are all included. To provide the best learning environment, the UI layer design should take into account the target audience's unique demands and preferences.

4.2 Augmented Reality Layer

The theoretical framework for customized augmented reality systems has two layers, the second of which is Augmented Reality (AR). It is in charge of establishing a virtual layer that supplements the physical one so that students can engage with digital content in a natural setting. The AR layer is essential for developing an immersive and engaging learning experience that enables students to interact more deeply with the instructional material.

The Augmented Reality layer is made up of the following elements:

1. Tracking: The tracking capabilities of the AR layer allow the system to keep track of the person's location and orientation in the actual world. There are various tracking technologies, including SLAM (Simultaneous Localization and Mapping) technology, marker-based tracking, and markerless tracking. While markerless tracking and SLAM technology use computer vision algorithms to track the user's location and orientation based on visual features in the real world, marker-based tracking involves using predefined markers to track the user's location and orientation.
2. Rendering: Another element of the AR layer that enables the system to superimpose digital content on the real world is rendering technology. Model-based and image-based rendering technologies are the two main categories. Instead of projecting digital content onto a real-world image captured by the device's camera, model-based rendering entails creating 3D models of the surrounding environment and superimposing digital content on these models.
3. Object Recognition: The AR layer also has object identification tools that let the system identify actual items and add digital material to them. In order for the system to give digital content that is contextually appropriate, object identification technology uses computer vision algorithms to recognize actual items in the real environment.
4. Calibration: The calibration technology incorporated in the AR layer allow the system to precisely align the virtual material with the actual world. In order to ensure that the virtual content appropriately corresponds with the real-world context, calibration technology entails altering the location, orientation, and scale of the virtual content.
5. Occlusion: Another feature of the AR layer is occlusion technologies, which allow the system to produce believable occlusions in which virtual things are concealed behind actual ones. By identifying real-world objects and producing realistic occlusions, occlusion technology increases the degree of authenticity of the augmented reality experience.

In conclusion, the Augmented Reality layer is essential for developing an immersive and dynamic learning environment that enables students to interact more deeply with the instructional material. It covers occlusion technologies, object recognition technologies, tracking technologies, rendering technologies, and technologies for object recognition. To provide the best learning environment, the AR layer should be designed with the target audience's particular needs and preferences in consideration.

4.3 Personalization Layer

The third layer in the conceptual framework for customized augmented reality systems is the Personalization layer. It is in charge of designing a personalized and flexible learning environment that takes into account the special requirements and preferences of every student. By adapting the educational content to each learner's unique knowledge level, learning style, and interests, the personalization layer is essential for improving learner engagement, motivation, and learning results.

The Personalization layer's components are as follows:

1. Learner Modeling: The Personalization layer contains technology for learner modeling, allowing the system to generate a model of the learner's preferences, abilities, and knowledge. Learner modeling technology builds a profile of the learner's knowledge level, learning style, and interests using information gathered from the learner's interactions with the system, such as their responses to questions and their browsing history [16].
2. Content Adaptation: The Personalization layer also contains technologies for content adaptation, which allow the system to modify the instructional materials in accordance with the needs and preferences of the learner. Using content adaptation technology, educational materials are chosen and delivered according to the learner's interests, learning preferences, and level of knowledge. To increase the learner's engagement and comprehension, the system can alter how the educational content is presented, such as changing the text size, color, and layout.
3. Feedback and Guidance: The Personalization layer also includes technology for feedback and guidance, which give students individualized input and direction on their academic achievement. Technology used for feedback and guiding involves giving students feedback on their performance, such as pointing out areas where they may improve or suggesting more reading. Additionally, the system can offer advice on the learning process, such as recommending study techniques that are specific to the learner's learning preferences.
4. Gamification: The Personalization layer also contains gamification technologies, which increase motivation and learner engagement by introducing game-like elements into the educational process. Gamification technology involves rewarding and incentivizing learners to meet learning objectives and goals by giving them points, badges, and leaderboards.
5. Social Interaction: The Personalization layer also provides tools for social interaction that let students communicate with one another as well as teachers. The use of social interaction technology entails giving students access to communication and collaboration tools including chat, forums, and video conferencing. Additionally, the system can support peer-to-peer learning and give students the chance to get feedback and direction from teachers.

In conclusion, the Personalization layer is essential for developing a personalized and adaptive learning environment that caters to the particular requirements and preferences of individual learners. It covers technology for learning modelers, content adapters, feedback and guiding systems, gamification systems, and social interaction systems. To provide the best learning environment, the Personalization layer's design needs to take into consideration the target audience's unique needs and preferences.

4.4 Content Layer

The theoretical framework for customized augmented reality systems has two layers, the second of which is the Content Layer. It is in charge of supplying the instructional materials that are displayed to the learner via the augmented reality interface. The content in this layer, which might include text, photographs, videos, and 3D models, is created to meet the system's learning objectives.

The Content Layer is made up of the following elements:

1. Content Creation: In the Content Layer, subject-matter experts create educational content. The system's learning objectives and target audience must be met by the material. Creating educational content might involve creating study tools like simulations, interactive tests, and instructional movies.
2. Content Curation: The collection of current instructional content is another aspect of the Content Layer. To enhance the instructional materials developed by subject matter specialists, the system can incorporate pre-existing educational content, such as textbooks, academic articles, and web resources. In the curation process, content is chosen and arranged according to its quality and relevance.
3. Content Management: The Content Layer involves the administration of the system's educational content. A content management system that permits the development, storage, and retrieval of instructional content is a requirement for the system. Version control, access control, and content backup functions must all be offered by the content management system.
4. Content Localization: The Content Layer also provides educational content localization to accommodate students from various cultural backgrounds. The educational material must be localized to meet the target audience's cultural, linguistic, and educational needs. This may entail customizing the material to local educational standards, translating it into several languages, and changing it to match regional traditions and customs.
5. Content Assessment: The Content Layer also contains an evaluation of the educational material to make sure it is successful in accomplishing the system's learning goals. The evaluation process may include gathering input from students and subject-matter experts, analyzing the performance of the learners utilizing the educational material, and performing user studies to determine how the educational content affects learning outcomes.

In conclusion, it is essential for the Content Layer to deliver the educational materials that are displayed to the learner through the augmented reality interface. It covers content administration, content localization, content assessment, and the generation and curation of instructional information. To create effective educational content that supports the learning process, the Content Layer should be designed with the system's learning objectives and target audience in mind.

4.5 Analytics Layer

The theoretical framework for customized augmented reality systems has four layers, the fourth of which is the analytics layer. Its function is to analyze the system's data in order to offer information about the learning process and to guide personalization choices. User interactions with the system, performance indicators, and contextual data are all collected in this layer. The Analytics Layer is made up of the following elements:

1. Data collection: The analytics layer involves data gathering from system-generated data. Users' interactions with the augmented reality interface, performance indicators like quiz results, and contextual data like the place and time of day can all be included in the data. Regulations governing data privacy must be followed, and user anonymity must be guaranteed.
2. Data Processing: The Analytics Layer processes the gathered data in order to produce insightful findings. Data extraction, transformation, and analysis utilizing statistical and machine learning methods are all possible components of data processing. The system's data processing needs to be built to handle massive amounts of data and deliver insights instantly.
3. Learning Analytics: In the Analytics Layer, learning analytics approaches are applied to the data gathered to offer insights into the learning process. Learning analytics is the process of using data analysis to identify patterns in learner behavior, evaluate motivation and engagement, and measure how well the instructional material is performing. Learning outcomes can be enhanced and customized choices can be made using the insights produced by learning analytics.
4. Personalization Analytics: Using personalization analytics techniques on the data obtained to direct personalization decisions is another aspect of the analytics layer. Analyzing data to determine learner preferences, strengths, and weaknesses and recommending personalized learning experiences is known as personalization analytics. The creation of individualized learning routes and the choice of educational content can both benefit from the insights produced by customization analytics.
5. Visualization: In order to provide meaningful and interactive representations of the insights produced by the system, the Analytics Layer additionally includes data visualization. Charts, graphs, and dashboards that let users explore the data and acquire understanding of the learning process can be included in the visualization.

In conclusion, the Analytics Layer is essential for assessing the data produced by the system to offer perceptions into the learning process and to drive personalization choices. Data processing, learning analytics, personalization analytics, and visualization are all included. To provide pertinent insights that enhance learning outcomes, the analytics layer should be designed with the system's learning objectives and target audience in mind.

4.6 Backend Layer

The fifth layer in the theoretical framework for customized augmented reality systems is the backend layer. It is in charge of running the system and managing all aspects of data management, including storage, retrieval, and security. The Backend Layer is made up of the following elements:

1. Data Management: The Backend Layer contains components for managing, storing, and retrieving data produced by the system. Storage of instructional materials, user

profiles, and system-generated information like activity logs and performance metrics are all included in this. Large amounts of data should be handled by the data management system, which should also provide real-time access and guarantee data security and privacy.
2. Application Server: The Backend Layer has an application server in charge of overseeing the business logic of the system and responding to user queries. The user interface layer, augmented reality layer, personalization layer, content layer, and analytics layer are all managed by the application server. The application server needs to be built to enable several user requests at once, real-time communication, and secure data transfer.
3. Authentication and Authorization: Controlling user access to system resources is the responsibility of the authorization and verification components found in the backend layer. By using credentials like usernames and passwords or third-party authentication services like Google or Facebook, the authentication system confirms user identification. On the basis of user roles and permissions, the authorization system grants access rights.
4. Security: The Backend Layer has security parts in charge of making sure the system is secure and guarding against security risks. This comprises access restrictions to prevent unwanted access to the system's services as well as data encryption and secure data transmission. The security elements should be created to adhere to legal and regulatory criteria for data security.
5. Integration: The Backend Layer has integration parts that link the system to other educational platforms like learning management systems and student information systems. Integration gives the system access to outside data sources and allows it to give users a smooth learning experience.

In conclusion, the Backend Layer is essential for controlling system activities and guaranteeing the security and dependability of the system. It consists of components for data management, application servers, security, integration, authentication, and authorization. The Backend Layer should be designed in accordance with data privacy and security regulations, while also taking into consideration the system's capacity, safety, and integration requirements.

5 Conclusions

In summary, the conceptual framework for personalized augmented reality systems provided in this study offers a thorough framework for creating educational systems that make use of augmented reality and personalization to improve the learning experience. User Interface Layer, Augmented Reality Layer, Personalization Layer, Content Layer, Analytics Layer, and Backend Layer are the five layers that make up the model.

The User Interface Layer offers a user-friendly interface that makes it easier for users to interact with the augmented reality parts of the system. The Augmented Reality Layer makes use of augmented reality technology's capabilities to offer a comprehensive and engaging learning environment. The Personalization Layer makes use of data-driven personalization approaches to offer a customized learning environment that takes into account each learner's interests and needs. The educational material offered by the Content Layer is specialized to the learner's ability level and learning goals. The Analytics Layer makes use of data analytics methods to offer perceptions into the educational process and guide customization choices. The Backend Layer is in charge of overseeing system operations and assuring the security and dependability of the system.

The combination of these layers offers a potent and successful strategy for creating educational systems which tackle the problems with conventional educational systems. The customized augmented reality system provides a dynamic and interesting learning environment that changes depending on the demands of particular students. Additionally, the system provides analytics, assessments, and real-time feedback so that students can track their development and pinpoint areas for improvement.

The successful development and use of the various layers is essential to the tailored augmented reality system's success. A thorough grasp of the educational objectives, the intended audience, and the available technology is necessary for designing a system that integrates these levels efficiently. Additionally, it necessitates a multidisciplinary approach involving specialists in system, augmented reality, personalization, education, and data analytics.

In conclusion, the suggested conceptual model provides a thorough framework for creating individualized augmented reality systems that have the potential to alter the way we learn. The customized augmented reality system has the ability to change the educational landscape by giving students a more interesting and productive learning experience with its realistic and interactive learning environment.

References

1. Papakostas, C., Troussas, C., Krouska, A., Sgouropoulou, C.: User acceptance of augmented reality welding simulator in engineering training. Educ. Inf. Techno.l (Dordr) **27**(1), 791–817 (2022). https://doi.org/10.1007/s10639-020-10418-7
2. Papakostas, C., Troussas, C., Krouska, A., Sgouropoulou, C.: Measuring user experience, usability and interactivity of a personalized mobile augmented reality training system. Sensors **21**(11), 3888 (2021). https://doi.org/10.3390/s21113888
3. Papakostas, C., Troussas, C., Krouska, A., Sgouropoulou, C.: Exploring users' behavioral intention to adopt mobile augmented reality in education through an extended technology acceptance model. Int. Hum. Comput. Interact. **39**, 1294–1302 (2022)
4. Muzyleva, I., Yazykova, L., Gorlach, A., Gorlach, Y.: Augmented and virtual reality technologies in education. In: 2021 1st International Conference on Technology Enhanced Learning in Higher Education (TELE), pp. 99–103. https://doi.org/10.1109/TELE52840.2021.9482568
5. Nugroho, A., Permanasari, A.E., Hantono, B.S.: Augmented reality implementation for human anatomy learning in medical education: a review. In: 2022 8th International Conference on Education and Technology (ICET), pp. 256–260. https://doi.org/10.1109/ICET56879.2022.9990829
6. Cholifah, P.S., Nuraini, N.L.S., Ferdiansyah, A.: Designing augmented reality book for improving practical skill in Scout Education Course. In: 2022 2nd International Conference on Information Technology and Education (ICIT&E), pp. 403–407. https://doi.org/10.1109/ICITE54466.2022.9759857
7. Kanetaki, Z., Stergiou, C.I., Bekas, G., Troussas, C., Sgouropoulou, C.: Creating a metamodel for predicting learners' satisfaction by utilizing an educational information system during COVID-19 pandemic. In: NiDS, pp. 127–136 (2021)

8. Kanetaki, Z., Stergiou, C., Troussas, C., Sgouropoulou, C., Sgouropoulou, C.: Development of an innovative learning methodology aiming to optimise learners' spatial conception in an online mechanical CAD module during COVID-19 pandemic, pp. 31–39 (2021). https://doi.org/10.3233/FAIA210072
9. Yan, H., Liu, W., Xia, X., Xu, Y., Ssong, T.: Design research of interactive picture books of cultural education based on augmented reality technology. In: 2021 16th International Conference on Computer Science & Education (ICCSE), pp. 958–962. https://doi.org/10.1109/ICCSE51940.2021.9569391
10. Singh, S., Kaur, A.: Amalgamation of 3-dimensions in education field using augmented reality technology. In: 2022 7th International Conference on Communication and Electronics Systems (ICCES), pp. 114–119. https://doi.org/10.1109/ICCES54183.2022.9835871
11. Kanetaki, Z., Stergiou, C., Bekas, G., Troussas, C., Sgouropoulou, C., Sgouropoulou, C.: Data mining for improving online higher education amidst COVID-19 pandemic: a case study in the assessment of engineering students, pp. 157–165 (2021). https://doi.org/10.3233/FAIA210088
12. Keller, J.M.: Development and use of the ARCS model of instructional design. J. Instr. Dev. **10**(3), 2–10 (1987). https://doi.org/10.1007/BF02905780
13. Wu, P.-H., Hwang, G.-J., Su, L.-H., Huang, Y.-M.: A context-aware mobile learning system for supporting cognitive apprenticeships in nursing skills training. Educ. Technol. Soc. **15**, 223–236 (2012)
14. Bacca-Acosta, J., Baldiris, S., Fabregat, R., Graf, S., Kinshuk, D.: Augmented reality trends in education: a systematic review of research and applications. Educ. Technol. Soc. **17**, 133–149 (2014)
15. Costa, M., Manso, A., Patrício, J.: Design of a Mobile Augmented Reality Platform with Game-Based Learning Purposes, vol. 11 (2020). https://doi.org/10.3390/info11030127
16. Papakostas, C., Troussas, C., Krouska, A., Sgouropoulou, C.: Modeling the knowledge of users in an augmented reality-based learning environment using fuzzy logic. In: Krouska, A., Troussas, C., Caro, J. (eds.) Novel and Intelligent Digital Systems: Proceedings of the 2nd International Conference (NiDS 2022), Springer International Publishing, Cham, pp. 113–123 (2023)

Identification of the Problem of Neural Network Stability in Breast Cancer Classification by Histological Micrographs

Dmitry Sasov, Yulia Orlova, Anastasia Donsckaia(✉), Alexander Zubkov, Anna Kuznetsova, and Victor Noskin

Software Engineering Department, Volgograd State Technical University, Lenin Ave, 28, Volgograd 400005, Russia
poas@vstu.ru , donsckaia.anastasiya@yandex.ru

Abstract. This paper discusses the actual neural network architectures used to classify breast cancer according to histological micropreparations. A comparative analysis of models and frequently used datasets was conducted. The main methods of augmentation and data preprocessing were also identified. The main goal of the study was to test the trained models on data other than the training set. ResNet 152, DenseNet121, Inception_resnet_v2 models, and the transfer learning approach were chosen for training. As a result of the testing, it was concluded that training on data without color normalization of images and with standard augmentation (rotation and flipping) makes the models vulnerable to changes in the input data. In further studies, we plan to improve the models' stability to changes in the color gamut of images using various methods of data preprocessing.

Keywords: classification of cancer · breast cancer · neural networks · histological micropreparations

1 Introduction

Breast cancer is the most common type of cancer in women worldwide. The incidence of this type of cancer has been increasing since 2008 [2]. In 2020, 2.3 million women worldwide were diagnosed with breast cancer and 685,000 died from it [2,15]. In Russia, the breast cancer mortality rate among women is 15.9%, and its prevalence is 21.2%. Therefore, the issue of fighting breast cancer is also relevant for Russia.

Early diagnosis and effective treatment of breast cancer are absolutely critical to survival. For patients who visited a doctor with a small breast cancer tumor at an early stage, the survival rate is higher. So a lot of new technologies are being developed to detect tumors and their further effective treatment [5].

Among numerous diagnostic methods, histopathological evaluation of a biopsy is extremely relevant for breast cancer diagnosis and treatment. Histological analysis is the standard for determining the cancer type [18]. However, this

K. Kabassi et al. (Eds.): NiDS 2023, LNNS 783, pp. 80–90, 2023.
https://doi.org/10.1007/978-3-031-44097-7_8

method is laborious and nontrivial, so there may be discrepancies in diagnoses even between two experienced specialists [8].

The purpose of this study is to test the models' stability. Its essence is to compare the metrics of models that were trained on a dataset with classes similar to the training dataset and on the training dataset with color correction and resolution changes of images.

2 Convolutional Neural Networks for Breast Micrograph Analysis

Artificial intelligence, in particular convolutional neural networks (CNNs), is actively used for medical image analysis [12]. Due to modern computer performance, it has become possible to train CNNs on images without first feature extraction [1]. CNNs show excellent results in medical image classification in terms of accuracy, sensitivity, AUC, and DSC. Currently, CNNs are presented as a black box, which cannot give a clear answer why such a diagnosis was made after image processing [11]. However, this problem is not considered in this study.

There are various different CNN architectures today: AlexNet, ResNet, Inception, DenseNet, Xception, VGG, MobileNet, etc. They are often used in medical image classification problems [3]. Inception, ResNet, EfficientNet, MobileNet, VGG, and DenseNet architectures are often used for histological micrograph classification and show good results in binary and multiclass classification [4,7,13,14,16]. In most studies, a certain frequently used dataset is selected to train several models on it. Then the results of these models are compared using various metrics. Transfer learning is usually chosen for model training because it allows to achieve high accuracy with a small amount of data.

There are some studies that are developing a new architecture for classifying breast cancer by histological micrographs based on existing architectures. For example, the study by Yun et al. [17] The authors proposed their own CNN architecture BHCNet (breast cancer histopathological image classification network). It consists of an input layer, a convolutional layer, SE-ResNet blocks, and a fully connected layer. The training results of the proposed model can be considered quite successful. The accuracy, recall, and f-measure in the binary classification ranged from 98.52% to 99.44%, and the AUC-score ranged from 99.66% to 99.93%. In the multiclass classification, accuracy, recall, and f-measure ranged from 90.71% to 95.55%.

In the study by Li et al. [10], the IDSNet architecture was proposed, which is a combination of the DenseNet121 model and the SENet module. The following metrics were used to assess the training quality: patient recognition rate (PRR) and image recognition rate (IRR). The SENet module addition improved the PRR in comparison with the basic DenseNet121 model by 2% at worst and by 9.4% at best, and the IRR by 1.2% at worst and by 7.3% at best.

Another direction in the breast histological micrograph classification is the ensemble construction from CNN models [8,9]. Kassani et al. [9] proposed an ensemble based on VGG19, MobileNetV2, and DenseNet201. With its help, accuracy improves from 1.4% to 5% on various datasets compared to the best accuracy for the same models separately. The study by Hameed et al. [8] considered ensembles of models VGG16 and VGG19, which also performed better than the same models separately.

During this study, attention will be focused on the methods of image preprocessing and augmentation. The Table 1 presents the sources and methods of image augmentation and preprocessing.

Table 1. Comparison of image augmentation and preprocessing methods

Source	Dataset	Augmentation Methods	Image Preprocessing
Xie et al. [16]	BreaKHis	Image counterclockwise rotation and displacement along the X and Y axes	Saturation change; the values of all channels conversion to the range [−1, 1]
Bagchi et al. [4]	ICIAR 2018	Image rotation, flipping, and shifting along the X and Y axes	Normalization of Matsenko coloring
Gupta et al. [7]	BreaKHis	No augmentation methods	Normalization of all channels' values (in the range [0, 1])
Voon et al. [13]	BreaKHis, BCG	Horizontal and vertical flipping, image rotation and magnification	Normalization of all channels' values (in the range [0, 1])
Wakili et al. [14]	BreaKHis	Image horizontal flipping, rotation, and scaling	Normalization of Matsenko coloring, normalization of all channels' values (in the range [0, 1])
Yun et al. [17]	BreaKHis	Image shifting and horizontal flipping	Zero-mean normalization
Li et al. [10]	BreaKHis	Image rotation and flipping	Image resolution change and normalization
Kassani et al. [9]	Patch Camelyon, ICIAR 2018, BreaKHis, Bioimaging 2015	Image flipping, magnification, rotation and contrast change	Normalization of Matsenko coloring, normalization of all channels' values (in the range [0, 1])
Hameed et al. [8]	Dataset by Hameed et al.	Image magnification, rotation, shifting, and flipping	No image preprocessing

It can be noticed that color normalization is not used everywhere, although some sources note that images may have different shades due to different coloring methods [4,9,14]. In the previous studies, the model was trained and tested on the same data set. But it is unclear how the trained model will behave on similar data from other datasets or on data from the same training dataset with minor adjustments.

3 Datasets Description

In this study, the dataset presented in the study by Borbat et al. [6] was chosen as a control dataset. The dataset contains images of breast micropreparations with different resolutions and magnifications. The Table 2 presents the image distribution in the dataset.

This dataset presents several classification types: the nature of the lesion, the morphological type of lesion, and the degree of differentiation. In our case, we decided to choose a classification according to the nature of the lesion (benign, in situ, invasive) because it is also presented in other datasets (for example, ICIAR 2018). Images with 300 × 300 pixels resolution and ×4 and ×10 magnification were chosen as training data. Examples of images are presented in Fig. 1.

Table 2. Image distribution by magnification and resolution in the dataset

Resolution/Magnification	×4	×10
300 × 300	6730	23235
500 × 500	3049	9890

For image preprocessing, standard normalization in PyTorch was used, in which all channels' values are converted to the range [0, 1].

There was class imbalance in the used dataset, so augmentation techniques were used to solve it: rotation of images by 90, 180, and 270° and flipping vertically and horizontally. The image distribution before and after augmentation is presented in Table 3.

For initial testing, 15% of the total number of images were allocated. The remaining images were divided into training data (85%) and validation data (15%). The testing data included only non-augmented images for benign and invasive tumors. For non-invasive ones, both augmented and non-augmented were included in rhe dataset because they were the fewest.

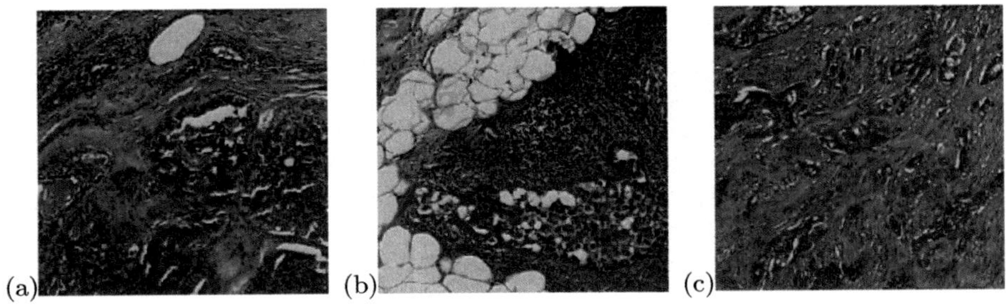

Fig. 1. Examples of histological breast micropreparation images: (a) benign tumor; (b) non-invasive tumor; (c) invasive tumor

The dataset was also tested on the same data, but in 500 × 500 pixels resolution with ×4 and ×10 magnification. However, the images were compressed to 300 × 300 pixels on the model's input. Another test was conducted on images with 300 × 300 pixels resolution with ×4 and ×10 magnification, and with a change in the color balance, contrast, and brightness. Examples of images with changed parameters are presented in Fig. 2. The PIL library for Python was used to change all these parameters. The following values of color balance, contrast, and brightness were chosen: 1.2, 0.8.

Table 3. Data distribution in the training dataset before and after augmentation

The nature of the lesion	Total	Test augmented	Fit + valid augmented	Total augmented
benign tumor	9 514	2 160	12 240	14 400
non-invasive tumor	899	2 160	12 224	14 384
invasive tumor	18 687	2 160	12 240	14 400

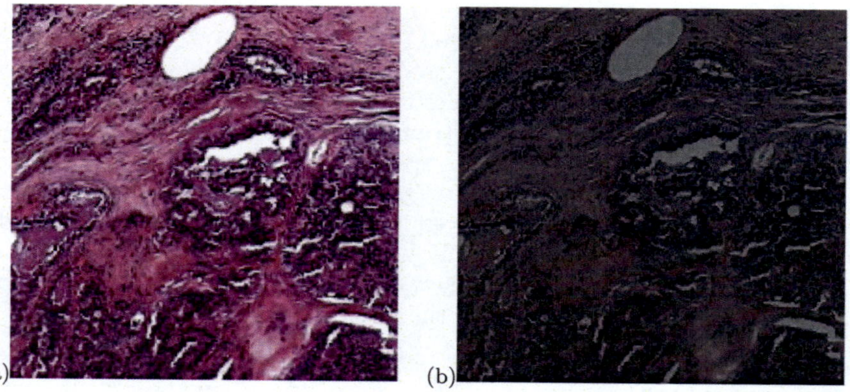
(a) (b)

Fig. 2. Examples of images with changed parameters

The ICIAR 2018 dataset was also chosen for testing the models because it contained the necessary image classification (according to the nature of the lesion). This dataset presents benign, invasive, and non-invasive lesions. Each class has 100 high-resolution images (2048 × 1536 px) with ×200 magnification. The images were compressed to 300 × 300 pixels because the models were trained on images at this resolution. Figure 3 presents images from the ICIAR 2018 dataset.

4 Model Training

This study considers three models related to architectures that have proven themselves well in other studies and have presented excellent results in breast histological micrograph analysis: Resnet 152, DenseNet 121, Inception_ResNet_v2.

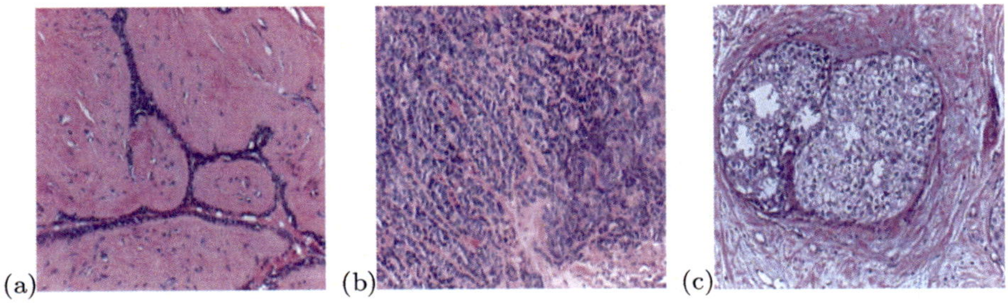

Fig. 3. Examples of images in ICIAR 2018 dataset: (a) benign tumor, (b) non-invasive tumor, (c) invasive tumor

All models have already been pre-trained on the ImageNet dataset, so we need to train the output layer. The transfer learning approach was chosen for training because it allows us to train the model quickly and get fairly high metrics.

For training, the PyTorch framework was used with calculations on an RTX 2060 graphics card with 6 GB of memory capacity. Table 4 presents the hyperparameter values. Adam was chosen as the optimizer, and cross-entropy loss was chosen as the error function.

Table 4. Hyperparameter values

Hyperparameter	Values
batch_size	8
epoches	20
learning_rate	0.001
gamma	0.1
step_size	8

During the training process, the values of metrics were recorded after each epoch: train and validation loss, accuracy, macro average precision (hereinafter macro avg precision), macro average recall (hereinafter macro avg recall), and macro average F1-score (hereinafter macro avg F1). The scikit-learn library was used to calculate the metrics. The best model weights were determined by the highest macro avg F1 score during training. Table 5 presents the best values of metrics based on the training results on the validation dataset.

Table 5. Metrics based on the training results

Metric	ResNet 152	Inception_resnet_v2	DenseNet 121
Accuracy	0.9857	0.9984	0.9944
Macro avg precision	0.9856	0. 9984	0.9944
Macro avg recall	0.9856	0.9984	0.9945
Macro avg F1	0.9856	0.9984	0.9944

5 Model Testing

During testing, the following metrics were evaluated: accuracy, macro avg precision, macro avg recall, macro avg F1, and average ROC-AUC score (hereinafter ROC-AUC score). The scikit-learn library was also used to calculate metrics during testing. The first testing was conducted on 15% of the control dataset data without any changes in it. The metrics based on the test results are presented in Table 6. In general, we can assume that the models successfully coped with the classification task on the control dataset since all metrics ranged from 0.8898 to 0.9437.

Table 6. Metric values for testing on the control dataset part

Metric	ResNet 152	Inception_resnet_v2	DenseNet 121
Accuracy	0.8898	0.9380	0.9022
Macro avg precision	0.9048	0.9437	0.9192
Macro avg recall	0.8898	0.9380	0.9022
Macro avg F1	0.8899	0.9377	0.9011
ROC-AUC score	0.9174	0.9535	0.9266
Error matrices	$\begin{bmatrix} 2010 & 0 & 150 \\ 86 & 1674 & 400 \\ 71 & 7 & 2082 \end{bmatrix}$	$\begin{bmatrix} 2114 & 0 & 46 \\ 57 & 1837 & 266 \\ 31 & 2 & 2127 \end{bmatrix}$	$\begin{bmatrix} 2100 & 4 & 56 \\ 42 & 1612 & 506 \\ 26 & 0 & 2134 \end{bmatrix}$

The next testing stage was conducted on data from the control dataset, but at a different resolution (500 × 500 pixels). On the model's input, all images were compressed to a resolution of 300 × 300 pixels. The testing results of the model on these data are presented in Tables 7.

Table 7. Metric values for testing on higher resolution images

Metric	ResNet 152	Inception_resnet_v2	DenseNet 121
Accuracy	0.9234	0.9501	0.9497
Macro avg precision	0.9001	0.9546	0.9518
Macro avg recall	0.8061	0.8639	0.8162
Macro avg F1	0.8401	0.8985	0.8616
ROC-AUC score	0.8784	0.9166	0.8908
Error matrices	$\begin{bmatrix} 4266 & 5 & 247 \\ 74 & 248 & 137 \\ 495 & 33 & 7434 \end{bmatrix}$	$\begin{bmatrix} 4426 & 0 & 92 \\ 56 & 304 & 99 \\ 393 & 6 & 7563 \end{bmatrix}$	$\begin{bmatrix} 4352 & 9 & 157 \\ 51 & 238 & 170 \\ 264 & 0 & 7698 \end{bmatrix}$

Table 8. Metric values for testing on images with increased and decreased brightness and contrast

Metric	ResNet 152		Inception_resnet_v2		DenseNet 121	
2-7	increased	decreased	increased	decreased	increased	decreased
Accuracy	0.1762	0.4111	0.0976	0.2649	0.2953	0.3426
Macro avg precision	0.3523	0.3265	0.2314	0.5719	0.2722	0.4829
Macro avg recall	0.4013	0.3285	0.3708	0.4750	0.4292	0.4295
Macro avg F1	0.1639	0.2602	0.1033	0.2985	0.2232	0.3343
ROC-AUC score	0.5338	0.4888	0.5083	0.6003	0.5449	0.5819

Table 9. Error matrices for testing on images with increased and decreased brightness and contrast

Model	Error matrices for increased	Error matrices for decreased
ResNet 152	$\begin{bmatrix} 3856 & 5431 & 227 \\ 195 & 690 & 14 \\ 8543 & 9563 & 581 \end{bmatrix}$	$\begin{bmatrix} 722 & 2356 & 6436 \\ 66 & 291 & 542 \\ 1357 & 6381 & 10949 \end{bmatrix}$
Inception_resnet_v2	$\begin{bmatrix} 1069 & 7312 & 1133 \\ 27 & 855 & 17 \\ 3955 & 13815 & 917 \end{bmatrix}$	$\begin{bmatrix} 3615 & 5024 & 875 \\ 29 & 780 & 90 \\ 356 & 15016 & 3315 \end{bmatrix}$
DenseNet 121	$\begin{bmatrix} 7669 & 1075 & 770 \\ 479 & 408 & 12 \\ 15125 & 3047 & 515 \end{bmatrix}$	$\begin{bmatrix} 5542 & 3090 & 882 \\ 143 & 443 & 313 \\ 2834 & 11867 & 3986 \end{bmatrix}$

According to the test results, it can be noticed that the macro avg recall, macro avg F1 and ROC-AUC score metrics for all three models deteriorated. Problems mainly arose in determining the non-invasive tumor class. The Inception_resnet_v2 model performed best because it had the highest metric values among the other models. Further testing was conducted on images from the control dataset with changes in contrast, color balance, and brightness. The test results are presented in Tables 8 and 9. The models clearly could not cope with the classification of such images because all models had bad metric results. It follows from this that the models relied on a specific color scheme when deciding on the class choice. It can also be noticed that when the values of the image parameters increased, the models classified most of the images as class 1 (for ResNet 152 and Inception_resnet_v2, most of them fell into the class of non-invasive tumor; for DenseNet 121, into the class of benign tumor). ResNet 152 and Inception_resnet_v2 models behaved similarly when the image parameters were reduced, and DenseNet 121 was able to correctly identify most of the images of benign and non-invasive tumors.

The next testing stage was conducted on the ICIAR 2018 dataset. Since the images in this dataset differed from the control dataset in terms of color gamut, magnification, and resolution, we should not expect high metrics from the models. Table 10 presents the test results. As expected, the models failed to achieve good results in the image classification from the ICIAR 2018 dataset. However, ResNet 152 performed better than other models (which simply assigned almost all images to one class) and was able to correctly classify almost half of the images.

Table 10. Metric values for testing on the control dataset part

Metric	ResNet 152	Inception_resnet_v2	DenseNet 121
Accuracy	0.8898	0.9380	0.9022
Macro avg precision	0.9048	0.9437	0.9192
Macro avg recall	0.8898	0.9380	0.9022
Macro avg F1	0.8899	0.9377	0.9011
ROC-AUC score	0.9174	0.9535	0.9266
Error matrices	$\begin{bmatrix} 2010 & 0 & 150 \\ 86 & 1674 & 400 \\ 71 & 7 & 2082 \end{bmatrix}$	$\begin{bmatrix} 2114 & 0 & 46 \\ 57 & 1837 & 266 \\ 31 & 2 & 2127 \end{bmatrix}$	$\begin{bmatrix} 2100 & 4 & 56 \\ 42 & 1612 & 506 \\ 26 & 0 & 2134 \end{bmatrix}$

6 Conclusion

During several tests of models on various data, a problem with the models' stability to changes in input data was identified. When training without image color range normalization, models do not classify images well when contrast, brightness, and color balance change. However, with a slight scaling of images, the values of metrics become worse not so much. When tested on a dataset with a similar classification, the models also showed poor results. All this leads to the conclusion that normalization of channel values in the range [0, 1] and augmentation using rotation and flipping are not enough for high-quality model training. Such models do not adapt well to changes in the color gamut and scale of images, which are quite common in real life.

In future research, we plan to increase the model stability through the use of various image preprocessing methods, which will help to cope with differences in the color gamut of images. The following approaches will be used to normalize breast histological micrographs: CycleGANs, StainCUT, and Macenko stain normalization. These approaches will be compared in terms of both performance and accuracy improvements.

Acknowledgments. The reported study was funded by VSTU according to the research project No. 60/478-22, 60/473-22.

References

1. Araújo, T., et al.: Classification of breast cancer histology images using convolutional neural networks. PLOS ONE **12**(6), 1–14 (06 2017). https://doi.org/10.1371/journal.pone.0177544
2. Breast cancer statistics and resources. https://www.bcrf.org/breast-cancer-statistics-and-resources. Accessed 01 May 2023
3. Different types of CNN models. https://iq.opengenus.org/different-types-of-cnn-models/. Accessed 05 Apr 2023
4. Bagchi, A., Pramanik, P., Sarkar, R.: A multi-stage approach to breast cancer classification using histopathology images. Diagnostics **13**, 126 (2022). https://doi.org/10.3390/diagnostics13010126
5. Bhushan, A., Gonsalves, A., Menon, J.U.: Current state of breast cancer diagnosis, treatment, and theranostics. Pharmaceutics **13**(5) (2021). https://doi.org/10.3390/pharmaceutics13050723, https://www.mdpi.com/1999-4923/13/5/723
6. Borbat, A., Lishchuk, S.: Pervyj rossijskij nabor dannyh gistologicheskih izobrazhenij patologicheskih processov molochnoj zhelezy. Vrach Inform. Tekhnol. **3**, 25–30 (2020). in Russ
7. Gupta, K., Chawla, N.: Analysis of histopathological images for prediction of breast cancer using traditional classifiers with pre-trained CNN. Procedia Comput. Sci. **167**, 878–889 (2020). https://doi.org/10.1016/j.procs.2020.03.427
8. Hameed, Z., Zahia, S., Garcia-Zapirain, B., Javier Aguirre, J., María Vanegas, A.: Breast cancer histopathology image classification using an ensemble of deep learning models. Sensors **20**(16) (2020). https://doi.org/10.3390/s20164373, https://www.mdpi.com/1424-8220/20/16/4373
9. Kassani, H.S., Hosseinzadeh Kassani, P., Wesolowski, M., Schneider, K., Deters, R.: Classification of histopathological biopsy images using ensemble of deep learning networks, p. 8 (2019)
10. Li, X., Shen, X., Zhou, Y., Wang, X., Li, T.Q.: Classification of breast cancer histopathological images using interleaved DenseNet with SENet (IDSNet). PLOS ONE **15**(5), 1–13 (05 2020). https://doi.org/10.1371/journal.pone.0232127
11. Sarvamangala, D.R., Kulkarni, R.V.: Convolutional neural networks in medical image understanding: a survey. Evol. Intell. **15**, 1–22 (11 2022). https://doi.org/10.1007/s12065-020-00540-3
12. Tang, X.: The role of artificial intelligence in medical imaging research. BJR|Open **2**(1), 20190031 (2020). https://doi.org/10.1259/bjro.20190031
13. Voon, W., et al.: Performance analysis of seven convolutional neural networks (CNNs) with transfer learning for invasive ductal carcinoma (IDC) grading in breast histopathological images. Sci. Rep. **12**, 19200 (2022). https://doi.org/10.1038/s41598-022-21848-3
14. Wakili, M.A., et al.: Classification of breast cancer histopathological images using DenseNet and transfer learning. Comput. Intell. Neurosci. **2022** (2022). https://doi.org/10.1155/2022/8904768
15. Wild, C., Weiderpass, E., Stewart, B. (eds.): World Cancer Report: Cancer Research for Cancer Prevention. International Agency for Research on Cancer, Lyon (2020)

16. Xie, J., Liu, R., Luttrell, J., Zhang, C.: Deep learning based analysis of histopathological images of breast cancer. Front. Genet. **10**, 80 (2019)
17. Yun, J., Chen, L., Zhang, H., Xiao, X.: Breast cancer histopathological image classification using convolutional neural networks with small SE-ResNet module. PLoS ONE **14** (2019). https://doi.org/10.1371/journal.pone.0214587
18. Zeiser, F.A., da Costa, C.A., Roehe, A.V., da Rosa Righi, R., Marques, N.M.C.: Breast cancer intelligent analysis of histopathological data: a systematic review. Appl. Soft Comput. **113**, 107886 (2021). https://doi.org/10.1016/j.asoc.2021.107886, https://www.sciencedirect.com/science/article/pii/S1568494621008085

A Web Tool for K-means Clustering

Konstantinos Gratsos[1], Stefanos Ougiaroglou[1(✉)], and Dionisis Margaris[2]

[1] Department of Information and Electronic Engineering, School of Engineering, International Hellenic University, 57400 Sindos, Thessaloniki, Greece
stoug@ihu.gr

[2] Department of Digital Systems, School of Economics and Technology, University of the Peloponnese, 23100 Sparta, Greece
margaris@uop.gr

Abstract. The K-Means clustering finds many applications in different domains. Researchers and practitioners utilize K-Means through specialized software or libraries of programming languages. This implies knowledge on these tools. This paper presents Web-K-Means, a user-friendly web application that simplifies the process of running the K-Means clustering and identifying the optimal number of clusters via the elbow method. Web-K-Means allows users to upload datasets, select relevant features, and finally execute the algorithm. The application displays the graph generated by elbow method, suggests a value for the number of clusters and presents the cluster assignments in a data table along with an exploratory data plot. Thus, the users can easily perform clustering analyses, without specialized software or programming skills. Additionally, Web-K-Means includes a REST API that allows users to run clustering tasks and retrieve the results programmatically. The usability of Web-K-Means was evaluated by the System Usability Scale (SUS) and experiments were conducted for the CPU times evaluation. The results indicate that Web-K-Means is a simple, efficient and user friendly tool for cluster analysis.

Keywords: Clustering · K-Means · Elbow method · Web application · Web service

1 Introduction

Clustering or unsupervised learning [1] is a common data analysis task that involves grouping instances with similar characteristics into clusters. A cluster is a set of instances in which each instance is closer (or most similar) to every instance in the cluster, rather than every instance outside the cluster. Nowadays, Clustering finds many applications in various domains, including market research, search engines, psychology and medicine, biology, etc.

Clustering was originated by Zubin and Tryon in psychology domain and Driver and Kroeber in anthropology domain in the 1930s. However, the development of clustering techniques was delayed due to computational difficulties. From the late 1950s on-wards, the rise of computing power led to the development of various clustering techniques

K. Kabassi et al. (Eds.): NiDS 2023, LNNS 783, pp. 91–101, 2023.
https://doi.org/10.1007/978-3-031-44097-7_9

widely used today. Therefore, computer science plays a crucial role in clustering as it is used to perform complex calculations and processing that are necessary to identify clusters.

K-Means [2] may be characterized as the most popular and widely-used clustering algorithm. It aims to group instances into k clusters based on their similarity. The algorithm is based on a iterative procedure that assigns instances to the nearest cluster centroid. A major issue that must be addressed by the user is the k parameter determination. The elbow method [3] is a popular technique for determining the optimal number of clusters. The K-Means clustering with or without the elbow method has been incorporated in many specialized stand-alone software (e.g. Matlab, Weka [4], SPSS, Orange [5], etc.) and in libraries of programming languages (e.g. scikit-learn [6]). The use of K-Means and elbow method through software and programming environments may imply software license, download and installation, specialized knowledge on the specialised software and programming skills. To the best of the authors' knowledge, there is no free web application for clustering purposes. This observation constitutes the motivation of the present work.

The contribution of the paper is Web-K-Means, a user-friendly web application that enables researchers and practitioners to easily perform K-Means clustering with the well-known K-Means++ initialization technique [7] via the web. The elbow method has been integrated in Web-K-Means and allows the user to identify the optimal number of clusters. Moreover, Web-K-Means provides an exploratory plot of the cluster assignments that enables users to understand the characteristics of each discovered cluster.

The paper is organized as follows: Sect. 1 provides an overview of the K-Means clustering. Section 2 reviews the elbow method. Section 3 presents Web-K-Means in detail. System evaluation through CPU time measurements and usability testing using the System Usability Scale (SUS) questionnaire are presented in Sect. 4, and finally Sect. 5 concludes the paper and outlines future work.

2 K-Means Clustering

K-Means clustering groups instances based on similarities. It involves iteratively assigning of each instance to a cluster, based on its distance from the centroid (mean of each cluster) and then computing the updated centroid, until the centroids no longer change. The algorithm aims to minimize the sum of squared distances between data instances and their assigned cluster centroids, known as the within-cluster sum of squares (WCSS). More formally, given a set of instances, $X = \{x_1, x_2,\ldots, x_n\}$, K-Means clustering aims to assign the n instances to k clusters $(k \leq n)$ $S = \{S_1, S_2,\ldots, S_k\}$ so as to minimize the following function(within-cluster sum of squares):

$$E = \sum_{j=1}^{k} \sum_{x_1 \in S_j} x_i - \mu_j^2 \quad (1)$$

where μ_j is the mean of cluster Sj and $\|x_i - \mu_j\|$ is the chosen distance metric between the data point x_i and the corresponding mean.

The algorithm starts by randomly initializing k centroids, which are used as the initial cluster centers. Then, each instance is assigned to the nearest centroid, and the centroid is moved to the center of the assigned instances. This process is repeated iteratively until the centroids no longer move. Algorithm 1 illustrates the pseudo-code for the K-Means clustering technique. Firstly, the algorithm takes a dataset $X = x_1, x_2,\ldots, x_n$ and forms k clusters *(C)*. The algorithm initially assigns each instance x_i to the cluster with the closest centroid *(mean)*, which is termed as the assignment step (lines 4–6). After all the instances have been assigned, the algorithm recalculates the new means, by averaging the corresponding instances of the clusters (lines 7–9), which can be expressed as

$$cj = \frac{1}{|Cj|} + \sum_{x_1 \in C_j} x_i \tag{2}$$

The algorithm then re-executes the assignment step by taking into account the new means. As a result, the K-Means are adjusted in each step. The algorithm terminates when the means (cluster centroids) do not change between two iterations (cluster consolidation) (line 10). Since each instance is closer to the centroid of the cluster it belongs to, the within-cluster sum of squares function is minimized.

Algorithm 1 k-means clustering

Require: $X = x_1, x_2, \ldots, x_n$ - dataset, k - number of clusters
Ensure: $C = C_1, C_2, \ldots, C_k$ - set of clusters
1: Select k initial centroids c1, c2, . . . , ck randomly from X
2: **repeat**
3:    Create k empty clusters C1,C2, . . . ,Ck
4:    **for** i = 1 to n **do**
5:       Assign xi to the cluster Cj with the closest centroid cj
6:    **end for**
7:    **for** j = 1 to k **do**
8:       Update centroid cj as the mean of all points in cluster Cj
9:    **end for**
10: **until** no more changes in the assignment of points to clusters

The resulted clusters depend on the randomly selected initial centroids. In effect, K-Means will discover different clusters by examining the same data but utilizing different initial cluster centroids. K-Means + + is a popular initialization technique. It improves the initial centroid selection by probabilistically choosing the initial centroids that are farthest away from each other, resulting in a more accurate and efficient clustering. The algorithm starts by selecting a random instance as the first centroid, and then selects each subsequent centroid based on the distance from the previous centroids, with a higher probability of choosing instances that are farther away. This helps to avoid the common issue of K-Means clustering getting stuck in sub-optimal solutions due to poor initialization.

K-Means is popular and therefore, many variants have been proposed. K-medians [8], k-modes and k-prototypes [9] are well-known variants. K-medians clustering replaces the mean calculation with the median calculation. The main advantage is that it is less sensitive to outliers, which can cause problems in K-Means. K-modes is designed to handle categorical data. In k-modes, instead of computing the mean or median, the mode (most frequent value) of each attribute in the cluster is used as the cluster center.

3 Elbow Method

The elbow method is a well-known technique for determining the optimal number of clusters in K-Means clustering. It works by plotting the within-cluster sum of squares (WCSS) for different values of clusters (k). The WCSS is the sum of the squared distances between each point and its assigned cluster centroid. Considering a specific k value, it is computed by summing up the squared distances for all instances within each cluster and then by finding the sum of the individual k cluster sums.

As k increases, the WCSS generally decreases, since each instance is closer to its assigned centroid. However, at a certain point, the marginal decrease in WCSS diminishes, creating a noticeable bend in the plot. This bend, or "elbow", represents the point of diminishing returns in terms of clustering performance, beyond which adding more clusters does not provide much additional benefit. The optimal number of clusters is often chosen at the elbow point on the plot. By this way, K-Means discovers compact and well-separated clusters.

In Fig. 1, the x-axis represents the number of clusters (k), while the y-axis represents the WCSS values. As k increases, the WCSS decreases, but at $k = 3$, there is a noticeable bend in the plot. This bend represents the "elbow" point, which indicates that adding more clusters beyond this point does not provide significant improvements in clustering performance. In this case, the optimal number of clusters might be chosen as 3.

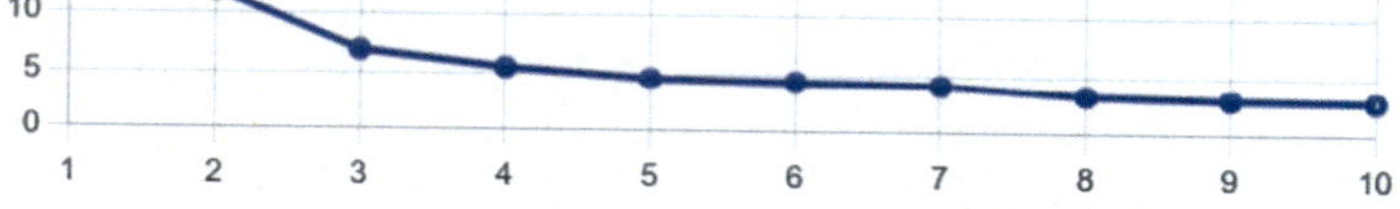

Fig. 1. Example of an Elbow Method Plot.

4 The Web-K-Means Application

4.1 Description

Web-K-Means is a user-friendly web application for cluster analysis. With Web-K-Means, users can upload datasets in csv, xls or xlsx format, choose the attributes that will be used in clustering, and finally execute the K-Means clustering by utilizing K-Means + + initialization method. Web-K-Means displays the "Elbow diagram", recommends the optimal cluster number, and presents the instances and their assigned clusters in an exploratory plot and a tabular. The elbow graph, the exploratory plot and the dataset with the assigned clusters can be downloaded for further use and processing. Furthermore, Web-K-Means provides a REST API that allows developers to submit their data, execute the elbow method and K-Means clustering and obtain the cluster assignments from their own applications by incorporating simple API calls into them. The REST API is an advantageous feature of Web-K-Means, expanding its usability beyond the web-based interface and allowing for more efficient and automated usage of the K-Means clustering.

Following the architecture of WebApriori [10], Web-K-Means includes three components: (i) modules for the elbow method and K-Means clustering, (ii) a modern and user-friendly web interface, (iii) the back-end and the REST API service. The application was developed using open-source technologies and Git, with its source code available on GitHub[1]. Python was used to code the modules of the first component. Web-K-Means uses the K-Means implementation which is available in the scikit-learn library. Also, Pandas [11] library was utilized for datasets manipulation, and the kneed library [12] was used to obtain the number of clusters from elbow graph. PHP was used for the development the back-end as well as the REST API. The composer library was used for package management and PHPMailer for sending confirmation emails. A MySQL database was designed to manage users and their access levels (privileges). For the development of the front-end and the web interface, Javascript with the jQuery library, AJAX, and the Bootstrap framework were used. Web-K-Means utilizes the server's file system to store the uploaded datasets. The REST API acts as the interface where all technologies can communicate with each other, as shown in Fig. 2.

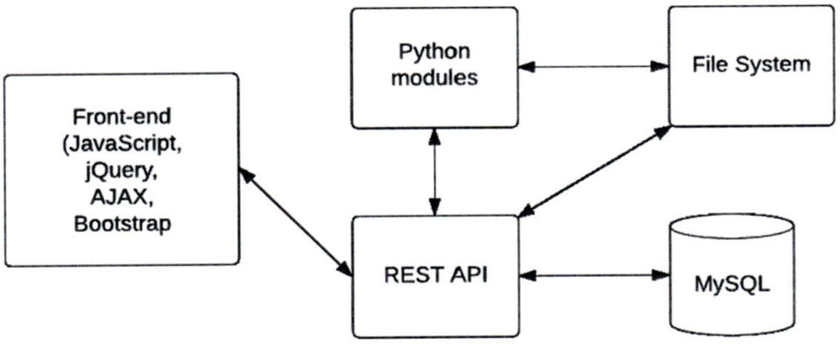

Fig. 2. Application architecture.

[1] https://github.com/KostisGrf/WebKmeans.

Each dataset can be either public or private. Private datasets are limited to the user who uploaded them, while public datasets are accessible to all registered users. However, only users with advanced privileges are able to upload public datasets. This feature is especially beneficial for educators aiming to share datasets with their students. The application offers three user roles: (i) simple user, (ii) public dataset creator, and (iii) administrators. Simple users are the users who can upload private dataset and use Web-K-Means on them. They can also use public datasets uploaded by a public dataset creator. Public dataset creators have the same privileges with simple users but they can also upload public datasets. Administrators have the ability to upgrade a user account from simple user to public dataset creator.

It is worth mentioning that Web-K-Means has been deployed in a web server at the Department of Information and Electronic Engineering of the International Hellenic University[2]. It is free and open source. Therefore, users can access the source code on Github and deploy it on their own server.

4.2 Web Interface

To use Web-K-Means, users must first register and confirm their email. Then, they can log in. The main page is divided into three distinct sections. The first section provides a web interface that enables users to either upload a new dataset on the Web-K-Means server or select an existing one (see Fig. 3(c)). The supported datasets formats are csv, xls, and xlsx. Once a dataset has been selected, its data will be displayed in a tabular format, and the user can download or delete it (assuming that the user has the necessary permissions). Also, the names of the numerical attributes are displayed in check-boxes. Note that K-Means can be applied to numerical data. The users can un-check the numerical attributes they want to be ignored in the K-Means clustering.

The second section concerns the elbow method. The users are able to enter the maximum number of clusters they want to examine and click "Get elbow chart". This generates the elbow graph and suggests an appropriate number of clusters in the range from 2 to the maximum number provided (see Fig. 3(b)).

The last section requires users to input their desired number of clusters. This can be either the number suggested from the previous step or chosen independently by the user. By clicking "Get cluster assignment", users will receive a table displaying the data along with the cluster assignments (see Fig. 3(a)). The cluster assignments can also be presented in an exploratory data plot which summarises the key characteristics of each cluster (see Fig. 3(d)). The "Download CSV" button allows users to download the table in CSV format. The elbow graph and the exploratory data plot can be also downloaded.

[2] https://webkmeans.iee.ihu.gr.

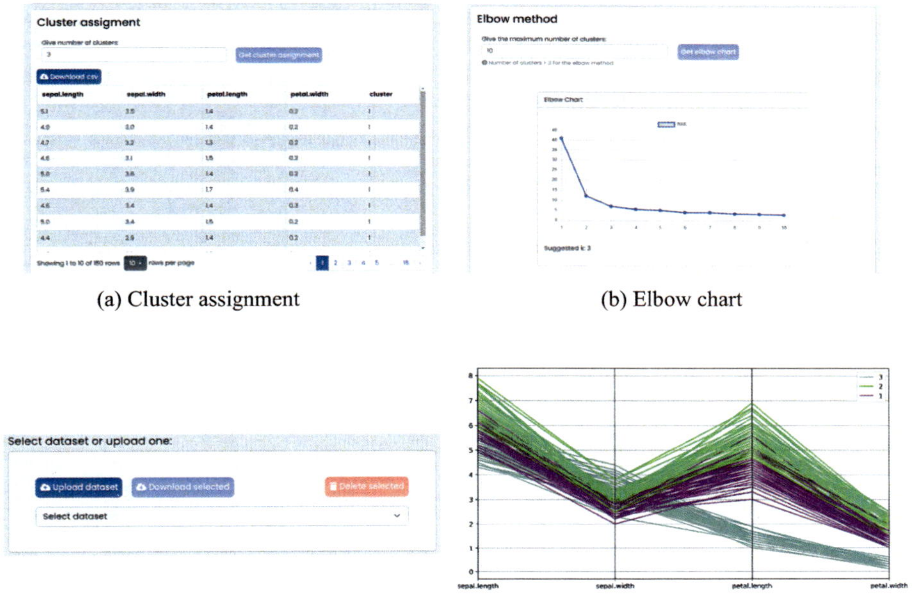

(a) Cluster assignment

(b) Elbow chart

(c) Dataset selection

(d) Exploratory plot

Fig. 3. Web-K-Means interface.

4.3 Web Service

The web service is designed to be a REST API with eleven endpoints which make the aforementioned functionalities available to other applications through HTTP requests. The users/programmers must sign up and acquire an API key to access the web service. Each HTTP requests must be accompanied by the user's API key. Four endpoints are dedicated to user account functions, such as sign-up, login, editing of account, and deletion. Another four endpoints are exclusively reserved for managing datasets: upload, accessing, deletion and retrieval of datasets name and characteristics. The remaining three endpoints are for executing K-Means and visualizing its results. More specifically, one of these triggers the elbow method and returns WCSS values along with the recommended number of clusters. Another endpoint triggers K-Means and returns cluster assignments. The last endpoint generates the exploratory plot of the clustered data. All endpoints return their results in JSON format. For example, below are depicted a JSON request and a JSON response of the endpoint that triggers the elbow method:

JSON request:

{“dataset”: “sample.csv”, “dataset-type”: “personal”, “clusters”: “10”, “columns”:[“Age”,"Salary”],"apikey”:"0a8366a07c0d8fccx48bab2e657f12d0”}.

JSON response:

{“sse”: [“41.166”, “12.128”, “6.982”, “5.5258”, “4.589”, “4.674”, “3.757”, “3.193”, “3.299”, “2.695”],

“suggested-k”:"3”}.

The web interface includes a web page with instructions for utilizing the eleven endpoints along with the user's API key. The web page presents examples of possible HTTP requests with the corresponding responses.

5 System Evaluation

5.1 CPU Time Measurements

Running the K-Means clustering can be intensive in terms of CPU and RAM usage. To tackle this issue, the Pandas library was employed to effectively manage the large datasets. Additionally, the scikit-learn library was utilized to enhance the algorithm's execution efficiency.

Table 1. CPU time measurements.

Dataset	Size (KB)	# of rows	# of columns	Time for elbow (k = 20)	Suggested K	Time for K-Means
penbased	538	10,992	16	1.06s	6	0.24s
letter	716	20,000	16	2,49s	7	0.35s
magic	1,462	19,020	10	1,67s	5	0.23s
texture	1,495	5,500	40	1,02s	4	0.45s
shuttle	1,559	57,999	9	3,08s	5	0.31s
poker	24,563	1,025,009	10	72s	6	23.25s

In order to measure the performance of Web-K-Means, we conducted an experimental study by using six datasets distributed by the keel dataset repository [13][3]. The experimental measurements are obtained by executing the elbow method for 20 clusters and the K-Means clustering using the suggested number of clusters through the web interface. The experimental results are presented in Table 1. As we expected, the results indicate that the execution times are directly affected by the size and the number of rows and columns of each dataset. It is worth mentioning that the execution of the elbow method is more time consuming than the execution of the K-Means clustering. This happens because the elbow method executes clustering several times (from $k = 1$ to $k = 20$ in the case of our experimentation) and compute a WCSS value for each clustering task. Poker is a quite large dataset. It contains over a million rows. The elbow method took more than a minute to run, while the K-Means clustering took only 24 s. The measurements obtained by using such large datasets can be improved by hosting Web-K-Means to a more powerful computer.

[3] https://sci2s.ugr.es/keel/datasets.php.

5.2 Usability Testing

The usability of Web-K-Means was evaluated using the System Usability Scale (SUS) questionnaire[4]. More specifically, SUS was used to measure the users overall satisfaction with Web-K-Means, as well as their perception of its effectiveness, efficiency, and ease of use. SUS is a widely used tool for measuring the usability of web applications. It consists of ten questions that participants complete to rate their experience. The questions of the questionnaire are presented bellow.

1. I think that I would like to use this website frequently
2. I found the website unnecessarily complex
3. I thought the website was easy to use
4. I think that I would need the support of a technical person to be able to use this website
5. I found the various functions in this website were well integrated
6. I thought there was too much inconsistency in this website
7. I would imagine that most people would learn to use this website very quickly
8. I found the website very cumbersome to use.
9. I felt very confident using the website
10. I needed to learn a lot of things before I could get going with this website

Each question is rated on a 5-point Likert scale ranging from "strongly disagree" (1) to "strongly agree" (5). To calculate the SUS score, for each of the odd numbered questions, 1 is subtracted from the score. For each of the even numbered questions, their value is subtracted from 5. The resulting scores are then added up and multiplied by 2.5 to give a final score between 0 and 100. A SUS score of 80 or above is considered excellent.

We requested people to complete the questionnaire, and 22 of them did so. Most of them were computer science undergraduate students who attend a data mining course. Table 2 presents the response count for each range. The SUS score is 83.4. Therefore the results of SUS illustrate that the users are satisfied with the experience of using Web-K-Means.

Table 2. Results of System Usability Scale (SUS) Questionnaire.

Question	1 (strongly disagree)	2	3	4	5 (strongly agree)
Q1	0	2	3	9	8
Q2	14	7	0	0	1
Q3	0	0	1	4	17
Q4	14	6	2	0	0

(continued)

[4] https://forms.gle/dUSecKe1pgES661z8.

Table 2. (*continued*)

Question	1 (strongly disagree)	2	3	4	5 (strongly agree)
Q5	0	0	3	7	12
Q6	18	4	0	0	0
Q7	1	0	4	4	13
Q8	6	8	8	0	0
Q9	0	0	1	7	14
Q10	6	4	9	2	1

6 Conclusions and Future Work

The paper presented Web-K-Means, a user-friendly web application that allows researchers and practitioners to easily perform K-Means cluster analysis with the kmeans ++ centroid initialization method. The elbow method has been integrated in order to give the user the ability to identify the optimal number of clusters. With Web-K-Means, users can upload datasets, choose the attributes that will be used in clustering, and conduct cluster analysis with K-Means. The application displays the elbow graph, recommends the optimal cluster number, and presents the clustering results in a tabular form and an exploratory data plot. Additionally, Web-K-Means provides a REST API that allows developers to submit their data, execute the elbow method and the K-Means clustering, and obtain the clustering results from their own applications by incorporating simple API calls into them.

In our future work, we plan to extend Web-K-Means by integrating k-prototypes. Also, we plan to integrate a mechanism for data pre-processing tasks, such as missing values imputation and normalization.

References

1. Aggarwal, C.C., Reddy, C.K.: Data Clustering: Algorithms and Applications. Chapman Hall/CRC, 1st edn. (2013)
2. MacQueen, J.B.: Some methods for classification and analysis of multivariate observations. Proce. Fifth Berkeley Symp. Mathe. Stati. Probab. **1**(14), 281–297 (1967)
3. Kodinariya, T.M., Makwana, P.R.: A review on the Elbow method in clustering. Int. J. Comp. Appli. **1**(6), 97–100 (2013)
4. Frank, E., Hall, M.A., Holmes, G., Kirkby, R., Pfahringer, B., Witten, I.H.: Weka: A machine learning workbench for data mining, pp. 1305–1314. Springer, Berlin (2005). http://researchcommons.waikato.ac.nz/handle/10289/1497
5. Demšar, J., et al.: Orange: Data mining toolbox in python. J. Mach. Lear. Res. **14**, 2349–2353 (2013)
6. Pedregosa, F., et al.: Scikit-learn: Machine learning in Python. J. Mach. Learn. Res. **12**, 2825–2830 (2011)
7. Arthur, D., Vassilvitskii, S.: K-Means++: The advantages of careful seeding (2007)

8. Sengupta, J.S., Auchter, R.F.: A k-medians clustering algorithm. Appl. Stat. **39**(1), 67–79 (1990). https://doi.org/10.2307/2347822
9. Huang, Z.: Extensions to the K-Means algorithm for clustering large data sets with categorical values. Data Min. Knowl. Disc. **2**(3), 283–304 (1998)
10. Malliaridis, K., Ougiaroglou, S., Dervos, D.A.: Webapriori: A web application for association rules mining. In: Kumar, V., Troussas, C. (eds.) Intelligent Tutoring Systems, pp. 371–377. Springer International Publishing, Cham (2020)
11. McKinney, W.: Data Structures for Statistical Computing in Python. In: der Walt, S., Millman, J. (eds.) Proceedings of the 9th Python in Science Conference. pp. 56–61 (2010). https://doi.org/10.25080/Majora-92bf1922-00a
12. Satopaa, V., Albrecht, J., Irwin, D., Raghavan, B.: Finding a "kneedle" in a haystack: Detecting knee points in system behavior. In: 2011 31st International Conference on Distributed Computing Systems Workshops, pp. 166–171 (2011)
13. Alcalà-Fdez, J., et al.: Keel data-mining software tool: Data set repository, integration of algorithms and experimental analysis framework. J. Multiple-Valued Logic & Soft Computing **17**(2–3), 255–287 (2011)

Debriefings on Prehospital Care Scenarios in MedDbriefer—A Tool to Support Peer Learning

Sandra Katz(✉), Pamela Jordan, Patricia Albacete, and Scott Silliman

University of Pittsburgh, Pittsburgh, PA 15260, USA
katz@pitt.edu

Abstract. Across the healthcare professions, many students don't get enough practice doing simulated clinical interactions during course labs to feel confident about passing certification exams and treating actual patients. To address this problem, we are developing MedDbriefer, a web-based tutoring system that runs on a tablet. MedDbriefer allows peers to engage in supplemental clinical scenarios on their own. With its current focus on paramedic training, one student "voice treats" a simulated patient as the leader of a mock emergency medical services team while a peer uses MedDbriefer's checklists to log the team leader's verbalized actions. The system then analyzes the event log and generates a debriefing, which highlights errors such as assessment actions and treatment interventions that the team leader missed or performed late. This paper focuses on how the system analyzes event logs to generate adaptive debriefings.

Keywords: Simulation-based Training · Debriefing · Healthcare Training

1 Introduction

Simulation-based training (SBT) provides students in the healthcare professions realistic clinical experiences without risk to actual patients [1]. Some exercises focus on psychomotor skills such as intubating a patient's airway, administering fluids through an intravenous line, and transferring a patient safely to an ambulance. Other exercises immerse students in realistic clinical scenarios that challenge them to apply psychomotor, clinical reasoning, and team coordination skills. Although simulation cannot replicate all benefits of interacting with actual patients, such as emotional engagement, its effectiveness for developing clinical knowledge and skills is well established [e.g., 2].

Students who struggle to acquire these skills can benefit from supplemental simulation-based practice, outside of their course labs. Unfortunately, instructors who are trained to facilitate simulation exercises are in short supply [e.g., 1, 3]. Many instructors are themselves active clinicians (e.g., doctors, nurses, paramedics), which limits the time they can devote to teaching. To address this problem, instructors often encourage peer learners to do clinical scenarios on their own, without an instructor present. However, peer-to-peer learning needs to be supported to be effective. Left unguided, it

K. Kabassi et al. (Eds.): NiDS 2023, LNNS 783, pp. 102–113, 2023.
https://doi.org/10.1007/978-3-031-44097-7_10

can become fraught with problems [4]. For example, students often can't find or invent clinical scenarios that are as challenging as those they will be tasked to perform during certification exams and on the job. With limited clinical knowledge and skills, especially during early stages of training, students sometimes call out unrealistic patient findings and other information their peers request while treating the scenario's patient(s). Most seriously, students typically can't provide helpful feedback on their peers' performance, and explanations that require a sufficient understanding of human anatomy, physiology, etc. to produce.

In response to these limitations of peer-to-peer simulation-based training, we are developing MedDbriefer—a web-based tutoring system that runs on a tablet. It allows two or more paramedic trainees to practice realistic prehospital care scenarios and immediately receive an automated debriefing on their performance. While one student treats a simulated patient as the leader of a mock emergency medical services (EMS) team, a peer uses the tablet's checklists to log the team leader's actions. The team leader may be assisted by one or more peers (see Fig. 1). Immediately after the scenario, the system analyzes the event log and generates an adaptive debriefing that highlights errors—for example, missing patient assessment steps and interventions, inappropriate interventions, and errors in how interventions were performed. Our ultimate goal is for MedDbriefer to be more scalable, to support peer-to-peer simulation-based training across the healthcare professions.

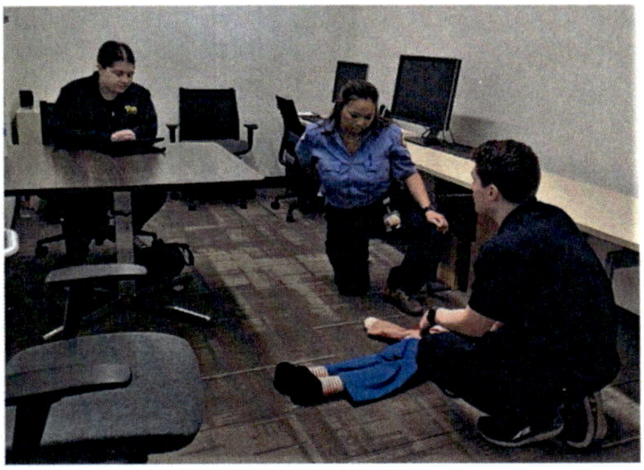

Fig. 1. MedDbriefer in use during a clinical scenario. Peer members of a mock EMS team treat a virtual patient (at right); session observer, at left, uses a tablet to log the team's actions.

The next section presents an overview of MedDbriefer. Previous papers provide more detailed descriptions of the two approaches to debriefing that MedDbriefer implements [5, 6]. At this writing, a randomized controlled trial to compare the effectiveness of these approaches is in progress. This paper focuses on how the system analyzes event logs of students' actions during scenarios to generate adaptive debriefings.

2 MedDbriefer

MedDbriefer focuses on developing paramedic trainees' clinical reasoning skills, which includes identifying clinical problems, determining which interventions to perform to manage these problems and how to perform them. For example, paramedics need to be able to recognize symptoms of hypovolemic shock such as significant blood loss; pale, cool, moist skin; hypotension; and rapid, shallow breathing. This diagnosis should trigger the decision to administer intravenous fluids unless other circumstances render fluids contraindicated (e.g., the patient is a near drowning victim). The paramedic also needs to decide how to perform this intervention—for example, which type of fluid to administer when more than one option is available, at what dosage, how large a catheter to use, etc.. Whereas psychomotor skills must be repeatedly rehearsed "hands on" to be mastered—for example, by starting an IV and administering fluids in patient manikins—clinical reasoning skills can be practiced by "voice treating" simulated patients, which can be anything tangible: a manikin (if available from a simulation lab), doll, peer, etc. Voice treating entails verbalizing the actions the "EMS team" would perform, how they would perform them, which actions the team leader would delegate to partners, etc. Students often mime actions and use readily available equipment (e.g., a stethoscope); however, costly simulation equipment is unnecessary.

Fig. 2. MedDbriefer's Observer Interface

Since MedDbriefer runs on a tablet, students will be able to use the fully developed version to do practice scenarios just about anywhere—for example, in a small meeting room or dorm room. As shown in Fig. 1, a peer who is neither the EMS team leader nor a team member plays the role of "session observer," using MedDbriefer's checklists to record the team leader's verbalized actions. MedDbriefer's *Observer Interface* (OI) provides two main checklists. The *assessment checklist* (Fig. 2, left) is patterned after one of the scorecards used to assess candidates during the National Registry of Emergency Medical Technicians' (NREMT) paramedic certification exam [7]. When the observer checks an assessment step, the system displays a finding to call out. For example, if the team leader states that he is checking breathing quality, the observer is cued to call out "gurgling," highlighted in yellow in Fig. 2. The *intervention checklist* (Fig. 2, right) includes treatments and other actions that EMS providers perform such as ventilating a patient with a bag-valve mask and securing a patient onto a longboard.

Interspersed throughout the checklist menus are prompts for the observer to issue if the team leader fails to provide sufficient detail while voice treating. For example, the right side of Fig. 2 displays an Airway Management intervention, bag-valve mask ventilation, along with questions the observer should ask a student who fails to specify the ventilation rate, oxygen flow rate, and/or target O_2 saturation level.

As noted previously, MedDbriefer implements two approaches to automated debriefing. The first is a step-by-step walkthrough of students' actions, color coded to signal correct (green) and incorrect (red) actions (see Table 1). This is the standard approach to debriefing taken in computer-based simulation systems such as vSim for Nursing [8]. The second approach adapts one of several protocols that have been developed to guide simulation instructors in conducting effective debriefings: DEBRIEF [5, 6, 9]. This acronym stands for **D**efine the debriefing rules; **E**xplain the learning objectives; specify the performance **B**enchmarks; **R**eview what was supposed to happen; **I**dentify what actually happened; **E**xamine why things happened as they did; and **F**ormalize the "take home" points. MedDbriefer uses the same procedures to analyze event logs, and presents identical feedback, across these two debriefing approaches.

Table 1 presents a sample of debriefing feedback in the first approach, a timestamped walkthrough of a mock EMS team's scenario solution. The scenario involves the near drowning of a four-year-old boy left unattended in a swimming pool. Note that the system flagged an error in how the team ventilated the patient (line 15), failure to check for external bleeding (line 17), and late assessment of the patient's pulse (line 21).

MedDbriefer's development was informed by extensive observations of actual simulation-based training sessions, facilitated by human EMS instructors; feedback from experienced EMS and nursing educators; and field trials of the system that included paramedic trainees as participants. As a first step, we videotaped, transcribed, and analyzed over 100 h of simulated scenarios that took place during the University of Pittsburgh's 2020–2021 Emergency Medical Services program. Analysis of session transcripts enabled us to identify common errors that EMS trainees make while assessing and treating patients and the feedback that instructors provide to address these errors. Guided by this analysis, a physician authored most of the feedback that MedDbriefer provides during debriefings, as part of her graduate research project. She also analyzed the data from an initial field trial [6].

This trial revealed several "bugs" to address, and unclear feedback to revise, before we could conduct the randomized trial currently in progress. For example, we discovered that several decision rules deployed during event log analysis had to be refined, to prevent students from receiving negative feedback for correct actions. To illustrate, one temporal constraint stated that students should complete all assessment steps in the Primary Survey, and address all identified life threats, before starting the Secondary Survey (see Fig. 2). However, paramedics often need to examine the patient's head, neck and/or chest to identify and manage respiratory problems. Hence, we revised this rule so that students would not be told that they assessed the patient's upper body too early, in scenarios that require them to identify and manage compromised breathing.

3 Related Work

Other tools to support peer-to-peer and self-directed simulation are available on the market. For example, nursing educators at the University of Stavenger, Norway, experimented with using Laerdal's® Simpad to support peer-centered simulation-based training [4]. Like MedDbriefer, SimPad and its successor, Simpad Plus, provide checklists on a tablet-based interface that a student can use to log their peers' actions during a scenario. However, unlike MedDbriefer, SimPad and Simpad Plus don't debrief students on their performance. Instead, they provide tools that instructors can use for debriefing, such as a Log Viewer that displays a history of students' actions during a simulation session. Including the human instructor in the loop may cause delays in students' receipt of feedback. MedDbriefer emulates the log analysis and debriefing skills of human instructors, so students can receive immediate, high-quality feedback.

MedDbriefer also emulates human instructors' tendency to prompt students for additional details while voice treating a simulated patient when students' verbalized actions are vague. These additional specifications allow the system to provide more detailed, adaptive debriefings than it otherwise could. To our knowledge, other systems do not probe for important missing specifications.

Several computer-based simulation platforms generate debriefings, such as the American Heart Association's Heartcode™ BLS and ACLS programs to train basic and advanced cardiac life support skills, respectively, and vSim for Nursing [8]. The latter engages nursing students in realistic clinical interactions with patients in a hospital setting. Like MedDbriefer, these tutoring systems analyze the log of students' actions immediately after a simulation session, in order to generate a debriefing. However, unlike MedDbriefer, they do not afford hands-on interaction with a tangible, 3D "patient." Preliminary research suggests that interaction with tangible simulated patients such as manikins, peers acting as patients, etc. may predict superior patient care performance than interaction with screen-based simulated patients [10].

Table 1. Excerpt from a step-by-step debriefing that followed a near-drowning scenario (slightly modified for improved readability)

ID	Timestamp	Action description	Feedback
Breathing			
15	01:56	ⓧ Performed bag-valve mask ventilation. **Purpose:** Oxygenation **Oxygen flow rate:** 15 liters/min or greater **Target O2 saturation rate:** 95-99% **Ventilation rate:** 10-12 BPM	*Incorrectly administered intervention.* You were incorrect about ventilation rate. This patient should have been ventilated at a rate of 20 breaths per minute, which is the recommended rate for a child aged 2-12. When performing BVM ventilation, you use 15LPM O2 in order to both oxygenate and ventilate the patient to a target SpO2 >90%.
Circulation			
16	02:01	☺ Checked skin condition (moisture). **Found:** *Wet*	
17		ⓧ Performed a gross blood sweep. **Found:** *No major external bleeding*	*Missing assessment step.* A gross blood sweep, which was negative in this patient, will help you identify any life-threatening bleeding you may have missed earlier when forming your general impression.
18	02:02	☺ Checked skin color. **Found:** *Cyanosed*	
19	02:02	☺ Checked skin temperature. **Found:** *Cool*	
20	02:09	☺ Which pulse? **Pulse to check:** carotid and radial	
21	02:09	⏰ Checked pulse rate. **Found:** *Slow*	*Mistimed assessment step.* All unconscious patients should have a pulse check before starting the ABCs. However, paramedics often check breathing while doing a pulse check. If the patient is pulseless, you will start high-quality CPR immediately…

4 Analyzing Scenario Logs to Generate Debriefings

4.1 Overview

Once a student completes a scenario as an EMS team leader, the event log is automatically analyzed to identify what was done well and what needs improvement. The event log (EL) includes the observer's checked-off actions and is analyzed in three phases, as described in this section. Each phase utilizes the event log, the assessment hierarchy (AH) and the management hierarchy (MH). These hierarchies represent knowledge specified offline by domain experts and stored in a database for use during analysis.

The AH is a downward branching tree whose parent node is the goal of completing a full patient assessment and branches are assessment phases and subphases. For example, a thorough trauma patient assessment consists of an initial *Scene Size-up* to determine whether the scene is safe, how many patients there are, the mechanism of injury, etc.; a *Primary Survey* to qualitatively assess the patient's airway, breathing, and circulation (e.g., Is his airway clear? Is he breathing? Does he have a pulse?); *History Taking,* which includes taking baseline vital signs, finding out as much as possible about what happened and the patient's medical history; a *Secondary Survey* or focused head-to-toe assessment to check for injury and anatomically specific conditions, such as jugular venous distension at the neck; and ongoing reassessment and management. Figure 2 at left shows the top two levels of the assessment hierarchy. Lower levels are displayed when the observer selects a menu item. For example, Fig. 2 shows the checklist that appears when the observer selects Breathing in the *Primary Survey* menu.

Like the Assessment Hierarchy, the Management Hierarchy (MH) is a downward branching tree whose parent node is the goal of managing the clinical problems identified during patient assessment, the children are separate problems (e.g., severe bleeding, hypovolemic shock), and grandchildren are interventions necessary to address these problems, including acceptable alternatives. For example, in the scenario that involves a child drowning, the main management goals are to control the child's compromised airway and breathing. Managing the airway requires suctioning and intubation. The latter, in turn, entails inserting one of several appropriate airway adjuncts.

4.2 Analysis Phase 1: Interpreting the Event Log

During the first phase of analysis, the observed events in the EL are interpreted by comparing them to two models: the expected patient assessment actions specified in the AH and solutions to clinical problems specified in the MH. In addition, the system scores any responses to the observer's request for additional details. In our current implementation, domain experts manually specify the MH for each scenario instead of generating these solutions automatically. Interventions (the leaf nodes) in the MH are designated as either "required" or "optional" and, as noted previously, there may be more than one acceptable alternative for required interventions. Interventions that are not part of any solution are simply designated as "not indicated."

Finer distinctions could be made within this "not indicated" category—for example, irrelevant vs. contraindicated—that is, the intervention doesn't apply to the current scenario or is potentially harmful, respectively. For example, in the near-drowning scenario referred to previously, tourniquet usage would be considered irrelevant because the patient is not bleeding. However, administering IV fluids would be considered contraindicated because a drowning victim likely has too much fluid already in their system. We chose not to make these distinctions at the representational level in the current prototype. However, they are addressed in the feedback that domain experts authored. While it usually suffices to point out that an irrelevant intervention is unnecessary—for example, "this patient is not bleeding, so a tourniquet is unnecessary"—contraindicated interventions often require more complex explanations—for example, how IV fluids could cause pulmonary edema in a drowning victim.

Some interventions must be performed in an expected order to be effective, whereas timing is less critical for other interventions. For example, in the child drowning scenario, it is important to suction the child's airway so that it is clear before intubating and ventilating him. In our system, we represent these temporal constraints as rules and use these rules to assess the ordering of interventions recorded in the event log.

There is usually a simple one-to-one mapping between assessment actions in the EL and the AH, and between interventions in the EL and the lowest levels and leaf nodes in the MH. However, there are two complex cases. First, some interventions can appear multiple times in the EL and/or the MH, because the same intervention could be used to address multiple instances of the same type of problem. For example, there might be multiple wounds on the patient's body, which could all be managed by applying sterile dressings. In the current implementation, the session observer is prompted to specify the body part(s) that the student applies dressings to. Future versions of the system will attempt to infer which wound(s) sterile dressings are being applied to.

The second challenging analysis case is when the same intervention could be performed to satisfy more than one management goal in the MH. For example, establishing an IV might be indicated by protocol (i.e., it is standard practice for a trauma patient) and by the need to administer fluids to address shock. The analysis system chooses the solution path with the best fit to the EL. The accumulated findings up to the point when an intervention occurs in the log could also provide clues about which goal the student intends to address. However, we have not yet implemented semantic relationships between findings and goals. Alternative interventions (e.g., alternative advanced airways) are handled similarly to multiple instances of the same intervention. The analysis system picks the solution path that best fits the EL.

By associating events in the EL with items in the AH and the MH, knowledge is gained about the possible role of each event, such as what to expect some time before or after a particular event and the purpose of that event. This information facilitates recognizing assessment sections and management goals that may not have been completed during one contiguous time frame—for example, the student interrupted an assessment section to start a different one and returns to the interrupted section later. It is also used as part of checking temporal constraints in the second analysis phase and organizing the final debriefing presentation in the third phase, as described presently.

The AI in Analysis Phase 1 is this matching process, a search to find the solution path that best explains the events logged. Similar approaches have been used in other intelligent tutoring systems such as Andes [11] and the Cognitive Tutors [e.g., 12]—that is, generate solutions and do plan recognition by matching observations of what the student did to possible solutions [13].

As noted previously, in addition to assessment and treatment actions, the EL includes observer prompts for additional details about how to perform these actions. For example, when ventilating a patient with a bag-valve mask, the team leader is prompted to state the ventilation rate, oxygen flow rate, and target O_2 saturation, if he doesn't volunteer these details (see Fig. 2, right). The observer interprets the team leader's responses and selects the multiple-choice items in the interface that best match what the student said. Because the observer is expected to be a peer, not an instructor, the system determines whether the selected responses are correct during Analysis Phase 1.

4.3 Analysis Phase 2: Applying Temporal Constraints

In the second phase of analysis, the identified actions, assessment sections and management goals from the EL are analyzed relative to a set of temporal constraints. Note that assessment sections and goals are a collection of actions, so we need to consider temporal intervals when checking constraints [14]. Temporal representations and constraints, and constraints in general, are part of problem solving and plan recognition and thus are important in reasoning [13–16].

Although the AH and MH imply orderings for actions, these suggestions are ignored during this phase because instructors allow flexibility when ordering is unimportant. For example, the relative ordering of actions within the *Secondary Survey* (see Fig. 2, left) is not an instructional priority. As a case in point, it is not critical for a student to check the patient's head for injury before checking the patient's neck or chest, although proceeding in a "head-to-toe" fashion is recommended to help ensure a thorough, systematic patient assessment. Thus, temporal constraints represent those orderings that are a priority for instructors. Most constraints focus on managing life threats identified during the *Primary Survey* before doing anything else. For example, if the patient has severe bleeding from an extremity, apply a tourniquet before taking vital signs, starting the *Secondary Survey*, etc..

Most temporal constraints apply globally, across scenarios. However, some constraints apply conditionally—that is, they depend on the patient's state. For example, by default, one should check the patient's airway before checking breathing and circulation, an "ABC" ordering. However, if the patient is assessed to be unconscious, then check the pulse prior to assessing airway and breathing, a "CAB" ordering. If the arguments (i.e., actions or intervals) for a constraint are present in the annotated EL from the first phase of analysis and a temporal constraint fails, then the argument that is "late" in the constraint representation is annotated as being misordered. For example, if the constraint, "Check an unconscious patient's pulse before checking airway and breathing" fails because the student checked the patient's airway and breathing before checking his pulse, but the student does (eventually) check the patient's pulse, then the action "Checks pulse" is marked as "late" (e.g., see Table 1, line 21).

4.4 Analysis Phase 3: Identifying and Marking Missing Actions

In the final phase of analysis, missing assessment actions are identified and inserted into the section in the annotated EL in which they best fit and are assigned a status of "missing." The suggested orderings implied by the AH and MH are utilized so that missing actions are inserted in the annotated debriefing log where they are inferred to be most appropriate (see Table 1, line 17). The insertion heuristic first tries to locate other events related to the same assessment phase or management goal and inserts the missing one relative to the ordering specified in the AH or MH. If a management goal is missing entirely from the student's solution, the missing intervention is inserted at the end of the assessment section in which the MH indicated it should appear.

For example, if the student doesn't check the patient's pulse at all—as opposed to checking the pulse late—"checks pulse" would be inserted in the *Primary Survey*/Circulation sub-phase of the debriefing narrative and tagged as a "missing assessment step" with a red X (e.g., see Table 1, line 17). Missing interventions are likewise identified and inserted into the annotated debriefing log based on the solutions specified for their management goal in the MH and relative to where they best fit in the student's solution (the EL). For example, if the student failed to administer oxygen to the near-drowning victim and eventually intubate him, these interventions would be inserted in the *Primary Survey*/Breathing and *Primary Survey*/Airway sub-phases, respectively.

5 Conclusion

Using the approach to analyzing event logs described in this paper, we developed a prototype tutoring system that can serve as a platform to compare alternative approaches to debriefing [5, 6]. Although we do not yet know which of the two approaches that MedDbriefer implements will predict higher learning gains, if either, students' feedback on a post-participation survey has been highly positive and constructive. For example, all participants in the randomized trial to date agreed with the statement that MedDbriefer, when fully developed, will be useful for EMS training. To illustrate:

> Yes, I believe this will be useful for EMS training. The only way to improve the skills is to apply them, and sometimes our lab sessions aren't enough practice. I can see this system being extremely useful outside of the classroom with friends or people who aren't as familiar with EMS, because they are still able to proctor the simulation…The instant feedback prompts me to incorporate the objectives I missed in the previous simulation.

Study participants have also pointed out bugs and limitations of the system that we plan to address. For example, in some scenarios, vital signs fail to improve after the student performs suitable interventions. In addition, the system needs to better accommodate variations in state EMS protocols. Recently, several students enrolled in a paramedic program in the state of California participated in the RCT. Their feedback highlights the need to enhance the rules and routines that drive MedDbriefer's analysis of event logs so that the system can provide students with feedback that reinforces their state's EMS protocols.

More work needs to be done to make MedDbriefer more scalable in other ways besides accommodating different groups of users—in particular, streamlining content development. As noted previously, EMS experts manually specify each scenario's management goal hierarchies. We are exploring the possibility of implementing a "problem-solver" that uses a scenario's findings to determine what clinical problems need to be addressed and, correspondingly, which interventions are indicated to address them. Since state protocols specify how to manage most clinical problems that EMS providers encounter in the field, they could drive development of an automated problem solver, supported by related work on solution generation [e.g., 11, 12, 17].

In addition to saving domain experts the time needed to manually enter possible solutions for each scenario, automated solution generation could enable instructors to quickly alter the findings in an existing scenario. This would afford students practice with managing variations of clinical problems. For example, although the student recognized and applied an intervention that was indicated in the current scenario (e.g., apply direct pressure), perhaps in a variant of this scenario the same intervention would not control the patient's bleeding, so other interventions should be tried (e.g., apply a tourniquet). Such versatility in scenario design and solution generation would enhance training across the healthcare professions.

Acknowledgements. This research is supported by grant 2016018 from the National Science Foundation. The ideas and opinions expressed are those of the authors and do not necessarily represent the views of the NSF. We thank John Gallagher, Karen Kornblum, Emily Miller, Collin O'Connor, Erin O'Meara, Thomas Platt, Stuart Prunty, Samuel Seitz, Emma Sennott, Keith Singleton, Zachary Smith, Marideth Tokarsky, and Tiffany Yang for their contributions.

References

1. McKenna, K.D., Carhart, E., Bercher, D., Spain, A., Todaro, J., Freel, J.: Simulation use in paramedic education research (SUPER): a descriptive study. Prehosp. Emerg. Care **19**(3), 432–440 (2015)
2. Zendejas, B., Brydges, R., Wang, A.T., Cook, D.A.: Patient outcomes in simulation-based medical education: A systematic review. J. Gen. Intern. Med. **28**, 1078–1089 (2013)
3. Boet, S., Bould, M.D., Bruppacher, H.R., Desjardins, F., Chandra, D.B., Naik, V.N.: Looking in the mirror: Self-debriefing versus instructor debriefing for simulated crises. Crit. Care Med. **39**(6), 1377–1381 (2011)
4. Haraldseid, C., Aase, K.: Variability among groups of nursing student's utilization of a technological learning tool for clinical skill training: An observational study. J Nurs Educ Pract **7**(7), 66–76 (2017)
5. Katz, S., et al.: Comparing alternative approaches to debriefing in a tool to support peer-led simulation-based training. In: Intelligent Tutoring Systems: 18th International Conference, ITS 2022, Bucharest, Romania, June 29-July 1, 2022, Proceedings, pp. 88–94. Springer International Publishing, Cham (2022)
6. Katz, S., Albacete, P., Jordan, P., Silliman, S., Yang, T.: MedDbriefer: A debriefing research platform and tool to support peer-led simulation-based training in healthcare. In: Blikstein, P., Van Aalst, J., Kizito, R., Brennan, K. (eds.), Proceedings of the Twenty-third International Conference of the Learning Sciences—ICLS 2023, Montreal, Canada, June 2023: International Society of the Learning Sciences (2023)

7. NREMT: Advanced level psychomotor examination (2020). https://www.nremt.org/getmedia/6302f735-b899-499e-8a51-26fefac999df/Patient-Assessment-Trauma-v2020_2.pdf
8. Laerdal Medical: vSim for Nursing: Building competence and confidence—anytime and anywhere (2020). https://www.youtube.com/watch?v=rXak70MxnAk
9. Sawyer, T.L., Deering, S.: Adaptation of the US Army's after-action review for simulation debriefing in healthcare. Simulation in Healthcare **8**(6), 388–397 (2013)
10. Haerling, K., Kmail, Z., Buckingham, A.: Contributing to evidence-based regulatory decisions: a comparison of traditional clinical experience, mannequin-based simulation, and screen-based virtual simulation. J. Nurs. Regul. **13**(4), 33–43 (2023)
11. VanLehn, K., et al.: The Andes physics tutoring system: Lessons learned. Int. J. Artif. Intell. Educ. **15**(3), 147–204 (2005)
12. Koedinger, K.R., Corbett, A.: Cognitive tutors: Technology bringing learning sciences to the classroom (2006)
13. Carberry, S.: Techniques for plan recognition. User Model. User-Adap. Inter. **11**, 31–48 (2001)
14. Allen, J.F.: Towards a general theory of action and time. Artif. Intell. **23**(2), 123–154 (1984)
15. Jordan, P.W., Mowery, D.L., Wiebe, J., Chapman, W.W.: Annotating conditions in clinical narratives to support temporal classification. In: Proc American Medical Informatics Association Symposium, Vol. 2010, p. 1005 (2010)
16. Weida, R., Litman, D.: Terminological reasoning with constraint networks and an application to plan recognition. In: Principles of Knowledge Representation and Reasoning: Proceedings of the Third International Conference (KR'92), pp. 282–293 (1992)
17. Saibene, A., Assale, M., Giltri, M.: Expert systems: Definitions, advantages and issues in medical field applications. Expert Syst. Appl. **177**, 114900 (2021)

Case Study of Organization of Decision-Making and Feedback Synthesis in Intelligent Tutoring Systems with a Cross-Cutting Approach

Viktor Uglev(✉)

Siberian Federal University, Zheleznogorsk, Russia
vauglev@sfu-kras.ru

Abstract. This paper describes an approach to support the decision making process in the Intelligent Tutoring Systems (ITS) and provide graphical presentation of the ITS decisions. The discussion focuses on the issue of explainability of the decisions made by the ITS and their explanation to a human learner. Highlighting the most significant aspects in the interpretation of the learning situation and their further use by the ITS intelligent scheduler is based on the mapping mechanism and the cross-cutting approach to switching between the maps. The feedback synthesis in the form of the dialogue is based on parametric maps and their visualization using CMKD notation. The maps are combined into an atlas, which is used as the basis for decision-making when switching from the combined map to the particular maps. An example of learning situation analysis using the cross-cutting approach in the experimental ITS is discussed in detail.

Keywords: decision-making · Intelligent Tutoring Systems · cognitive visualization · decision explanation · feedback · cross-cutting approach

1 Introduction

Intelligent Tutoring Systems (ITS) used in modern e-learning are focused on individual work with a human learner. Decision making by the ITS intelligent scheduler and feedback synthesis in such systems should allow for verification of their adequacy. This can be done through natural language dialogue, implementing the concepts of XAI [2] and Humanistic AI [6]. But this would require a flexible system of representation and analysis of the learning situation. Let's consider this issue in more detail.

The interaction of a learner with the ITS, in the limit, occurs throughout the entire learning cycle starting from registration in the system at the beginning of the educational process to the final assessment of all the disciplines studied at the current level. The typical sequence of the learner's work with the e-course (discipline) consists of many portions of the learning material (didactic units).

K. Kabassi et al. (Eds.): NiDS 2023, LNNS 783, pp. 114–124, 2023.
https://doi.org/10.1007/978-3-031-44097-7_11

The set of the course material forms a semantic structure (ontology) which is the initial subject substrate for tutorial structuring (e-course creation) and further work of the tutoring system algorithms. This is how the data hierarchy is built from didactic units (nano-level) to the whole level of education (meta-level) [19].

A hierarchy of the disciplines, the learning material of which is associated with the target skills and competencies allows generating an individual model of the learning situation (relative to the current semester). Therefore, decisions regarding an individual discipline should correlate both with the individual learner's needs and with the strategic learning goals (curriculum level). It is much more difficult to implement this functionality because it involves working with the learning situation of all disciplines at the same time. But the flexibility of such an approach promises greater effectiveness in the development of pedagogical influences by ITS.

The processing of the learning situation shall cover, if possible, the following types of P parameters (see [20]): for details): the scope of the learning situation (P^M), levels of decision-making (P^L), aspects of consideration (P^A), dynamics in time (P^B). In addition, it is necessary to record the type of decision-making task relative to the moment of its occurrence (Ω). By default, events are recorded in ITS protocols (a digital footprint is generated), hypotheses are tested, and the learner model is updated [9]. In addition to the learner model the advanced ITS offer the models of the teacher and the tutor, allowing to form a "point of view" on the current learning situation [10]. A set of virtual models of "stakeholders" incorporated into the ITS is denoted by Q.

A particular decision can only be justified when it is possible not only to highlight the problematic aspects in the space of analyzed factors, but also to focus attention on those of them that are most essential for achieving the goals set. The same is also true for the explanations: synthesis of several phrases is sometimes enough to justify decisions based on the relatively small amount of data (e.g., trajectory of task solution), but if there is a lot of data (academic performance in all disciplines from the digital educational footprint), then one cannot do without attention focusing tools (text is focused on key aspects and supplemented with cognitive visualization tools) [3,5,7,15,18].

The purpose of this paper is to demonstrate the application of the cross-cutting approach to the synthesis of ITS decisions and their reasoning in the form of the dialogue text (feedback generation). For this purpose, in Sect. 2 we will give an explanation of some aspects of the cross-cutting approach concerning decision making and explanation (the approach itself is described in [19]) based on parametric maps and their visualization. In Sect. 3 we will describe a case study on implementation of decisions on the basis of the proposed approach.

2 Method

2.1 Generalization of the Input Data in Cross-Cutting Approach

Extended representation of data about the learning situation includes not only data about the immediate actions of a learner but also the data about the learning environment. To order the data, we use the Pospelov's square developed in applied semiotics [13]. Let us divide the substrate of primary data about the learning situation into three parts: the data forming the digital educational footprint in ITS protocols (activity, pragmatics, R); the data this footprint remains on (structural blocks, syntactics, W); and the data specifying the semantic and target structure of the learning process (semantics, Z), i.e. the "reconstructed" image of the one who left the footprint. All three data blocks intersect in such an entity as a didactic unit: any action, dialogue, or plan is described at the level of operations with one or more elements of a learning course. For this reason, it is possible to form a complex knowledge structure that comprehensively describes the learning situation in the form of graph G, as a set of subgraphs $<G_w \cup G_z \cup G_r>$.

The G_r configuration will define the existing learning situation at the time δ. One of the models included in Q, will act as the denotation, interpreting the learning situation in relation to one of the solved tasks included in Ω. If we limit ourselves to those points of call of the ITS intelligent solver (E events) that occur within an individual learning course, they can include the following [18]: enrolling in a course, solving an individual task, passing a test, the occurrence of an event in the presentation of theoretical material, interaction with a dialogue form. In fact, the combination of Ω and E largely determine the nature of the hypotheses that will test and justify the models from the set Q (control style).

The general approach to data transformation can be as described in [17]. At the first stage statistical concentration is performed which is the process of generalization of set of facts and their parameters from ITS protocols G_r and parameters of the learner and e-course models, using statistical methods of processing (e.g., calculation of average grades for each learning module). The second stage involves metric concentration which is formation of structural and functional configuration of entities in the form of invariant set of maps (personalized template of metric space in various aspects P, collected into an atlas), specified by G_w and G_z hierarchies (e.g., synthesis of a Cognitive Map of Knowledge Diagnosis in the ITS memory). During the third stage there is semantic concentration which is selection of maps from the atlas and emphasizing those aspects (according to G_r), which will be most significant in the current learning environment at the given type of task from Ω and active model from Q (e.g., to develop recommendations for the repetition of the learning material). At the fourth and final stage (logical concentration) takes place which involves extraction of arguments from maps by the ITS recommendation system to justify decisions made (relying on the knowledge base, direct display) and their implementation with subsequent demonstration to a human learner (e.g., synthesis and display of the text of the prompt message accompanied with the simplified cognitive map).

Structural map projections are used to display the learning situation in relation to the hierarchy of the learning material description (see levels shown in Fig. 1a from [19]). Their main elements are encoded entities in the actual level of detail (scale), supplemented by semantic links. As an illustration, let us consider an example of a map for an educational semester of the master's degree program in Informatics and Computer Science (nodes represent academic modules and edges represent semantic links) presented in the notation of Cognitive Maps of Knowledge Diagnosis (CMKD [19], Fig. 1a). It shows how course topics (syntactics, circular elements) interact with causal relationships (semantics, arrows) in a structural projection. If the map is specified for a particular learner and overlaid with data from the digital footprint (pragmatics, colors, shapes, fonts, etc.), it will describe a particular state of the learning process (Fig. 1b and Fig. 1c show the examples of the subject discipline and competency aspects, respectively).

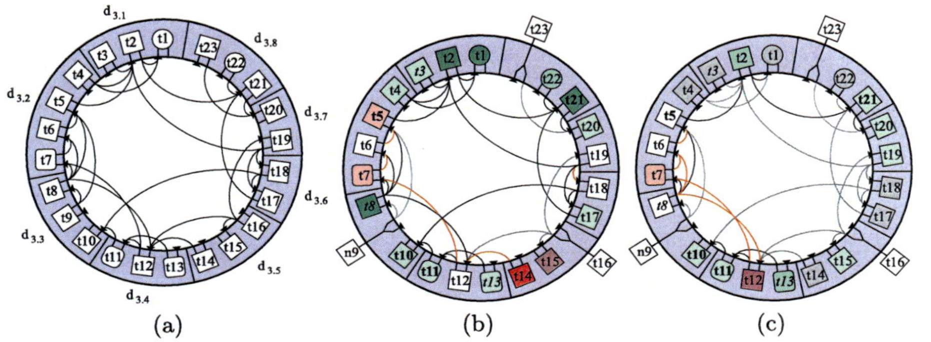

Fig. 1. Example of the CMKD for academic disciplines of one semester (a), the map with overlaid data of one of the students in the subject discipline (b, assessment of progress in points) and competency (c, detailed assessment of the level of mastery of one of the competencies) aspects

When there is a dialogue or repeated hypothesis testing for different elements of the course, the map (substrate) remains unchanged, and only the real-time data regarding Q is updated. This not only speeds up the process of analyzing the learning situation (minimizing repeated statistical and metric data concentration), but also provides an invariant description of the primary data for matching decisions of the teacher, learner, and tutor models [19]. A generalized description of the algorithm for generation the current analysis configuration can be written according to the following sequence:

1. an atlas entry point corresponding to the task specificity $\Omega(E)$ by scale m is determined;
2. the most significant projection type (structural or functional) is selected;
3. data from G, corresponding to the decision-making level for the learning goals is selected (if necessary, the map is expanded without changing scale);
4. the most significant aspect from K is evaluated and the corresponding learner activity data from the educational footprint is overlaid on the map;
5. additional notations reflecting the dynamics of the process b are introduced into the map;

6. the emphasis is shifted from the graphic view depending on the specifics of the point of view of the current model from Q or the direction of analysis.

At steps 5 and 6 the key positive and negative points are highlighted on the maps, allowing forming reasoning for the decision to be made. There is a possibility to vary the point of focus in the decision-making process, choosing the desired scale of coverage of the learning situation and aspects in factor space.

2.2 Cross-Cutting Approach in Decision-Making

The set of Q-models included in the ITS is responsible for the possibility to simulate several points of view on the learning situation (teacher, leaner and tutor). After the solver has assembled a personalized set of maps (the metric concentration stage), data from the digital educational footprint are overlaid on them and presented to each of the models for hypothesis testing. To do this, a blackboard metaphor [8] is used according to the following sequence of actions:

1. a hypothesis about the state of the learning situation is made (the blackboard is empty);
2. a learner model is chosen from Q;
3. maps from the atlas are looked through and evaluated for each aspect P, and the most illustrative representations of data which contribute to decision-making by the model are noted;
4. the model generates a decision based on the relevant part of the knowledge base (the decision complements/changes the state of the blackboard, including recording key arguments);
5. steps 3 and 4 are repeated for the teacher model;
6. individual decisions (from the learner model and the teacher model) are transferred to the tutor model to find a trade-off;
7. evaluation of the final decision with the blackboard by the ITS intellectual core, its planning and implementation;
8. an explanation is generated.

As a rule, at the last step not only the decision is presented to a learner, but also he is given a possibility to receive the explanation in the natural language (see [1,4,11,16,18]) and graphic form (see [7,14,19]). If the object of the decision is the configuration of the learning material, then the corresponding map is put on the blackboard and it becomes the object of modification by Q-models.

2.3 Feedback Synthesis

The feedback generated by the ITS can take two forms: non-verbal and verbal. The first refers to the selection and implementation of the decision by the system, which is reflected in changes in the learner's capabilities in the educational environment (changes in the composition of interface elements, available functions, graphic images). It is accompanied by logging and is not always correctly understood by a learner. The verbal form, as a rule, represents interaction of a learner with dialogue boxes [12]. To generate the stimulus, it is necessary to use the appropriate knowledge base block Kb_{Ω}, which defines a general scheme of reasoning. In general, it can be written in accordance with the following sequence:

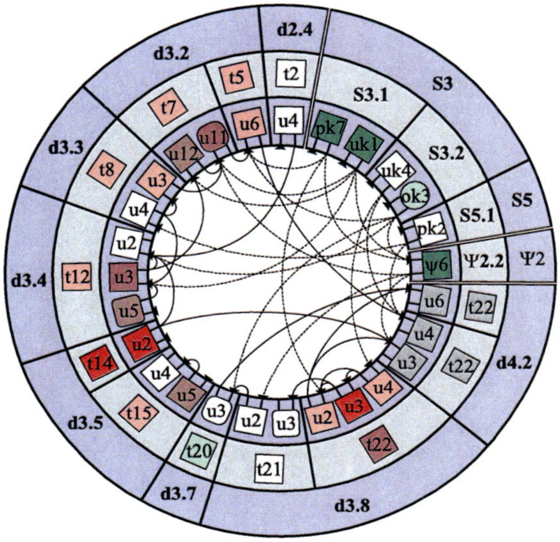

Fig. 2. An example of a combined map for a learning situation by the example of a master's degree program in Informatics and Computer Science (third academic term)

1. loading of all generalized data about the situation Ω, generalized on the atlas maps corresponding to the current scale M_{Ω} and forming a combined map (see Fig. 2) in the structural aspect;
2. iterative testing of a number of intermediate hypotheses on key structural and functional entities related to the scale M_{Ω};
3. testing hypotheses about the nature of learner behavior and his/her difficulties, taking into account the targets from the learner model;
4. selecting from the maps the most significant parameters reflecting the essence of the current situation and making hypothesis about selection of y decision;
5. confirmation of the hypothesis y about the effectiveness of application of the decision in a "dialogue" between Q-models or justification of an alternative hypothesis;
6. application of changes in the learning space, corresponding to decision made;
7. display of an appropriate description of the system reaction to a learner with the possibility of an explanatory dialogue $\gamma(y)$.

A combined map is a map in the CMKD notation, which combines two or more entities of analysis in a single image (ignoring the specificity of aspects) and defines the links reflecting the specificity of the learning situation. It displays those objects that, according to the assessment results and learner behavior, have a significant negative impact on the academic performance. Figure 2 shows an example of a simplified combined CMKD, which combines the structural hierarchy of the learning material (micro-meso-level), functional hierarchy (levels of competencies, key skills and their groups S), learner's expectations related to the field of study (academic program Ψ) and semantic links reflecting interdependence between the objects.

The dialogue focuses on two components at once: a natural-language-oriented description of the situation and its results (verbal form) and a visual complement (nonverbal form). The text of the dialogue is synthesized using templates, displaying key information in accordance with the emphases revealed with the help of maps [18]. The visual support of the dialogue is realized in the form of a simplified map focusing a learner's attention to the emphases selected by the ITS planner as the most significant. The maps are interactive and allow requesting tutorial clarification to any of the displayed entities. The main emphasis is made on synthesizing answers to the "how?" and "why?" questions (including personal motivation). The idea of the cross-cutting approach is realized here through the ability to move between scales (vertical movement) and to shift the focus from one element P to another (horizontal movement). The ITS records the trajectory and nature of the dialogue (the G_r footprint component).

3 Case Study and Discussion

3.1 Input Data

Let us consider examples of the analysis of a learning situation with the generation of a stimulus and the accompanying dialogue. As the basic program we will choose the master's degree program in Informatics and Computer Science at Siberian Federal University. For the current group of students (2021) the academic program tree will be as shown in Fig. 1b from [19] (macro-level and meso-level). Each discipline d_j corresponds to a set of learning topics (nodes t_i in the graph), divided into didactic units supplemented with the assessment and test material. When forming a personal learning trajectory, the individual elements of the learning material (didactic units, topics and even entire courses) may not be included in the program. In addition, there are the results of general and subject questionnaires filled out by a learner that contain his/her preferences for goals, material, competencies, practical skills and technologies.

During ITS-based learning, a group of students solves tasks and tests that assess both residual knowledge (subject discipline aspect) and the level of competency mastery. Combining the current grades, results of the previous tests and student preferences (taken from the questionnaires), the ITS solver makes decisions, activating a chain of steps shown in Fig. 1b from [19]. A description of the decision-making and feedback generation process (through dialogue) is given below.

3.2 Learning Situation Analysis

Step 0. System Display. Let us have a personalized configuration of the learning trajectory, the structural representation of which (meso-level for the third academic term) is shown in the CMKD notation in Fig. 1a. The student has passed the interim assessment and needs to get a comprehensive recommendation. To do this, there is a statistical concentration of the data from the digital educational footprint and its representation in the form of a set of maps (metric concentration). On the basis of the atlas maps a combined map (Fig. 2, semantic concentration).

The combined map shows that the most problematic is the topic "Pseudo Random Distribution" ($t7$) of the discipline "Simulation Modeling" ($d3.2$). As the entry point into the atlas, the ITS scheduler will select the CMKD from Fig. 1b (configuration of map parameters $P = <3, 2, 1, 1>$). We can see that according to the results of the interim assessment for the subject topics $t5$, $t7$, $t14$ and $t15$ there is insufficient mastering of the learning material.

The topic $t5$ is emphasized according to the data from the map shown in Fig. 1c, which reflects the data on competency uk1 "Can critically analyze problem situations on the basis of a systematic approach" from the current supra-subject competency profile [20]. And since the competency $uk1$ was indicated in the target map as a personal priority (green color from Fig. 2 on $uk1$ block), its mastery is traced through the chain of topics $t7$ ($u11$ and $u12$) and $t12$ ($u3$).

The following goal of academic action is formed: to improve the mastery as a result of convincing the student of the importance of restudy the material, which will improve the general mastery level in several directions at once, i.e. the element $t5$. This forces not only to analyze the learning situation from the teacher model position, but also to take into account the requests of the learner model, selecting the most personally meaningful system of reasoning (for example, take into account the links with $uk1$ and $\psi6$ reflected on the combined map).

As a result, the user is shown the following recommendation (fragment of text): "Your academic achievements are insufficient for the transition to interim assessment... To improve the results you are recommended first to study the topic "Factor Space Metrics" ($t14$) of the discipline "Data Mining" ($d3.5$); the topic "Pseudo Random Distribution" ($t5$) of the discipline "Simulation Modeling" ($d3.2$); and the topic "Structural Optimization (practice)" ($t12$) of the discipline "Foundation of Systems Engineering" ($d3.4$) ...".) ...". The feedback is accompanied with a map shown in Fig. 1b. The interface contains elements to initiate an explanatory dialogue.

Step 1. Learner's Choice. The learner selected the item $t12$ from the list of topics associated with the question "Why...?" and asked for further clarification.

System Display. The ITS displays the following (simplified) text as the feedback: "The topic $t12$ is one of the key topics in the $d3.4$ course, which you indicated in the questionnaire as very important. In addition, you have declared your intention to professionally master the competency $uk1$, which depends largely on the successful mastery of the $d3.2$ course topics". The feedback is accompanied with a structural map with the parameters $P = <3, 2, 2(uk1), 1>$ (see Fig. 1c).

Step 2. Learner's Choice. The learner selected the item $t12$ associated with the question "Why is it so important...?" (with the parameter "curriculum scale") and asked for further clarification.

System Display. To justify the interdisciplinary links, the map shown in Fig. 1c is supplemented with data from other semesters is supplemented with data from other semesters $d3.2$, but also the link with the upcoming topics of the fourth semester are highlighted in color. The following dialogue text is displayed: "Competency $uk1$ is not only mandatory within this program, but also lies at the core of educational program profiling ψ, affecting the group of key skills $S3$, i.e. "Mathematical and algorithmic methods of data processing", which you have indicated as the object of advanced study to be used in the your own dissertation research". The map clearly shows how $t12$ is influenced by $t7$, which in turn depends on $t6$ and $t5$.

Step 3. Learner's Choice. The learner selected a question "What is the place of... in the learning process...?" (with the $uk1$ parameter) and asked for further clarification.

System Display. The ITS solver shifts the focus from the structural projection to the functional projection and goes to the higher scale level. The anthropomorphic image in the UGVA notation shows the data on the contribution of competency $uk1$, taking into account the current assessments of the level of competency mastery. Colors from red (grades tend to zero) to green (grades tend to 100) with an uncertainty zone of 50 (white) are used as heat map scale values. Thus, $uk1$ is associated with the forearm of the right hand (skill group zone $S3$) and is insufficiently mastered (the grade is 37.3 points out of 100, which corresponds to the dark pink color). The dialog text with this data is displayed to a learner in addition to the image in UGVA notation (see [19] for more details on the topology of anthropomorphic images and its analysis.

Step 4. Learner's Choice. Having read the answer, the student returns to the previous dialogue box and selects a question "What exactly is needed...?" (with the parameter $t5$) and asks for further clarification.

System Display. The solver again turns to the structural projection, but to answer the question it retrieves the map from the micro-level (see Fig. 3a, which has the configuration $P = <2, 2, 2(uk1), 1>$). Its analysis shows that for the best effect the didactic units $u7$ and $u10$ should be studied first. This will give the basis for repeating the material $u11$ and $u12$ in the next step. The learner is shown the map from Fig. 3a, and the dialog phrase includes arguments from the competency and subject discipline levels. The learner can also click to the elements of the map to clarify individual indicators by reading the pop-up messages and changing aspects of the map drawing.

Figure 3b shows the diagram of transitions in the factor space for steps 0–4 and the emphases from the maps. It reflects the cross-cutting nature of moving the feedback focus not only relative to scale and projection, but also to other parameters from P (red arrows), allowing for reflection of emphases in clarification.

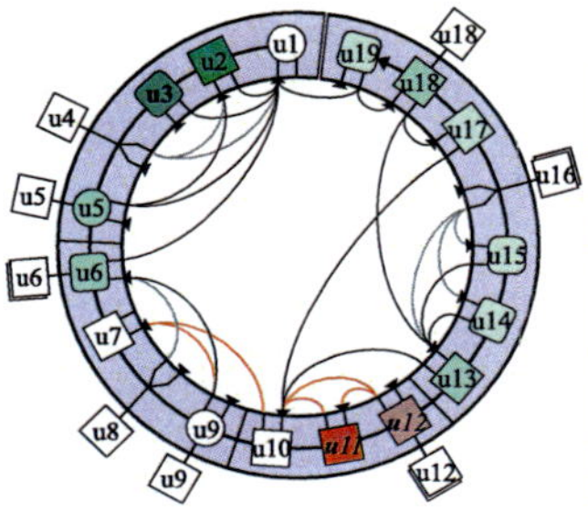

(a)

Step #	proje-ction	*m*	*l*	*a*	*b*	nota-tion
0	struct.	micro-macro	1-3	1-4	1-2	CMKD
0	struct.	meso	2	1	1	CMKD
1	struct.	meso	2	2(uk1)	1	CMKD
2	struct.	meso+	2	2(uk1)	1	CMKD
3	funct.	macro	3	2(uk1)	1	UGVA
4	struct.	micro	2	2(uk1)	1	CMKD

(b)

Fig. 3. The CMKD for the micro-level (a), the discipline "Simulation Modeling", the subject discipline aspect) and the cross-cutting moving between the maps (b).

4 Conclusion

The logic of the cross-cutting approach used for making decisions in ITS as well as the accompanying text and graphic images (feedback) are based on the parametric maps. The proposed approach allows switching from the particular indicators to the combined parametric map and then displaying the data again in the form of simplified particular maps. This makes the arguments for explaining the decisions more convincing, which increases the level of confidence, i.e. it brings the ITS functionality closer to the implementation of XAI ideas.

A key difficulty in implementing the cross-cutting approach is that the analysis of CMKD is performed by the mechanism of expert systems. Developing knowledge bases with the expertise of teachers and psychologists is the most debated and most promising part of our research.

At the current stage of research, we conduct a comprehensive experiment demonstrating the influence of the proposed approaches on the level of learners' satisfaction with ITS decision explanations and the effect of their presentation on the interaction with the tutoring system (change in confidence level).

References

1. Agarwal, A., Mishra, D.S., Kolekar, S.V.: Knowledge-based recommendation system using semantic web rules based on learning styles for MOOCs. Cogent Eng. **9**(1), 2022568 (2022). https://doi.org/10.1080/23311916.2021.2022568
2. Arrieta, A.B., et al.: Explainable Artificial Intelligence (XAI): concepts, taxonomies, opportunities and challenges toward responsible AI. Inf. Fusion **58**, 82–115 (2020)
3. Brusilovsky, P., Rus, V.: Social navigation for self-improving intelligent educational systems, pp. 131–145. Army Research Laboratory (2019). https://www.pitt.edu/peterb/papers/SocNav4SIS.pdf
4. Ezaldeen, H., Misra, R., Bisoy, S.K., Alatrash, R., Priyadarshini, R.: A hybrid e-learning recommendation integrating adaptive profiling and sentiment analysis. J. Web Semant. **72**, 100700 (2022)

5. Grann, J., Bushway, D.: Competency map: Visualizing student learning to promote student success. In: Proceedings of the Fourth International Conference on Learning Analytics and Knowledge, pp. 168–172 (2014)
6. Gruber, T.: Humanistic AI. Tom Gruber (2023). https://tomgruber.org/humanistic-ai
7. Ilves, K., Leinonen, J., Hellas, A.: Supporting self-regulated learning with visualizations in online learning environments. In: Proceedings of the 49th ACM Technical Symposium on Computer Science Education, SIGCSE 2018, pp. 257–262. Association for Computing Machinery, New York (2018)
8. Jackson, P.: Introduction to Expert Systems. Addison-Wesley Pub. Co., Reading (1999)
9. Karpenko, A., Dobryakov, A.: Model for automated training systems. Overview, science and education. Sci. Educ. **7**, 1–63 (2011). https://doi.org/10.7463/0715.0193116. (in Russian)
10. Kossiakoff, A., Sweet, W., Seymour, S., Biemer, S.: Systems Engineering Principles and Practice. Wiley-Interscience (2011)
11. Kuo, J.Y., Lin, H.C., Wang, P.F., Nie, Z.G.: A feedback system supporting students approaching a high-level programming course. Appl. Sci. **12**(14) (2022). https://doi.org/10.3390/app12147064
12. Mashbitz, E., Andrievskays, V., Komissarova, E.: Dialog in a tutoring system. Higher school, Kiev (1989). (in Russian)
13. Pospelov, D., Osipov, G.: Applied semiotics. News Artif. Intell. **1**, 9–35 (1999). (in Russian)
14. Sinatra, A., Graesser, A.C., Hu, X., Goldberg, B., Hampton, A.J.: Design Recommendations for Intelligent Tutoring Systems: Volume 8-Data Visualization. A Book in the Adaptive Tutoring Series. US Army Combat Capabilities Development Command-Soldier Center (2020)
15. Takada, S., et al.: Toward the visual understanding of computing curricula. Educ. Inf. Technol. **25**, 4231–4270 (2020). https://doi.org/10.1007/s10639-020-10127-1
16. Troussas, C., Papakostas, C., Krouska, A., Mylonas, P., Sgouropoulou, C.: Personalized feedback using natural language processing in intelligent tutoring systems. In: Frasson, C., Mylonas, P., Troussas, C. (eds.) ITS 2023. LNCS, vol. 13891, pp. 667–677. Springer, Cham (2023). https://doi.org/10.1007/978-3-031-32883-1_58
17. Uglev, V.: Implementation of decision-making methods in intelligent automated educational system focused on complete individualization in learning. AASRI Procedia **6**, 66–72 (2014). https://doi.org/10.1016/j.aasri.2014.05.010
18. Uglev, V.: Explanatory didactic dialogue in the intelligent tutoring systems based on the cross-cutting approach. In: Frasson, C., Mylonas, P., Troussas, C. (eds.) ITS 2023. LNCS, vol. 13891, pp. 371–380. Springer, Cham (2023). https://doi.org/10.1007/978-3-031-32883-1_34
19. Uglev, V., Gavrilova, T.: Cross-cutting visual support of decision making for forming personalized learning spaces. In: Krouska, A., Troussas, C., Caro, J. (eds.) NiDS 2022. LNNS, vol. 556, pp. 3–12. Springer, Cham (2022). https://doi.org/10.1007/978-3-031-17601-2_1
20. Uglev, V., Sychev, O., Gavrilova, T.: Cross-cutting support of making and explaining decisions in Intelligent Tutoring Systems using Cognitive Maps of Knowledge Diagnosis. In: Crossley, S., Popescu, E. (eds.) ITS 2022. LNCS, vol. 13284, pp. 51–64. Springer, Cham (2022). https://doi.org/10.1007/978-3-031-09680-8_5

Rescue Under-Motivated Learners Who Studied Through MOOCs by Prediction and Intervention

Hadjer Mosbah[1,2(✉)], Karima Boussaha[3], and Samia Drissi[4]

[1] Department of Mathematics and Computer Science, Faculty of Science and Technology, Mohamed-Cherif Messaadia University, Souk Ahras, Algeria
hadjer.mosbah@hotmail.com

[2] University of Constantine, 3 - Salah Boubnider, Constantine, Algeria

[3] Department of Mathematics and Computer Science, University of Oum El Bouaghi, Oum El Bouaghi, Algeria
karima.boussaha@univ-oeb.dz

[4] Department of Mathematics and Computer Science, LIM Laboratory, Mohamed-Cherif Messaadia University, Souk Ahras, Algeria
s.drici@univ-soukahras.dz

Abstract. Under certain circumstances, E-learning can prevent students from interrupting their educational process about what happened during the lockdown, which caused education facilities from different levels to close. As a result, traditional learning has been replaced by online learning and the popularity of MOOCs is increasing and rapidly spreading all over the globe. Even after the crisis, the conventional form of learning is being complimented by online education nowadays. However, MOOCs struggle with low completion rates and high dropouts even though a large number of students have joined the courses. A variety of factors may be contributing to the low completion rates, especially the lack of interaction. As a result, predicting MOOC dropouts is an interesting research topic. The problem we intend to address in our research is to develop a system that enables the delivery of customized interventions based on the classification of students who could have motivational barriers.

Keywords: Deep Learning · MOOCs · Sentiment Analysis · Learners' Motivation · Learners at Risk

1 Introduction

E-learning platforms are widely available that institutions can adjust to improve educational procedures [1]. There are no barriers to accessing e-learning platforms [2]. E-learning can complement or replace traditional education [3]. With Learning Management Systems (LMS) and Massive Open Online Courses (MOOCs), all universities adopted online learning as the only solution during the COVID-19 crisis. E-learning was fruitful in guaranteeing the continuity of education during the pandemic [4].

K. Kabassi et al. (Eds.): NiDS 2023, LNNS 783, pp. 125–129, 2023.
https://doi.org/10.1007/978-3-031-44097-7_12

Despite the important part that MOOCs play in the continuity of learning, the absence of interaction between the teacher and the student leads to the student's lack of motivation. It causes students to drop out before completing their courses [5]. There has been a lot of research on predicting student dropouts and the low completion rates of MOOCs in recent years [6]. In addition to predictions, interventions are the next step. To improve learning, researchers should not skip the intervention step [7].

Therefore, we aim to predict the unmotivated students in MOOCs based on their profiles, behaviors, and needs. The goal is to improve student interaction with e-learning platforms by incorporating Smart tutors. To accomplish this, many research questions should be addressed that require in-depth exploration among them:

- How to identify the students at risk of lack of motivation? What techniques should be used for that purpose?
- Using behavior analysis, how can the students at risk of motivation be grouped into clusters?
- How to implement customized tutoring for each cluster?
- How does customized systems implementation affect motivation-deficient students' learning?

The rest of this paper is organized as follows: Sect. 2 presents an outline of objectives. In Sect. 3, a brief for related works is described. Section 4 is reserved for the methodology and the expected outcomes. Finally, the general conclusion and future works are presented in Sect. 5.

2 Outline of Objectives

There has been a dramatic shift in the academic environment due to online education, especially Massive Open Online Courses (MOOCs) [8]. MOOCs have high enrollment, but over 90% of enrollees never complete their courses [5]. Several factors can prevent learners from completing the course. These factors include lack of time, engagement, and motivation, difficulty with course content, learners' interest, and lack of support [9]. Learners' motivation plays a significant role in MOOC dropout rates because motivation is directly linked to self-discipline, participation, performance, and satisfaction in educational practices [5].

Our work aims to predict learners at risk of lack of motivation in their learning environment. By explaining and clustering the factors contributing to students' lack of motivation, we aim to design a model to classify and predict students' motivation. It is possible to use such classifications and predictions for a variety of purposes. In our case, we aim to provide tutoring for students based on their classes during MOOCs through recommendation systems. By doing so, the completion and success rates can be improved (Fig. 1).

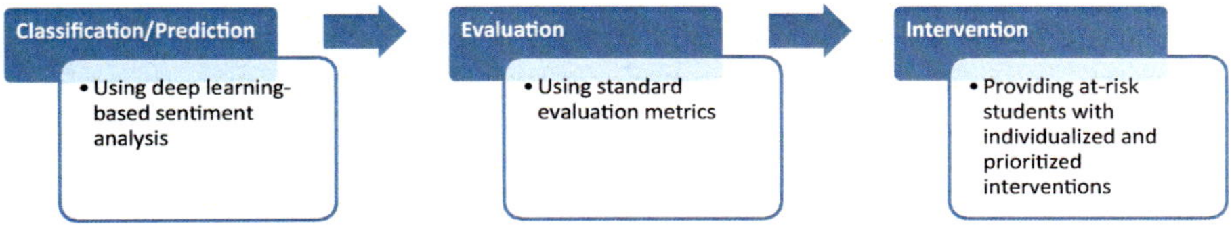

Fig. 1. Contribution, and the outlines

3 Related Works

The prediction of MOOC dropouts is an intriguing research topic considering the poor completion rate [10]. Various machine learning algorithms are adopted to tackle the problems of MOOC dropout [7], regression and support vector machines stand out among the many predictive models employed in different searches.

Meanwhile, in recent works using deep learning architectures in the prediction model, the authors combined Convolutional Neural Network (CNN), Bidirectional Long Short-Term Memory (Bi-LSTM), and static attention in the work presented in [11]. The authors presented a CNN and LSTM hyper-model in [12] to predict whether a student would withdraw from a class or fulfill it. A CNN model and Bi-LSTM were paired in another study by the authors in the work presented in [6] to build a fusion deep dropout prediction model. To predict and classify the students' behavior, the authors in the work presented in [13] used the RNN algorithm with three different architectures. Both CNN and modern RNN approaches could help reduce the complexity of the feature selection and processing [14].

Within the domain of education, sentiment analysis has been used for different task types [15]. Its new research insights could include sentiment analysis on MOOC evaluation. In this vein, the study presented by the author in [16] shows that deep learning-based architectures outperform ensemble learning methods and machine learning methods for sentiment analysis on educational data. The study presented by the authors in [17] used sentiment analysis and deep learning by adopting the RCNN model to identify the most essential factors frequently mentioned by learners in MOOCs that can impact their learning satisfaction. To address the dropout problem, the authors in [18] perform sentiment analysis on data from forum posts within a MOOC using a BNN model (Bayesian Neural Network).

4 Methodology and Expected Outcome

By proposing new models based on deep learning, our research aims to predict MOOC students at risk of lacking motivation. The following methodology is suggested as an initial proposal to achieve the underlined goal.

The dataset could be gathered from well-known MOOC platforms or accessible online datasets. Then, advanced deep learning-based sentiment analysis models will be used to extract features from raw input instead of conventional feature engineering techniques.

We aim to identify learners at risk of lack of motivation and divide students into groups based on their motivation. Because lower categories of students are more likely to drop out, they will be taken into account to ensure a customized intervention (Fig. 2).

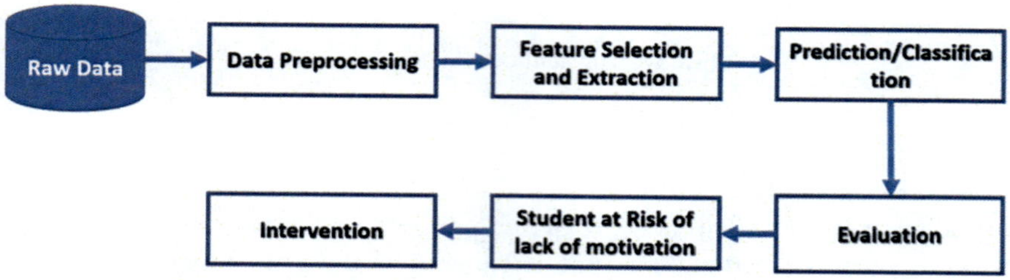

Fig. 2. The proposed methodology to predict students at risk of lacking motivation

5 Ongoing Works and Future Directions of the Search

Learners come to MOOCs for various reasons, from academic to professional motives [5]. The MOOC model represents a way to open up educational opportunities to all individuals regardless of their financial circumstances, social background, and stage of their professional development [19]. The significance and benefits of online learning make the study's topic particularly intriguing. Online education has an excellent opportunity to attract future learners since it provides an alternative study format to traditional face-to-face classes. In addition, it can be used to support existing in-class education and help students who need further assistance [8]. Also, a dropout from online courses is a major concern in the educational sector since it prevents institutions from satisfying even the smallest expectations and wastes financial, social, and academic resources [20]. The first step in our research is to understand the search field. The next step is to develop novel techniques to predict learners' lack of motivation risk.

References

1. Vora, M.: E-learning systems and MOOCs - a review. Int. J. Res. Appl. Sci. Eng. Technol. **8**, 636–641 (2020). https://doi.org/10.22214/ijraset.2020.31532
2. Denan, Z., Munir, Z.A., Razak, R.A., Kamaruddin, K., Sundram, V.P.K.: Adoption of technology on e-learning effectiveness. Bulletin of Electrical Engineering and Informatics **9**, 1121–1126 (2020). https://doi.org/10.11591/eei.v9i3.1717
3. Kumar Basak, S., Wotto, M., Bélanger, P.: E-learning, M-learning and D-learning: Conceptual definition and comparative analysis. E-Learning and Digital Media. **15**, 191–216 (2018). https://doi.org/10.1177/2042753018785180
4. Qazi, A., et al.: Adaption of distance learning to continue the academic year amid COVID-19 lockdown. Child Youth Serv Rev. **126**, (2021). https://doi.org/10.1016/j.childyouth.2021.106038
5. Badali, M., Hatami, J., Banihashem, S.K., Rahimi, E., Noroozi, O., Eslami, Z.: The role of motivation in MOOCs' retention rates: a systematic literature review. Res Pract. Technol. Enhanc. Learn. **17** (2022). https://doi.org/10.1186/s41039-022-00181-3

6. Dalipi, F., Imran, A.S., Kastrati, Z.: MOOC dropout prediction using machine learning techniques: review and research challenges. In: IEEE Global Engineering Education Conference, EDUCON, pp. 1007–1014. IEEE Computer Society (2018). https://doi.org/10.1109/EDUCON.2018.8363340
7. Moreno-Marcos, P.M., Alario-Hoyos, C., Munoz-Merino, P.J., Kloos, C.D.: Prediction in MOOCs: a review and future research directions. IEEE Trans. Learn. Technol. **12**, 384–401 (2019). https://doi.org/10.1109/TLT.2018.2856808
8. Schlögl, S., Ploder, C., Spieß, T., Schöffer, F.: Let's digitize it: investigating challenges of online education. In: Communications in Computer and Information Science, pp. 224–233. Springer Verlag (2019). https://doi.org/10.1007/978-3-030-20798-4_20
9. Kumari, P., Breslow, L., Pritchard, D.E., DeBoer, J., Stump, G.S., Ho, A.D.: Sambodhi digital learning through MOOCs: Advantages, Outcomes & Challenges. Gautam Buddh Nagar. 43
10. Feng, W., Tang, J., Liu, T.X.: Understanding Dropouts in MOOCs (2019)
11. Fu, Q., Gao, Z., Zhou, J., Zheng, Y.: CLSA: a novel deep learning model for MOOC dropout prediction. Computers and Electrical Engineering **94**, (2021). https://doi.org/10.1016/j.compeleceng.2021.107315
12. Mubarak, A.A., Cao, H., Hezam, I.M.: Deep analytic model for student dropout prediction in massive open online courses. Comput. Electr. Eng. **93**, 107271 (2021). https://doi.org/10.1016/j.compeleceng.2021.107271
13. Fotso, J.E.M., Batchakui, B., Nkambou, R., Okereke, G.: Algorithms for the development of deep learning models for classification and prediction of learner behaviour in MOOCs. In: Studies in Computational Intelligence, pp. 41–73. Springer Science and Business Media Deutschland GmbH (2022). https://doi.org/10.1007/978-3-030-92245-0_3
14. Sun, Z., Harit, A., Yu, J., Cristea, A.I., Shi, L.: A brief survey of deep learning approaches for learning analytics on MOOCs. In: Lecture Notes in Computer Science (including subseries Lecture Notes in Artificial Intelligence and Lecture Notes in Bioinformatics), pp. 28–37. Springer Science and Business Media Deutschland GmbH (2021). https://doi.org/10.1007/978-3-030-80421-3_4
15. Dolianiti, F.S., Iakovakis, D., Dias, S.B., Hadjileontiadou, S., Diniz, J.A., Hadjileontiadis, L.: Sentiment analysis techniques and applications in education: a survey. In: Communications in Computer and Information Science, pp. 412–427. Springer Verlag (2019). https://doi.org/10.1007/978-3-030-20954-4_31
16. Onan, A.: Sentiment analysis on massive open online course evaluations: a text mining and deep learning approach. Comput. Appl. Eng. Educ. **29**, 572–589 (2021). https://doi.org/10.1002/cae.22253
17. Chen, X., Wang, F.L., Cheng, G., Chow, M.K., Xie, H.: Understanding Learners' Perception of MOOCs Based on Review Data Analysis Using Deep Learning and Sentiment Analysis. Future Internet **14** (2022). https://doi.org/10.3390/fi14080218
18. Mrhar, K., Benhiba, L., Bourekkache, S., Abik, M.: A bayesian CNN-LSTM model for sentiment analysis in massive open online courses MOOCs. Int. J. Emerg. Technol. Learn. **16**, 216–232 (2021). https://doi.org/10.3991/ijet.v16i23.24457
19. Voudoukis, N., Pagiatakis, G.: Massive Open Online Courses (MOOCs): Practices, Trends, and Challenges for the Higher Education. European J. Edu. Pedago. **3**, 288–295 (2022). https://doi.org/10.24018/ejedu.2022.3.3.365
20. Mehrabi, M., Safarpour, A.R., Keshtkar, A.A.: Massive Open Online Courses (MOOCs) Dropout Rate in the World: A Protocol for Systematic Review and Meta-analysis. Interdiscip J Virtual Learn Med Sci. **13**, 86 (2022). https://doi.org/10.30476/IJVLMS.2022.94572.1138

Ontological Model of Knowledge Representation for Assessing the City Visual Environment Quality

Polina Galyanina[1], Natalya Sadovnikova[1](✉), Tatiana Smirnova[2], Artyom Zalinyan[3], and Ekaterina Baranova[3]

[1] Volgograd State Technical University, 28 Lenina Avenue, 400005 Volgograd, Russia
npsn1@ya.ru

[2] Volgograd State Pedagogical University, 27 Lenina Avenue, 400005 Volgograd, Russia

[3] Volgograd State Technical University, 1 Akademicheskaya Street, 400074 Volgograd, Russia

Abstract. Visual ecology is one of the new interdisciplinary directions, the development of which is necessary for the formation of the principles of modern urban planning. This is a section of sensory ecology that is responsible for the quality of the human environment, perhaps no less than clean air. The article examines the current state of issues related to the analysis of the quality of the visual environment of the city. The factors influencing visual ecology are considered: the monotony of the surrounding space, increased visual noise level, color harmony, etc. And a system of criteria for the formation of a knowledge model for this subject area is proposed. Ontology is chosen as a formal model of knowledge representation, which provides integration of heterogeneous information and the possibility of automating its processing. The "Protege" editor was used to implement the ontology. A technology based on SWRL rules was used to formalize the rules.

Keywords: Visual Ecology · Ontological Model · Environment Quality · Visual Noise Factors · Visual Environment Quality Criteria

1 Introduction

The visual environment of the city consists of various kinds of objects, both of natural origin (vegetation, various landforms, water bodies, etc.) and of anthropogenic origin (various buildings and structures, infrastructure facilities, vehicles). This is one of the important components of human life support that determine the quality of the urban environment.

The visual environment has become the object of studying video ecology, a new scientific direction, the foundations of which were laid in research conducted at the intersection of humanities and natural sciences. On the part of the humanities, the main position is occupied by the part of Vernadsky's noosphere teaching "ecology of culture", which was introduced into scientific circulation by D. S. Likhachev (1906–1999) in 1979, the object of which is world culture and which studies creative, engineering and

K. Kabassi et al. (Eds.): NiDS 2023, LNNS 783, pp. 130–139, 2023.
https://doi.org/10.1007/978-3-031-44097-7_13

economic relations in society with their impact on the human environment [1]. On the part of natural science disciplines, visual ecology has assumed a strict methodological apparatus, in the form of quantitative methods for measuring pollution of the human environment. Visual ecology can be characterized as a science that studies and explains the nature of the mechanisms of vision, as well as studies the theoretical foundations for the development of environmental principles for the organization and placement of various objects, in accordance with the comfortable perception of the human material environment [2].

For the first time, the term "video ecology" was used by Filin V.A., who studied the influence of visual images on mental and physical health of a person [3]. This direction has been developed in a variety of scientific disciplines – visual anthropology, landscape design, psychophysiology of visual perception, humanitarian ophthalmology [4, 5].

The object of the study of the visual ecology of the city is the spatial structure of visual environments: visual images of urban space, theoretical models and conceptual apparatus, reconstructed on the basis of modern theory of urban space, visual experience and ecological approach to visual perception [5].

The visual environment of the city is present in the human consciousness in the form of a network of perception and memory images functioning both at the conscious and unconscious levels of the psyche and there is reason to assert that it is responsible for the formation of patterns of processing visual information received as a result of contact with reality [6]. With the development of cities, the problem of degradation of the visual environment arises, and, unlike the natural visible environment, to which, over a very long period of time, in the process of evolution, our vision has adapted, modern ways of transforming the visual environment of a person, in the form of buildings, landscape changes, etc., depress human perception, which undoubtedly affects on the overall quality of the human environment [7].

The works [8, 9] present approaches to analyzing the quality of the visual environment of the city and emphasize the need to develop general requirements that will help reduce the negative impact on humans. The article [10] examines the issues of automated data collection for solving problems of visual ecology.

Despite the fact that interest in visual ecology is constantly growing, there are still no regulatory requirements governing the formation of the visual environment and uniform criteria for its assessment. In this regard, there is a need to create technologies for analyzing the visual environment of the city and recommendation systems to reduce the negative impacts of the visual environment [11]. The purpose of the research is to systematize the factors affecting the quality of the visual environment of the city and to develop a knowledge model based on ontology for use in decision support systems when designing urban space objects.

2 Systematization of Factors Affecting the Quality of the Visual Environment of the City

2.1 Increased Sealing

Asymmetric forms are proportional to nature and man, and not pure geometric ones, respectively, the modern excessively geometrized environment and material objects cannot but oppress a person.

2.2 The Monotony of the Surrounding Space

It is the predominance of dark, gray colors of buildings, coupled with their static nature and the presence of large planes, as well as a low percentage of green spaces within the city. That creates the very excessive monotony of the surrounding space.

2.3 Increased Visual Noise Level

Visual noise is the visual parameters of the urban environment, most of which can be perceived and evaluated by their observers as unpleasant, annoying [12].

Visual noise is created by combinations of various factors that can be attributed to various objects of the urban environment (Table 1).

2.4 Homogeneity of the Visible Medium

This phenomenon occurs due to the multitude of surfaces on which there are few or no visual elements at all. The ends of buildings, asphalt concrete roadbed, concrete fences, roofs of houses can serve as an example of such objects [13].

2.5 Aggressiveness of the Visible Environment

This factor is formed by visual fields that have many identical elements located uniformly relative to each other. These include ordinary buildings that have many windows, it can also be panels that are finished with monotonous tiles or brickwork, corrugated aluminum, grilles on the windows. To identify an object as "aggressive" for human perception, it is enough that this building has only nine repeating elements [14].

2.6 Color Harmony

Color has a special effect on a person, creating some kind of mood and image, affects the further actions of a person. It has been established that a color-appropriate visual environment attracts a person, creates a certain atmosphere, brings a sense of calm and contributes to improving relations between people.

Coloristics is the science of color, which gives color to the visual environment, being the most active direction in the field of design of urbanized landscapes of various functional purposes and their constituent elements, as well as artistic decoration of public spaces (streets, highways, squares) [15].

Table 1. Factors of visual noise formation.

Urban environment object	Visual noise generation factor
Advertisement	- location on a historical or cultural site; - incommensurability with the placement object; - color or light contrast; - improper condition of the advertising object; - restricted view of the terrain/scenic view
Buildings and structures	- the number of floors and the length of the building does not fit into the environment; - the building is part of a complex with the same building and color palette; - the color of the building stands out strikingly against the background of the range of surrounding objects; the building was subjected to acts of vandalism; - the facade of the building has a mirror finish, which leads to excessive reflection of light from the mirror panels; - the building is in a dilapidated condition, has not been repaired for a long time
Supporting infrastructure	- the object is located in the cultural-historical or "green" space of the city and stands out excessively; - careless placement of electrically conductive elements, which leads to the loss of aesthetic properties of the territory or buildings on which they are placed; - pipes, garbage cans, collectors, transformer booths and other gray infrastructure objects designed using outdated technologies; - chaotic parking spaces
Landscape areas	- ruderal vegetation; - dead wood and unkempt plants; - cluttered areas of the territory

A comfortable environment, from the point of view of coloristics, can be called such an artificial visual environment, which in its color scheme is as close as possible to the natural environment, simply put, one that is close to natural colors.

There are a number of approaches to the formation of harmonious combinations both in the interior and in the palette for the urban environment. The approaches are based on color schemes [16]:

- Complementary combination – a combination of colors in a circle located strictly opposite each other, which allows you to achieve contrast (orange-blue, red-green, etc.);
- Classical triad – based on a combination of three colors that should be located at the same distance from each other on the color circle (red-yellow-blue, purple-orange-green, etc.);
- A similar combination – combines from two to five colors located side by side within a circle (purple-red-purple, red-orange, orange);

- Separate-complimentary combination – allows you to achieve brightness and activity, while refinement of the image, by dividing the additional color into two adjacent ones, without using the additional color itself (red-blue-green-yellow-green);
- Tetrad – based on a combination of four tones located at the vertices of a rectangle on a color circle (green-blue-red-orange);
- Square – consists of tones located at the vertices of the square (blue, yellow-green, orange, red-purple).

2.7 Harmonious Combination of Architectural Forms

One of the main conditions for the harmony of buildings with space is the preservation and development of the plastic properties of the site. There are 5 conditions for creating a harmonious connection of architectural forms with the landscape [17]:

1. Preservation of natural "containers" – it is permissible to introduce as many new architectural volumes into the landscape as is permissible from the point of view of preserving the size and configuration of the space.
2. Preservation of the scale of the visual spatial unit of the landscape: taking into account the ratio of the scale height of buildings to the height of visual barriers.
3. Maintaining the closeness of the visual unit of the landscape.
4. Preservation of the natural configuration of visual objects.
5. Preservation of visual foci (points of special interest for the review).

2.8 Uniform Rules for the Organization of Visual Space

An important element in the formation of the urban environment is the presence of a design code. The design code is a set of rules and recommendations for designing a stylistically unified, comfortable and safe urban environment. Its task is to form a harmonious appearance of the city, emphasizing its uniqueness [18].

The design code specifies the requirements for the visual design of buildings, limits the dimensions of the elements of buildings and advertising and information structures, the materials of their execution [19].

The design code allows you to create a positive image of the city and ensure the perception of the surrounding space as a whole.

It is necessary to implement the following sequence of actions in order to solve the problem of matching the type of buildings with the design code [20]:

1. Determine the location zone of the building, in accordance with the established design code.
2. Determine the type of building.
3. Determine the types of advertising and information structures used.
4. Prepare a single integrated solution for the facade or entrance group.
5. Coordination with the authorized authorities.

Table 2 presents a system of criteria for assessing the quality of the visual environment of the city.

Table 2. A system of criteria for assessing the quality of the visual environment of the city.

Criteria	Description	Measurement method	Typical examples
Homogeneity	The same structure, "visual hunger"	Saturation of elements [20]	Concrete build. Without elem., solid glazing, asph
Monotony	Constant repetition of the same thing	Number of identical elements	Typical building, simplification of facades, grilles, partitions
Geometry	Aggressive shapes, lack of smooth lines, lack of proportions	Number of right angles and other factors affecting the aggression degree [20]	Typical building, simplification of facades
Scale to a person	Height of buildings and spaces isolation	Altitude, visibility of free spaces, sky	Mass constr. of "human hives", Point construct
Color harmony	Violation of the rules of color composition	Pres. of err. in the application and combin. of colors	Lack of design code
Information noise	Excess of advert. And sign., chaot. Placem	Existence yes/no	Advertisement
Visual images of ecological distress	Visible manifestations of environmental problems	Existence yes/no	Garbage, smoke, dirt, dry/diseased trees, weeds, lack of landsc
Visual imag. of industrial	Facilities of the support. Infrastruct	Existence yes/no	Antenna, cables, industrial zones

3 Development of an Ontological Model

The solution to the problem of evaluating the visual environment of a city can be implemented using an ontological model by developing and incorporating relevant rules and constraints into the ontology. The formal model of the ontology can be represented as follows:

$$O = <X, R, F>, \quad (1)$$

where X is a finite set of concepts in the domain, R is a finite set of relationships between concepts, F is a finite set of interpretation rules defined on concepts and/or relationships.

Formalized models, including ontologies, are used for a uniform presentation of knowledge from different subject areas, combining data from different sources and increasing the effectiveness of the use of information. Ontology is understood as a system of concepts of a certain subject area, which is represented as a set of entities connected by various relationships. Ontologies are used for the formal specification of concepts and relations that characterize a certain area of knowledge.

The advantage of ontologies as a way of representing knowledge is their formal structure, which simplifies their computer processing. Ontologies allow you to fix and unify the terminology of the subject area, and also allow you to organize navigation through the concepts of the subject area. Using the ontology, you can support case-based and rule-based systems. Figure 1 shows the classes of the developed ontological model (screenshot from Protege).

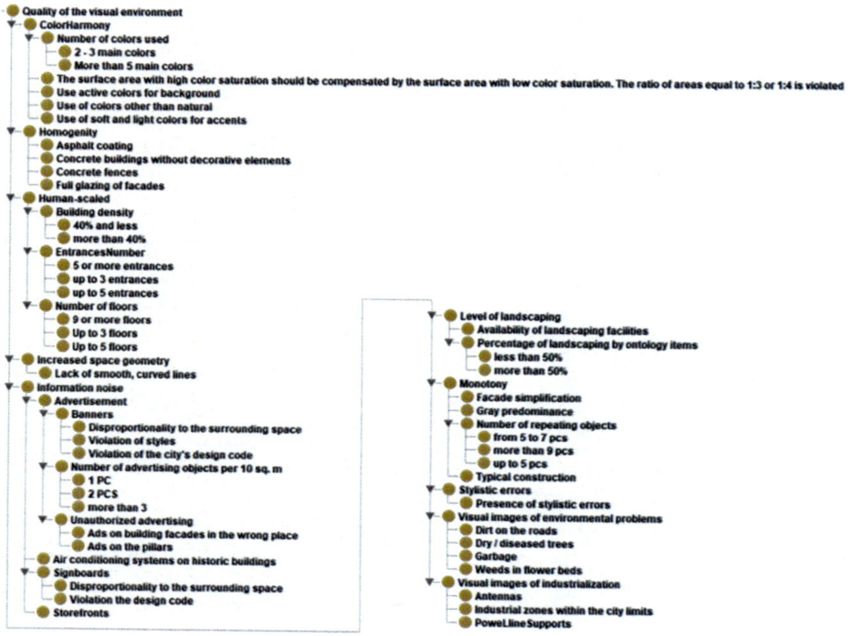

Fig. 1. Ontology model example.

Systems based on rule-based conclusions offer a simplified model of knowledge representation for both domain experts and programmers. It is usually easier for experts to express knowledge in a format similar to rules, and programmers usually find that rule-based programming is easier to understand and manipulate knowledge by unleashing calculations from rules. The former is executed by rules, whereas the latter is determined by the rule mechanism itself, that is, when and how to apply the rules. Thus, it is easier to add new rules or data, especially in constantly changing conditions. Rule–based reasoning (RBR – rule based reasoning) [21] is one of the most popular reasoning paradigms used in artificial intelligence. The architecture of systems based on rules-based reasoning has two main components: a knowledge base (usually consisting of a set of rules "IF… THEN…", representing domain knowledge) and an inference mechanism (usually containing some domain-independent inference mechanisms, such as a forward/reverse chain). The knowledge base in our case is the ontology of the subject area, namely the ontology of the visual environment of the city. There are several formal languages for describing ontology, but mainly the following are distinguished: RDF (RDFS) and

OWL [22]. RDF (Resource Description Framework) is a model for describing related data that allows Semantic Web technology to interpret information. The RDFS (RDF Schema) construction is based on the RDF dictionary and includes classes, properties, and auxiliary properties. In turn, this leads to the fact that RDF can now express the relationship between properties (property-subproperty) and classes (class-subclass). This advantage makes it possible to compose more flexible queries for extracting information [23]. OWL (Web Ontology Language) is a language for representing ontologies in the Web; a dictionary that expands the set of terms defined by RDFS. OWL ontologies contain descriptions of classes, properties, and their instances. The W3C Consortium recommends the OWL ontology description language [24], because it allows:

- to apply the logic of the first strand to describe axioms on ontology concepts through the description of descriptive logic constructs [25];
- to use existing OWL ontology inference machines that allow reasoning on the rules of descriptive logic [26];
- to use existing freely distributed ontology design tools in the OWL language.

Since 5 main city-like entities ("Buildings", "Private sector", "Schools", "Shopping centers", "Parks") were identified for the task to be solved, the rules were created taking into account their features. A set of criteria was formed based on the selected object for evaluation. The rule for the "Building" object example is shown in Fig. 2.

```
    IF(EXISTS {?obj a :DesignCodeViolation}, 0, 1) +
    IF(IF(?num_ads_less_than_2 = 1, 1, 0) = 0, 1, 0) +
    IF(EXISTS {?obj a :HistorialBuildingsWithAC}, 0, 1) +
    IF(EXISTS {?obj a :Trash}, 0, 1) +
    IF(EXISTS {?obj a :Antenna}, 0, 1) +
    IF(EXISTS {?obj a :PowerLines}, 0, 1) +
    IF(EXISTS {?obj a :UnkemptGreenery}, 0, 1)) as ?total_score
WHERE {
 ?obj a :AnyObject .
 OPTIONAL {?obj :FloorAmount ?num_floors .}

 SELECT COUNT(?color_obj) as ?num_colors_less_than_5 WHERE {
  ?obj a :AnyObject .
  ?obj :Color ?color_obj .
 } GROUP BY ?obj HAVING (COUNT(?color_obj) <= 5)

 SELECT IF(COUNT(?greenery_obj) > 0, 1, 0) as ?has_greenery,
     IF(EXISTS {?greenery_obj :perc ?percent . FILTER(?percent > 50)}, 1, 0) as ?greenery_percent_above_50
 WHERE {
  ?obj a :AnyObject .
  OPTIONAL {?greenery_obj a :Green .}
  FILTER(sameTerm(?greenery_obj, ?obj))
 }

 SELECT COUNT(?ad_obj) as ?num_ads_less_than_2 WHERE {
  ?obj a :AnyObject .
  ?obj :Ad ?ad_obj .
 } GROUP BY ?obj HAVING (COUNT(?ad_obj) <= 2)
}
GROUP BY ?total_score.
```

Fig. 2. Rule example.

IF the building object, residential building, microdistrict AND the number of 1 >= 20 AND <25 AND the number of duplicate objects <= 5 AND the number of floors <= 3 AND the number of entrances <= 3 AND the building density is less than 40% AND the number of colors used = 2 OR = 3 AND the proportion of landscaping is greater 50% AND the number of advertising = 1, THEN A good visual environment.

The example of executing the request is shown in Fig. 3.

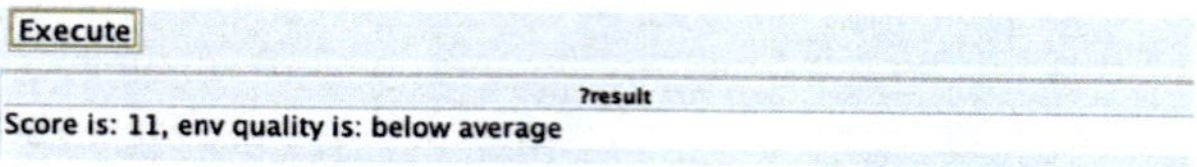

Fig. 3. The example of executing the request.

4 Conclusion

Visual images of the surrounding space can affect in different ways, increasing or decreasing the quality of the urban environment. A comprehensive analysis of the visual environment of the city is necessary both at the design stage in order to select optimal design solutions, and in the process of solving landscaping problems.

The analysis of factors affecting the quality of the visual environment of the city is carried out and a system of criteria is formed.

The developed ontological model contains criteria for assessing the quality of the visual environment. The model allows for logical inference based on rules for evaluating objects of the urban environment. OWL-DL was chosen as the language for describing ontologies. To describe the rules of logical inference for the search for a waste management strategy, the SWRL language was chosen, which is supported by a variety of tools and inference machines, is widely used in ontological modeling. SWRL extends the OWL-DL language. It is based on the rules of first-order predicate logic and allows you to describe the rules of logical inference in the form of a set of Horn clauses.

The ontological knowledge base of the intelligent decision support system for the selection of solutions for the transformation of the urban environment has been developed.

Acknowledgments. The study has been supported by the grant from the Russian Science Foundation (RSF) and the Administration of the Volgograd Oblast (Russia) No. 22-11-20024, https://rscf.ru/en/project/22-11-20024/. The authors express gratitude to colleagues from the Department of Digital Technologies for Urban Studies, Architecture and Civil Engineering, VSTU involved in the development of the project.

References

1. Chernyshev, S.N.: Culture in ecology — a part of the noosphere, the ideological base in reconstruction. Vestnik MGSU **12**, 123–130 (2013). (In Russian)
2. Grigoirev, E., Kuznetsova, G.: Visual ecology: a sight on structural form creation by joining positions of «deep ecology» and esthetics. Vestnik MGOU. Series "Linguistics" **1**, 184–194 (2011). (In Russian)
3. Filin, V.: Visible Environment in Urban Conditions as An Environmental Factor. Urban ecology. Science, Moscow (1990). (In Russian)
4. Visual Environment Pollution and Visual Ecology. https://culturolog.ru/content/view/3143. Last accessed 21 Mar 2023. (In Russian)
5. Savchuk, V.: Visual Ecology: Formation of Discipline. Publisher RCHA, Saint-Petersburg (2015). (In Russian)
6. Makarov, A.: Degradation of the visual environment of a modern city (on the example of Volgograd). Urban Sociol. **4**, 15–19 (2010). (In Russian)

7. Parygin, D., Sadovnikova, N., Kravets, A., Gnedkova, E.: Cognitive and ontological modeling for decision support in the tasks of the urban transportation system development management. In: Proceedings of the 6th International Conference on Information, Intelligence, Systems and Applications (IISA), pp. 1–5. Corfu, Greece (2015)
8. Fedosova, S.: Ecological and Technological Foundations of The Formation of The Visual Environment of a Large City. Russia, Bryansk (2008). (In Russian)
9. Korovina, V., Sidorenko, V.: Substantiation and development of methods of ecological evaluation of the visual city environment. Urban Sociol. **1**, 58–64 (2020). (In Russian)
10. Dubov, I., Rashevskiy, N., Yanin, K., Galyanina, P.: Approaches to collecting information to form a knowledge model of visual ecology. Eng. Constr. Vestnik Caspian Sea **2**, 115–120 (2022). (In Russian)
11. Ather, D., Rashevskiy, N., Parygin, D., Gurtyakov, A., Katerinina, S.: Intelligent assessment of the visual ecology of the urban environment. In: Proceedings of the 2nd International Conference on Technological Advancements in Computational Sciences (ICTACS), pp. 361–366. Tashkent, Uzbekistan (2022)
12. Goroshko, Y.: Criteria for assessing visual noise in an urban environment. Questions Student Sci. **7**, 41–47 (2022). (In Russian)
13. Tolstaya, E., Kozelko, N.: Visual environment as an important ecological factor. J. Belarusian State Univ. Ecol. **2**, 13–20 (2019). (In Russian)
14. Leontieva, I.G., Kormilitsyna, A.P.: Pollution of the visual environment. In: Proceedings of the VIII International Scientific and Practical Conference, pp. 323–328. Omsk, Russia (2021). (In Russian)
15. Voskresenskaya, A.I.: Coloristic organization of open spaces as a means of creating a visual comfortable urban environment. Forest. Bull. **5**, 66–70 (2015). (In Russian)
16. Six color schemes that always work. https://rerooms.ru/blog/posts/shest-tsvetovyh-shem-kotorye-rabotayut-vsegda. Last accessed 23 Mar 2023. (In Russian)
17. Formation of a comfortable urban environment, https://ecoportal.su/public/industry/view/795.html. Last accessed 24 Mar 2023. (In Russian)
18. Parygin, D., Sadovnikova, N., Gamidullaeva, L., Finogeev, A., Rashevskiy, N.: Tools and technologies for sustainable territorial development in the context of a quadruple innovation helix. Sustainability **14**(15), 9086 (2022)
19. Ryazan. Design code. https://designcode-rzn.ru. Last accessed 24 Mar 2023
20. Gorodkov, A.V.: Ecology of the Visual Environment: a Textbook, 2nd edn. Lan, Saint-Petersburg (2022). (In Russian)
21. Buchanan, B.G., Shortliffe, E.H.: Rule-Based Expert Systems: The MYCIN Experiments of the Stanford Heuristic Programming Project. Addison-Wesley, Boston (1984)
22. Lapshin, V.A.: Ontology in Computer Systems. RSDN MAGAZINE(4), 61–67 (2010). (In Russian)
23. Allemang, D., Hendler, J.: Semantic Web for the Working Ontologist: Effective Modeling in RDFS and OWL, 2nd edn. Morgan Kaufmann, Burlington, Massachusetts (2011)
24. OWL 2 Web Ontology Language. Document Overview, Second edn., https://www.w3.org/TR/owl2-overview. Last accessed 15 May 2017
25. Birger Hjørland. Semantic relations (meaning relations). http://www.iva.dk/bh/Lifeboat_KO/CONCEPTS/semantic_relations.html. Last accessed 15 May 2017
26. Sadovnikova, N., Matohina, A., Shabalina, O., Shirmanova, D., Romanova A.: Ontology-based urban transport system modelling for dynamic goal setting & decision making. In 7th International Conference on Information, Intelligence, Systems and Applications (IISA), pp. 1–5. Chalkidiki, Greece (2016)

Mental Disorders Prediction from Twitter Data: Application to Syndromic Surveillance Systems

Lamia Bendebane[1,2](✉), Zakaria Laboudi[1], and Asma Saighi[1,3]

[1] Mathematics and Computer Sciences Department, University of Oum El Bouaghi, Oum El Bouaghi, Algeria
lamia.bendebane@univ-oeb.dz

[2] ReLa(CS)2 Laboratory, University of Oum El Bouaghi, Oum El Bouaghi, Algeria

[3] LIAOA Laboratory, University of Oum El Bouaghi, Oum El Bouaghi, Algeria

Abstract. The widespread use of Twitter social media to express opinions, ideas, and desires related to various topics, including public health, makes it a rich source of relevant information. This is because it allows extracting a huge amount of data, more specifically in textual form which represents the largest part. The aim of this short paper is to give an outline about the research-works carried out as part of our ongoing researches (doctoral studies). The overall objective is to develop new effective syndromic surveillance systems that contribute to the improvement of healthcare systems, with a particular focus on mental disorders. To this end, we first analyze and compare some recent researches that employ Twitter data for healthcare purposes, especially those used within syndromic surveillance systems. Then, we sketch the general methodology that we adopt namely sentiment analysis of social media data. Finally, we highlight our ongoing research-works that focus on analyzing Twitter social data for detecting mental disorders, more specifically depression and anxiety. To do so, we propose to perform a classification task using multi-labeled deep-learning models. The idea is to adopt simple and hybrid three-class classifiers instead of the binary classification used in many researches in the literature. As a partial result, some of the used models could achieve an accuracy about of 89%.

Keywords: social media · sentiment analysis · healthcare · mental disorders · syndromic surveillance

1 Introduction

Syndromic surveillance refers to real-time (or near real-time) collection, analysis, interpretation, and dissemination of health-related data to enable the early identification of the impact of potential human or veterinary public health threats that require effective public health action. The systems collect health data in real or near real-time to track trends in the occurrence of disease conditions of public health importance in a defined population [1, 2].

On another side, the expansion in digital technology and increasing access to online user-generated content has provided a relevant source of data for healthcare syndromic

K. Kabassi et al. (Eds.): NiDS 2023, LNNS 783, pp. 140–145, 2023.
https://doi.org/10.1007/978-3-031-44097-7_14

surveillance [1]. In this regard, social media data can significantly help develop such systems. Indeed, social media has become an essential need for human existence as every day, millions of people share their thoughts, feelings, and experiences through Facebook, Instagram, Twitter, and so on. This large use of social media has resulted in the generation of giant volumes of data, which has encouraged conducting many researchers to analyze their data through efficient methods.

In this short paper, we are interested in studying the ways we can exploit social media data to build efficient syndromic surveillance systems for mental disorders detection, with a special focus on Twitter data. Indeed, prior to psychiatry confirmation of mental illness, ill persons may exhibit behavioral patterns, symptoms, signs, or laboratory findings that can be tracked through social data. Thus, these posts provide a rich source of data since people are ready to share their sentiments and problems on social media, hoping that they find solutions to their problems due to the vast spread of social networks. That is why it is believed that social media can significantly help us in such cases. To achieve this goal, we have to deal with two main challenges:

1. The first concern is about the determination of which social data will be processed; this is a crucial issue that depends on the goals of the applications.
2. The second concern is about the way in which the social data will be used in useful tasks leading to contributing to the improvement of health systems to protect and monitor the health of individuals through efficient syndromic surveillance systems.

2 Related Works

The Twitter platform represents an excellent source of information for health-state prediction. It provides a broad population of users who share their posts (known as tweets) generated with an average of 6.000 tweets per second. In particular, tweets are very useful for analysis since they are usually emitted during daily activities and events. Such data allows extracting behavioral attributes that are relevant to an individual's thinking and activities, especially as Twitter provides them for open source.

Involving Twitter data to build syndromic surveillance systems is an active research field; for instance, one can observe the increasing number of related research-works. Table 1 gives a summarized analysis of some recent studies in the public health domain that use Twitter data to detect mental disorders [6, 10, 13–15]. In these researches, only a few machine/deep learning based models are developed and deployed in syndromic surveillance systems for various purposes.

3 Research Methodology

In order to propose relevant contributions related to building efficient syndromic surveillance systems for mental disorders detection, we rely on sentiment analysis, which has emerged as one of the most important techniques to analyze the contents of social media for healthcare purposes. To do so, we involve Twitter data to build an efficient syndromic surveillance system for mental disorders detection, as shown in Fig. 1.

The process consists of a three-stage procedure. First, relevant data are collected using Twitter API, according to different parameters such as geographical location,

Table 1. Comparison between some studies involving Twitter data to predict mental disorders.

Ref	Overview	Year	Learning algorithms	Deployed in syndromic surveillance systems	Results (%)
[6]	Models for detecting depression levels using demographic characteristics and sentiment analysis from tweets	2022	SVM, DT, NB, RF, DL, ANOVA, SVM-FRE,	Yes	F1-measure: RF = 74.35 ANOVA = 76.57 SVM-RFE = 76.46
[10]	Stress detection in tweets via two models. The first model is used to check the existence of in tweets. The second model is based on logistic regression to predict stress levels incorporated in the content of previous tweets	2021	LR, SVM, RF, NTSD	No	NTSD with neighborhood = 2, Saracasm_Level attribute used D1: Accuracy = 79.51 F1-score = 73 used D2: Accuracy = 80.26 F1-score = 80
[13]	A gated recurrent neural network for depression analysis using Bangla social media data	2019	GRU	No	Accuracy = 75.7
[14]	Machine learning classifiers applied to Twitter data for detecting depressive cases if a user's tweet expresses some depression signs	2020	LSTM, CNN, LR, RF, NB, Linear SVC	No	CNN: Accuracy = 95
[15]	Hybrid deep learning classifier for predicting depressive cases from Twitter users	2022	CNN, RNN, CNN-BiLSTM	No	CNN-BiLSTM: Accuracy = 94.28 F1-score = 94.78

List of abbreviations
SVM: Support Vector Machine, **DT**: Decision Tree, NB: Naïve Bayes, **RF**: Random Forest, **DL**: Deep Learning, **ANOVA**: Analysis Of Variance, **SVM-FRE**: Support Vector Machine-Recursive Feature Elimination, **LR**: Logistic Regression, **NTSD**: Neighborhood tweet-based stress detection, **GRU**: Gated Recurrent Unit, **LSTM**: Long Short Term Memory, **CNN**: Convolutional Neural Network, **Linear SVC**: Linear Support Vector Classifier, **RNN**: Recurrent Neural Network, **Bi**: Bidirectional.

period and search keywords. Then, these data are passed to different deep learning models for training; the aim is to find the well-performing models. Finally, the selected deep learning models are deployed within a syndromic surveillance system for mental disorders detection.

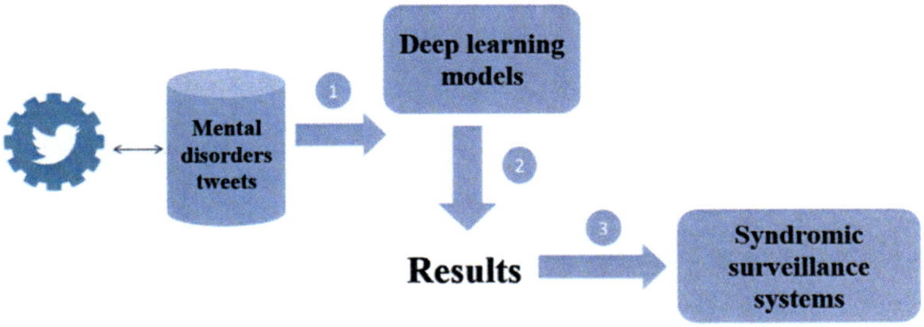

Fig. 1. Twitter-based syndromic surveillance system for mental disorders detection.

4 Work in Progress

Currently, we are working on applying sentiment analysis techniques using deep learning techniques to predict mental disorders from Twitter social data (see Fig. 1). Several unstructured free-text tweets related to healthcare are shared on Twitter, which is becoming a popular area for doing research. Most of related-researches are based on people discussing their healthcare cases on Twitter, sharing their illnesses, indications, and drugs; all these experiences are covered in sentiment analysis, which is in fact the power source of that kind of analysis. From this standpoint, many researchers have focused their efforts on the analysis of tweets in order to detect mental disorders such as depression, anxiety, bipolar, stress, and schizophrenia [7–12]. Although these researches have been able to achieve good results, they nevertheless rely on binary classification models by treating each mental disorder separately. Indeed, it would be better if we manage to develop systems capable of dealing with several mental disorders. In particular, we should deal with two main challenges:

- The first challenge concerns the methods adopted to extract data from a big data environment (Twitter platform) so that they lead to relevant information regarding the objectives set.
- The second challenge concerns the methods adopted to devise well-performing deep learning models (e.g. simple, hybrid, and transfer learning mechanisms).

Hence, we propose to rely on multi-class models for detecting depression and anxiety through tweets classification. The idea consists in testing some hybrid variants of deep learning models to examine their strengths and weaknesses. In what follows, we provide some partial results we obtained using some hybrid models for mental disorders prediction, hoping that they lead to higher performances. For that purpose, we used Twitter API to collect and label data as '0' for the "normal" class, '1' for the "depression" class, and '2' for the "anxiety" class. Then, we built and compared hybrid classifiers such as CNN-RNN, CNN-LSTM, CNN-GRU, CNN-Bi-RNN, CNN-Bi-LSTM, and CNN-Bi-GRU. Figure 2 shows the performances of the best hybrid model (CNN-Bi-LSTM). As a result, the best accuracy achieved was about 88.93% recorded by the CNN-Bi-LSTM model. Figure 3 shows that the best F1-scores for normal, depression and anxiety classes are about 86%, 90%, and 91%, respectively.

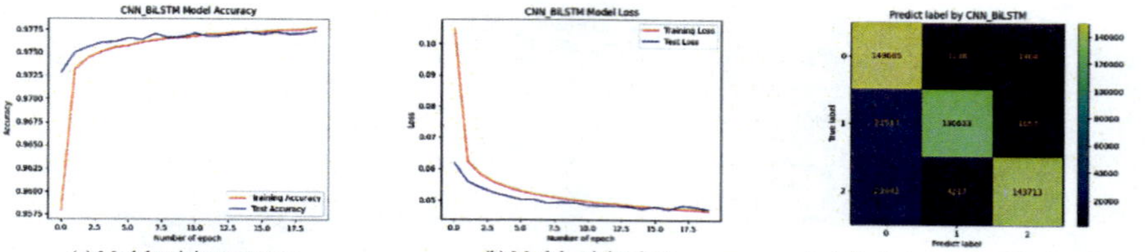

(a) Model training accuracy

(b) Model training loss

(c) Confusion matrix of the best hybrid model (CNN-Bi-LSTM)

Fig. 2. The performances of the best hybrid model (CNN-Bi-LSTM).

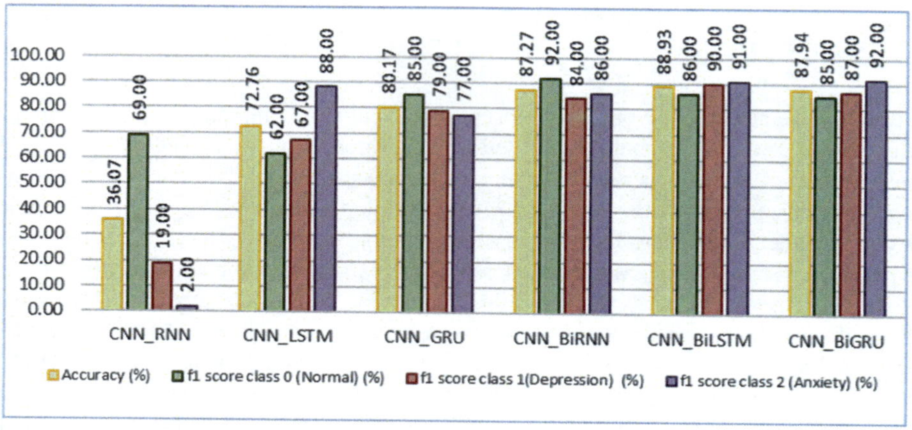

Fig. 3. Variation of Accuracy and F1-score of normal, depressive, and anxious tweets for the hybrid models tested.

5 Conclusion

The aim of this paper was to provide details about our ongoing researches as well as the expected outcomes. The global objective we target is to contribute to improving healthcare systems using effective syndromic surveillance systems for mental disorder detection. Thus, we first analyzed some relevant researches involving Twitter data for healthcare purposes, especially those relating to syndromic surveillance systems. Then, we presented our methodology outline that relies on sentiment analysis of Twitter data. Finally, we highlighted our works in progress, in which we focus on analyzing Twitter data for predicting mental disorders using hybrid deep-learning models.

References

1. Edo-Osagie, O., Lake, I., Edeghere, O., De La Iglesia, B.: Attention-based recurrent neural networks (RNNs) for short text classification: an application in public health monitoring. In: Rojas, I., Joya, G., Catala, A. (eds.) IWANN. LNCS, vol. 11506, pp. 895–911. Springer, Cham (2019)

2. Edo-Osagie, O., Iglesia, B.D.L., Lake, I., Edeghere, O.: An evolutionary approach to automatic keyword selection for Twitter data analysis. In: de la Cal, E.A., Villar Flecha, J.R., Quintián, H., Corchado, E. (eds.) HAIS 2020. LNCS (LNAI), vol. 12344, pp. 160–171. Springer, Cham (2020). https://doi.org/10.1007/978-3-030-61705-9_14
3. Lake, I.R., Colon-Gonzalez, F.J., Barker, G.C., Morbey, R.A., Smith, G.E., Elliot, A.J.: Machine learning to refine decision making within a syndromic surveillance service. BMC Public Health **19**(1), 1–12 (2019)
4. Bollig, N., Clarke, L., Elsmo, E., Craven, M.: Machine learning for syndromic surveillance using veterinary necropsy reports. PLoS ONE **15**(2), e0228105 (2020)
5. Oliveira, L.: BioInfo@ UAVR at eRisk 2020: On the use of psycholinguistics features and machine learning for the classification and quantification of mental diseases. In: Proceedings of the CEUR Workshop Proceedings, Thessaloniki, Greece, pp. 22–25 (2020)
6. Angskun, J., Tipprasert, S., Angskun, T.: Big data analytics on social networks for real-time depression detection. J. Big Data **9**, 69 (2022). https://doi.org/10.1186/s40537-022-00622-2
7. Skaik, R., Inkpen D.: Using Twitter social media for depression detection in the canadian population. In: ACM ISBN 978-1-4503-8883-2/20/12. AICC Kyoto Japan (2020)
8. Bhattacharya, C., et al.: The nature, cause, and consequence of COVID-19 panic among social media users in India. Soc. Netw. Anal. Min. **11**(1), 53 (2021)
9. Huang, Y.H., Wei, L.H., Chen, Y.S.: Detection of the prodromal phase of bipolar disorder from psychological and phonological aspects in social media. arXiv preprint arXiv:1712.09183 (2017)
10. Prashanth, K.V.T.K.N., Ramakrishnudu, T.: A novel method for detecting psychological stress at tweet level using neighborhood tweets. J. King Saud Univ.-Compu. Inform. Sci. **34**(9), 6663–6680 (2022)
11. Jilka, S., et al.: Identifying schizophrenia stigma on Twitter: a proof of principle model using service user supervised machine learning. Schizophrenia **8**(1), 1 (2022)
12. Dheeraj, K., Ramakrishnudu, T.: Negative emotions detection on online mental-health related patients texts using the deep learning with MHA-BCNN model. Expert Syst. Appl. **182**, 115265 (2021)
13. Uddin, A.H., Bapery, D., Arif, A.S.M.: Depression analysis of bangla social media data using gated recurrent neural network. In: 2019 1st International Conference on Advances in Science, Engineering and Robotics Technology (ICASERT), pp. 1–6. IEEE (2019)
14. Shetty, N.P., Muniyal, B., Anand, A., Kumar, S., Prabhu, S.: Predicting depression using deep learning and ensemble algorithms on raw twitter data. Int. J. Electr. Comput. Eng. **10**(4), 3751 (2020)
15. Kour, H., Gupta, M.K.: An hybrid deep learning approach for depression prediction from user tweets using feature-rich CNN and bi-directional LSTM. Multimed. Tools Appl. **81**(17), 23649–23685 (2022)

Check for updates

Internet of Wearable Things Systems: Comprehensive Review

Sabrina Mehdi[1,2](✉), Sofia Kouah[2], and Asma Saighi[2]

[1] Abdelhamid Mehri, Faculty of New Technologies of Information and Communication, University Constantine 2, Ali Mendjeli, Constantine, Algeria
sabrina.mehdi@univ-oeb.dz

[2] Department of Mathematics and Computer Sciences, University of Oum El Bouaghi, route de Constantine, BP 321, 04000 Oum El-Bouaghi, DZ, Algeria
{sofia.kouah,saighi.asma}@univ-oeb.dz

Abstract. The Internet of Wearable Things (**IoWT**) has emerged as a revolutionary technology in recent years, with the potential to impact various fields including health, sports and entertainment. The Smart wearable devices are designed to be worn on the user's body. They can take many forms, including accessories, medical devices, clothing or parts of clothing, and are capable to collect a large amount of data, analyze it, make smart decisions and provide personalized responses to users. Wearable systems can be broadly defined as mobile electronic devices that can be unobtrusively embedded in the user's outfit as part of the clothing or an accessory. In particular, unlike conventional mobile systems, they can be operational and accessed without or with very little hindrance to user activity. To this end they are able to model and recognize user activity, state, and the surrounding situation: a property, referred to as context sensitivity. Thus, **IoWT** offers a convenient platform to ensure these functionalities and exchange vital information with dedicated servers. It also provides outstanding features to monitor body biometrics while using non-invasive means. Nevertheless, several challenges arise, such as: decision-making, interoperability, energy consumption of portable devices, confidentiality and data security, which makes their development difficult. Therefore, our doctoral research aims to propose an approach to develop **IoWT** systems. The current work aims to study **IoWT** systems, extract their characteristics, components and important issues and trends that should be addressed during development. Thus, the study focuses on investigating recent and relevant literature.

Keywords: Internet of Wearable Things · Design Patterns · Artificial Intelligence · Software Engineering · Learning

1 Introduction

"Wearable technology" and "wearable devices" are expressions that describe electronics and computers that are integrated into clothing and other accessories that can be worn comfortably on the body. Examples of wearable devices include glasses, watches,

K. Kabassi et al. (Eds.): NiDS 2023, LNNS 783, pp. 146–151, 2023.
https://doi.org/10.1007/978-3-031-44097-7_15

headbands, and jewelry. Although these devices show great influence in fashion and entertainment, they have the largest impact in the areas of health, medicine, and fitness. Librarians are also exploring wearable technology's potential for enhancing services and expanding outreach to their organizations [1]. The **IoWT** seeks to enhance the people's daily lives quality. This promising technology encompasses the utilization of sensors installed in wearable devices, monitoring the individual's activities, health features and other relevant information. Data collected from the **IoWT** can be fed into medical infrastructure, allowing clinicians to remotely access their patient data in real time throughout their daily lives. Building on the Internet of Things (IoT) architecture [2], a new integration framework for the **IoWT** is being developed. **IoWT** is a breakthrough technology that has the potential to change the healthcare industry by creating an ecosystem for automated telehealth treatments [3].

Our work focuses on the development of **IoWT** systems. These systems are characterized by several features and requirements, such as the miniaturization and autonomous operation of portable devices, the use of various hardware and software technologies, the diversity of models, platforms, tools and services utilized for the development, as well as the variation of application fields which present particular specificities. Moreover, the combination of several paradigms and methods of artificial intelligence, such as machine and deep learning algorithms, is often necessary to assist the decision-making process. To provide a comprehensive understanding of **IoWT** systems, our paper is organized as follows: Section 2 introduces an overview of wearable technology, including its different applications, **IoWT** systems and the **IoWT** architecture. Section 3 reviews some relevant related works highpoints the main **IoWT** challenges, features and specificities. In Sect. 4, we state the problem and objectives of our study. Finally, Sect. 5 concludes the paper and highlights future research direction to be undertaken in the future.

2 Internet of Wearable Things Systems

The advent of the Internet of Things era has propelled the development of wearable technology [4]. Wearable devices are widely used in medical, healthcare, sports, and safety applications, bringing more convenience to the living environment. Wearable technology refers to a large class of electronic and mobile devices that are integrated into implants, accessories, or outfits and can be worn on the body as portable measurement tools [5]. This technology also includes the software applications with analytic algorithms embedded in wearable devices. These technologies can connect to communication infrastructure; sense, store, and transmit data; and perform some computations. Due to their small size and ease of usage, these wearable technologies have increasingly been applied to provide smarter solutions in healthcare-related fields. Wearable systems can be broadly defined as mobile electronic devices that can be unobtrusively embedded in the user's outfit as part of the clothing or an accessory. In particular, unlike conventional mobile systems, they can be operational and accessed without or with very little hindrance to user activity. To this end they are able to model and recognize user activity, state, and the surrounding situation: a property, referred to as context sensitivity. Wearable systems range from micro sensors seamlessly integrated in textiles through consumer electronics embedded in fashionable clothes and computerized watches to belt worn PCs with a head mounted display. The wearable computing concept is part of a broader framework of ubiquitous computing that aims at invisibly enhancing our environment with smart electronic devices [6].

In recent years, wearable systems have become increasingly popular due to advances in sensors, wireless technology, and data analytics. They are often used in healthcare, fitness, sports, and military applications, among others [6]. The development of **IoWT** systems is a topic of great importance due to the rapid rise of wearable technology and the Internet of Things (IoT) [7]. Therefore, research on **IoWT** system development methodology is essential to overcome these challenges and enable the deployment of reliable and efficient systems (Fig. 1).

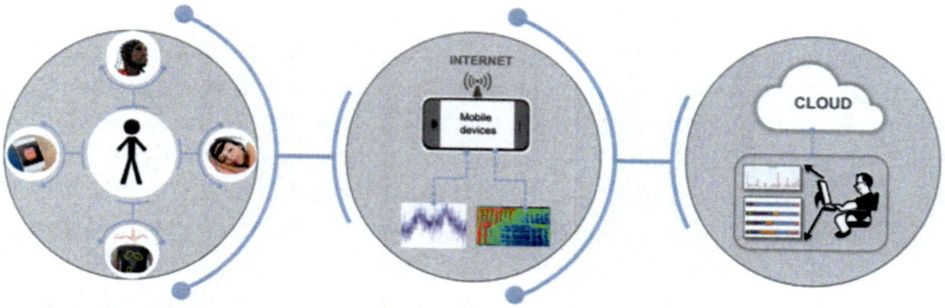

Fig. 1. Internet of Wearable Things [8]

3 Literature Review

There have been many previous studies on wearable IoT systems. Some of these have focused on the security challenges associated with using these devices, examining vulnerabilities and potential threats to data security. Others have focused on the development of technologies for connected objects, the optimization of real-time data analysis algorithms and the use of artificial intelligence in this area.

Among the works, we consider the following: Authors in [9] have provided a concise study of IoT-based devices such as wearable assistive devices for AD (Alzheimer Disease) patients and the adopted approaches for wearable device applications in healthcare to improve life quality of AD patients. In [10], the authors have proposed semantic-based service framework, Sensor-based Semantic (SeSem) framework, in which an end-user can subscribe for the intended physiological parameter of interest, for better understanding and monitoring patient's health status. In [11], the need for using wearables in healthcare and how it can change the future of medical systems have been discussed. A case study on how wearables can help patients as well as doctors is presented. In [12], the author has presented Web-based intelligent healthcare management system (CIHMS), which provides computer (if data are already available), or doctor, medical feedback/assistance for the patient. Adequate knowledge regarding the patient's condition will be collected from the correct sensor.

There are also many works for the development of **IoWT**. In [13], the authors have presented energy efficient approaches for minimizing power alleviation issue in **IoWT**. In [14], the authors have provided a design of a wearable medical monitoring system based on the Android platform of Iot, which can help users understand their physical conditions in real-time and assist doctors in completing diagnosis, thereby providing more references for extracting relevant disease characteristics. In [15], the authors have proposed a fault-tolerant data processing method handle uneven sensor data. The non-uniform and unsynchronized observation interval-based sensor data is analyzed for its impact on healthcare recommendations. In [16], the authors have presented a wearable sensor network system for IoT connected safety and health applications. In [17] the authors have addressed latest developments in Smart Health applications, focusing on technical challenges from both system and service perspectives in the context of wearable and medical IoT. Their findings show that the development faces a lack of system standardization and that data processing moves from servers or cloud locations toward the devices. In [18] the authors have proposed an IoT platform for smart maternal healthcare services with wearable devices and cloud computing and the key technologies.

The presented works demonstrate the complexity nature behind developing such systems, which are applicable to multiple domains and highlight the challenges inherent from IoT and using various technologies and paradigms. On the other hand, to the best of our knowledge, the previous works do not focus on methodological and design aspects. Therefore, the development of Smart Design Platforms, models and methodologies is of utmost importance.

4 Problematic, Objective and Outcomes

Despite the numerous advantages offered by **IoWT** systems, they present several challenges. These challenges include ensuring secure data transmission and storage, managing power consumption and battery life, dealing with the vast amounts of data generated by Smart wearable devices, and guaranteeing interoperability between different devices and systems. Other challenges include addressing privacy concerns, ensuring user acceptance and adoption, and addressing governing and legal issues. Nonetheless, by addressing these challenges, **IoWT** systems can be developed to provide significant benefits to individuals and society as a whole.

The problem we are addressing relates to the development of **IoWT** systems. Our PhD research aims to contribute to the challenges related to the design and development of **IoWT** systems. In this regard, we should propose solutions that includes models, architectures, approaches, etc. for a rigorous development of such systems by focusing on several complementary aspects. These aspects comprise in particular the combination of several development paradigms meeting the characteristics and requirements of specification, development, implementation and deployment of **IoWT** systems, as well as taking into account software quality criteria.

In this context, we believe that the use of artificial intelligence and software engineering paradigms to create **IoWT** system design patterns that meet their intrinsic requirements and support reuse, could lead to a promise solution. To achieve this, an alternative way consists of utilizing Design Patterns, Multi agents Systems, and machine and deep learning algorithms.

5 Conclusion

The emergence of the IoT has led to remarkable growth in the number of smart devices participating in all areas of human life. It is expected that this trend will retain its increasing pace in the coming years toward a connected society [19, 20]. Such a paradigm defines a novel subclass of the **IoT**, namely **IoWT**. **IoWT** is wearable system which provide people with a convenient platform to perceive, store, process, and exchange information with the Internet. The paper presented an overview of this technology and thoroughly explored its main challenges. After that, it highlighted the aim and the importance of our PhD research, which intends to provide a Smart Design framework, encompassing methodologies, models and patterns for the development of **IoWT** systems. An ongoing work is dedicated to provide an in-depth analysis of existing models, frameworks, and **IoWT** systems behavior, in order to identify the most effective approach for designing, developing, implementing, and deploying such systems.

References

1. Wright, R., Keith, L.: Wearable technology: if the tech fits, wear it. J. Electr. Res. Med. Libr. **11**(4), 204–216 (2014)
2. Sofia, K., Ilham, K.: Multi-layer agent based architecture for internet of things systems. J. Inform. Technol. Res. **11**(4), 32–52 (2018)
3. Seneviratne, S., et al.: Une enquête sur les appareils portables et les défis. IEEE Commun. Surv. Tuteur. **19**, 2573–2620 (2017)
4. Rahmani, A.M., et al.: The internet of things for applications in wearable technology. IEEE Access **10**, 123579–123594 (2022)
5. Madanian, S., Nguyen, H.H., Mirza, F.: Wearable Technology. In: Encyclopedia of gerontology and population aging, pp. 5388–5395. Springer International Publishing, Cham (2022)
6. Lukowicz, P., Kirstein, T., Tröster, G.: Wearable systems for health care applications. Methods Inform. Med. **43**(03), 232–238 (2004)
7. Kouah, S., Ilham, K.: Internet of things agents diagnosis architecture: application to healthcare IoT system. In: 3rd International Conference on Advanced Aspects of Software Engineering (ICAASE'18). Constantine 2 University, 01–02 Dec 2018
8. Hiremath, S., Yang, G., Mankodiya, K.: Wearable Internet of Things: concept, architectural components and promises for person-centered healthcare. In: 2014 4th International Conference on Wireless Mobile Communication and Healthcare-Transforming Healthcare Through Innovations in Mobile and Wireless Technologies (MOBIHEALTH). IEEE (2014)
9. Salehi, W., et al.: IoT-based wearable devices for patients suffering from Alzheimer disease. Contrast Med. Mol. Imaging **2022**, 1–15 (2022)
10. Radhika, R., Bhuvaneswari, A., Kalpana, G.: An intelligent semanticification rules enabled user-specific healthcare framework using IoT and deep learning techniques. Wireless Pers. Commun. **122**, 2859–2883 (2022)
11. Sivani, T., Mishra, S.: Wearable devices: evolution and usage in remote patient monitoring system. In: Connected e-Health: Integrated IoT and Cloud Computing, pp. 311–332. Springer International Publishing, Cham (2022). https://doi.org/10.1007/978-3-030-97929-4_14
12. Srinivasan, S., Prasanna Bharathi, S., Chamundeeswari, G., Muthu Kannan, P.: Intelligent healthcare monitoring system using cloud computing. In: Kumar, R., Mishra, B.K., Pattnaik, P.K. (eds.) Next Generation of Internet of Things. LNNS, vol. 201, pp. 343–351. Springer, Singapore (2021). https://doi.org/10.1007/978-981-16-0666-3_27

13. Ray, P.P., Dash, D.: A review on internet of wearable things for pervasive e-health care: energy efficiency and prospects. In: Kumar, A., Ghinea, G., Merugu, S., Hashimoto, T. (eds.) Proceedings of the International Conference on Cognitive and Intelligent Computing. Cognitive Science and Technology. Springer, Singapore (2023). https://doi.org/10.1007/978-981-19-2358-6_7
14. Li, Z., Lian, L., Pei, J., et al.: Design and implementation of wearable medical monitoring system on the internet of things. J. Ambient Intell. Human. Comput. (2021). https://doi.org/10.1007/s12652-021-03257-y
15. Pradhan, M.R., Mago, B., Ateeq, K.: A classification-based sensor data processing method for the internet of things assimilated wearable sensor technology. Cluster Comput. **26**, 807–822 (2023)
16. Wu, F., Wu, T., Yuce, M.R.: Design and implementation of a wearable sensor network system for IoT-connected safety and health applications. In: 2019 IEEE 5th World Forum on Internet of Things (WF-IoT), Limerick, Ireland, pp. 87–90 (2019). https://doi.org/10.1109/WF-IoT.2019.8767280
17. Schnell, B., et al.: Challenges in smart health applications using wearable medical Internet-of-Things—A review. In: Yang, X.-S., Sherratt, S., Dey, N., Joshi, A. (eds.) Proceedings of Sixth International Congress on Information and Communication Technology. LNNS, vol. 216, pp. 283–296. Springer, Singapore (2022). https://doi.org/10.1007/978-981-16-1781-2_27
18. Li, X., et al.: Building the Internet of Things platform for smart maternal healthcare services with wearable devices and cloud computing. Future Gener. Comput. Syst. **118**, 282–296 (2021)
19. Nouri, P., et al.: Machine-type wireless communications enablers for beyond 5G: enabling URLLC via diversity under hard deadlines. Comput. Netw. **174**, 107227 (2020). https://doi.org/10.1016/j.comnet.2020.107227
20. Xie, X., Rong, B., Kadoch, M.: Explaining 6G spectrum THz, mmWave, sub 6, and low-band. In: Xie, X., Rong, B., Kadoch, M. (eds.) 6G Wireless Communications and Mobile Networking, pp. 1–22. Bentham Science Publishers (2021). https://doi.org/10.2174/9781681087962121010003

Introducing a Biomimetic Rig for Simulating Human Gait Cycles and Its Potential Applications

Christos Kampouris[1,2(✉)], Philip Azariadis[2], and Vasilis Moulianitis[1]

[1] University of the Aegean, Konstantinoupoleos 1, Ermoupolis, 84100 Syros, Greece
cab@aegean.gr

[2] University of West Attica, Ag. Spyridonos 12243, Athens, Greece

Abstract. The human gait cycle is extensively studied. Huge effort for its reproduction is made, especially for identifying and compensating abnormal gait phases on humans with motorial problems. This paper reviews the main characteristics of an implemented biomimetic test rig, emphasizes on combinatorial experimental capabilities due to the rig's design and structure and notes on its motion envelope. Furthermore, we present a setup example with a pair of passive leg models for stance and gait reproduction and we illustrate an easily applicable optical method to assess the rig's repeatability. Finally, we discuss on applications and fields of usage of such a biomimetic rig platform.

Keywords: Gait Analysis · Biomimetic Rig · Knee Joint · Optical Gait Tracking

1 Introduction

Specialized test rigs have been significant tools for the field of gait analysis and relevant studies, since the previous century. Among the pioneers, Dowson described the first knee simulator [1] for testing total knee replacements (TKRs) in 1977, in accordance with hip simulators whose development had preceded. Dowson's single station hydraulic machine emulated the cyclic loading of walking on the joint under test in order to measure its wear. Progress followed with knee simulators from M.I.T., Syracuse, Munich University among several others. Notable multi-station simulators appeared in 1997, such as the 6 station, displacement-control Durham [2] and the 4 station, force-controlled Stanmore [3]. Nowadays, the AMT VIVO [4] with 6 degrees of freedom (DOF), both force and displacement-controlled, considered one of the most realistic joint simulators, can test knee, or hip, or other joints. Beyond single joints, other test rig designs support whole leg models including hip, knee, ankle joints. Such is the Bristol-Bath robotic platform [5] with a ball screw actuator mimicking the quadricep-hamstring antagonist pair in an inverted (upside-down) leg placement. It is designed for the application of ISO 10328:2016 standards on lower-limb prosthetics testing. More typical structures, such as Ricther's [6], or Haider's [7], rely on a 2 DOF arrangement for sagittal studies, with a rotary motor for the hip angle and a linear stage for the hip's height. To test semi-active

K. Kabassi et al. (Eds.): NiDS 2023, LNNS 783, pp. 152–163, 2023.
https://doi.org/10.1007/978-3-031-44097-7_16

prostheses, Jung-Hoon and Jun-Ho [8] used a 3 DOF leg-simulator (2 linear axes plus hip rotation) in an interesting arc structure. In the quite common sled-type design e.g., Oxford, Purdue, Kansas, Gent [9] the ankle is "pinned to the floor" while the hip is mounted on a vertically moving sled. The moving force may be applied on artificial quadriceps by pneumatic pistons, linear stages, wires and pulleys, while the knee joint may be loaded in multiple directions, by dead weights and actuators on hip and ankle. Industrial robotic arms [10] and hexapods have also been proposed [11] in order to offer 6 DOF, or more (e.g., 6 DOF at the hip plus 6 DOF at the ankle), in order to assess the knee joint compliance. The extents and multidisciplinary character of posture & gait analysis [12] led to the formation of specialized institutions/organizations with diverse applications in health, sports, entertainment, education, ergonomics, exoskeletons, robotics, security.

The main objective of our research is the design and construction of a bipedal, flexible, easy to use platform, for reproducing gait cycles, other exercises and daily activities (such as deep knee bending, jogging, mild running, small jumps, stair climb) in a reliable, controlled manner, keeping the complexity and cost at low levels [4]. This research has led to the design of a biomimetic rig that bears two horizontal and two vertical linear axes (see Fig. 1).

2 The Rig

The most notable specifications of the presented Rig are:

- axes combining capability: 1 to 4 axes can be used in the same experiment, or 1 to 4 experiments can be conducted concurrently,
- leg model flexibility: it can drive passive models, or support active and semi-active models, or combinations like passive models with orthotics/exosuits, etc..
- dual model capacity: supports comparative studies of different joints, or same joints at different planes, of leg models with/without orthotics, different instrument sets, different marker placement protocols, etc.,
- vertical axes placement versatility: the axes can be mounted at various positions,
- vertical axes tilting possibility: a variety of activities can be simulated.

2.1 Physical Characteristics and Main Parts of the Rig

Overall dimensions are 1900 × 950 × 1650 mm (L × W × H), including control unit, adjustable feet and transportation wheels. The useful motion envelope is 1465 × 650 × 850 mm. Utilizing a form-follows-function approach, a rigid aluminum frame surrounds the limbs motion envelope, carries the horizontal and vertical linear motion axes, protects users from moving parts and bears the necessary measuring equipment and geometrical patterns for optical calibration at the walls and floor. The control box (see Fig. 1) carries the main processor unit, sensors, driver cards, power supplies, EMI filters, emergency stop switch according to EN60204, joystick and Go button which starts/pauses at stance/continues the execution of the gait cycle. Custom software on android tablet provides the machine's interface. The rig includes 4 linear motion axes. Two axes (X1, X2) are permanently arranged across the horizontal (X) direction, at the bottom, long edges of the cage, as shown in Fig. 1. The other 2 axes (Y1, Y2), as independent modules, are predominantly mounted vertically on auxiliary horizontal beams, but can also be tilted (e.g., at 70°) if required for certain experiments. With minimum setup time, Y1 and Y2 can be easily moved inside the useful envelope, placed at any one of the 4 vertical sides of the cage, or removed when not required. Any specific limb test arrangement can be achieved with Y1, Y2 assembled at either of 2 opposite sides, facilitating the placement of cameras and measuring equipment.

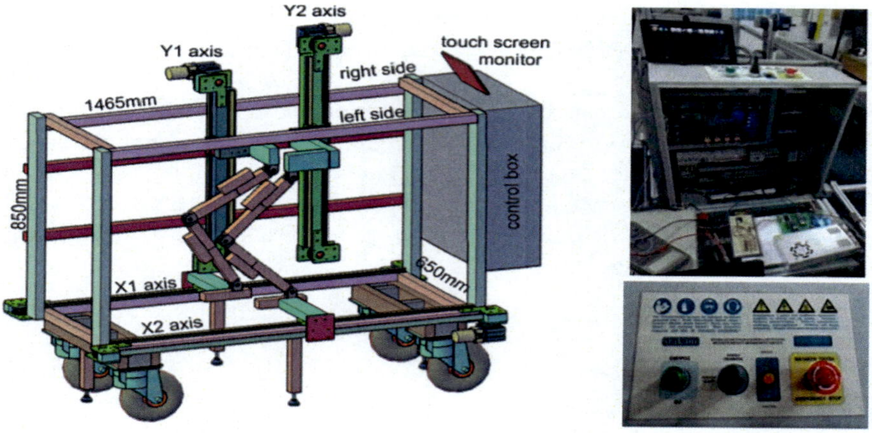

Fig. 1. Left: The rig with 2 passively driven limb models mounted for sagittal study. Right: inner and top view of the control box with the electronic circuits, buttons and joystick.

Each axis comprises of a steel linear guideway, motor with reducer, timing belt and pulleys (see Fig. 2). Each steel guideway is mounted on an aluminum beam and bears its own heavy load carriage where the model of a limb is attached via removable plates or supports. Carriages are equipped with circulating ball rollers and offer smooth motion with low friction coefficient. Four stepping motors are used, with resolution of 1.8 degrees/step, coupled to worm-gear reducers, which are directly attached to the driving pulleys via wedged shafts. Besides increasing the motor torque and resolution, the reducers serve as brakes when no power is applied, assisting the process of mounting the limb models and enhancing overall safety. Since Y1 and Y2 axes are – mostly – positioned vertically to the ground (see Fig. 2), if no locking reducer was applied, the carriages would slide down to their lowest position in case of a power outage, a failure, or during an emergency stop, posing a risk of injury to the personnel, or of damage to the leg models. Polyurethane timing belts reinforced with steel strands pass through pairs of timing pulleys (a driving and a driven), mounted with taper bushes at each end of the linear axes. The belt slack is minimized through pretension at the side of the driven pulley.

The rig can host passive and active models of limbs, as well as their combinations with orthotic devices or exoskeletons. Passive models do not include motors or actuators, contrary to the active. In the simplest cases, 2 axes (e.g., X1 and Y1) are assigned to a passive model of a knee joint, which also includes femur, tibia, hip and ankle joints, as discussed below. Depending on the required study, the limbs may be mounted with their sagittal or coronal plane corresponding to the X1-Y1 plane (see Fig. 3, right). If the model's sagittal plane is not mounted parallel or perpendicular to the X1-Y1 plane, then compound knee joint motions can be studied, provided that hip and ankle joints allow compound rotations (e.g., spherical joints) or that elastically deformable supports are included. In 2-axes-per-model arrangements, 2 independent models can be tested concurrently in the same or different planes. Alternatively, models can be mounted upside-down (hip attached to X1, ankle to Y1), or more than two axes can be assigned to one model, utilizing dedicated sub-assemblies.

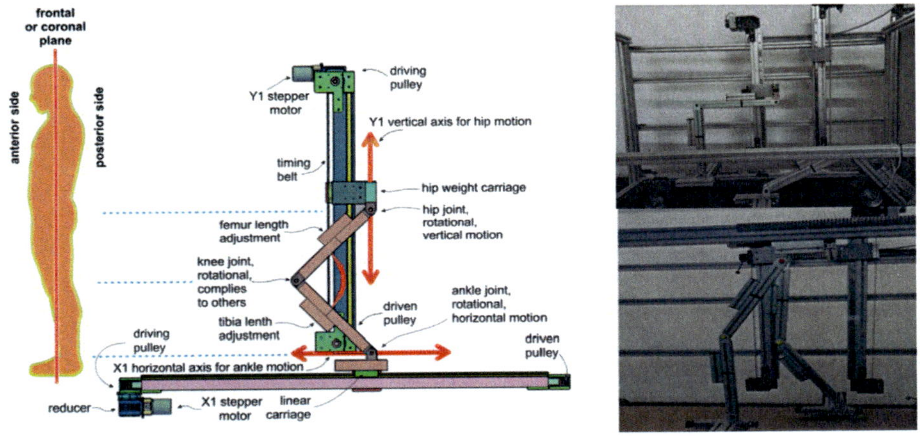

Fig. 2. Left: human figure for rough reference, anatomical planes terminology, side view of a leg model mounted for sagittal study. Dotted lines indicate the corresponding joints. Right: two independent passive models mounted for sagittal study, with Y1, Y2 attached at the same side.

2.2 Passive Models of Limbs

Figure 1 illustrates a passive limb model mounted for sagittal study on 1 horizontal axis (e.g., X1) and 1 vertical axis (e.g., Y1). The adjustable length of model's femur and tibia enable the artificial limbs to match a human subject's dimensions. Sensors and optical recognition markers can be attached with high repeatability and can be adjusted with ease, due to the special extruded industrial aluminum profiles used. The marker placement repeatability is a particularly important factor for gait studies, since even slight misplacement could lead to angular errors of several degrees [13]. There are 3 rotational joints (see Fig. 2), for the hip, knee and ankle, on the same parasagittal plane, without medial or lateral angle ability (neither inner, nor outer). Typically, the joint under study is the knee joint (with angular range from -2 up to 180°), while the hip joint serves merely for mounting the model to the Y linear carriage. Likewise, the ankle joint mounts to the X carriage. Consequently, the hip performs a vertical linear motion and the ankle a horizontal. This model does not include a "foot" (talus and calcaneus).

When humans walk or run, hips' distance to ground fluctuates far less than the ankles' distance. Hence, the above model set-up with the ankle at fixed (zero) height, although counter-intuitive/strange at first sight, offers a significant advantage: extra weights can be easily added to the limbs/joints, to emulate a particular body type and build, or to increase the stress on the joints, e.g., for fatigue tests of mechanisms. The play of the rotational joints is minimal (practically zero), i.e., an order of magnitude below 1/8″ in any plane, in accordance to Mennel's threshold for the definition of involuntary joint movements. In essence, the 2 limbs and the virtual hip-ankle segment behave as an ideal Euclidean triangle, forcing the knee joint angle to comply.

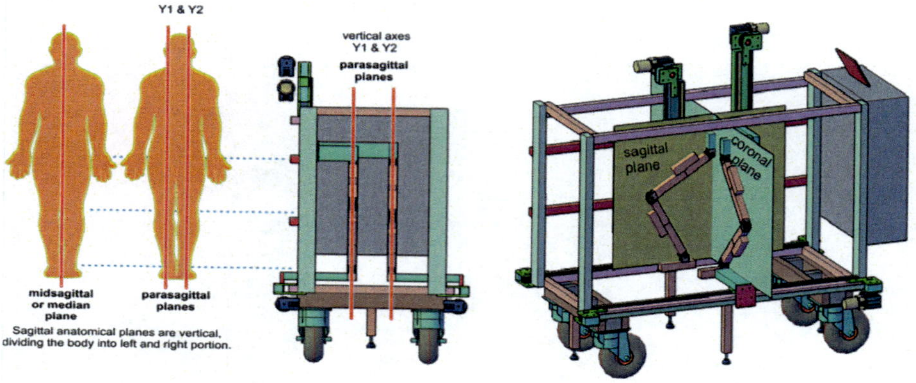

Fig. 3. Left: anatomical planes terminology. Centre: frontal view of the rig with 2 leg models, corresponding to the set-up of Fig. 2. Right: Set-up with one passive limb model mounted for sagittal study and one for coronal study. Notice that the planes are perpendicular.

This results in exceptional repeatability at the reproduction of gait cycles and sequences of cycles: certain x,y coordinates will always produce the same knee joint angle θ (where x: ankle's horizontal distance from hip, and y: hip's vertical distance from ankle). The specific limb model implementation offers a number of advantages:

- Robustness of construction and ease of initial placement.
- Mounting (on the rig) ease and fast tuning.
- Adjustability. Scales or tape measures inside the profile slots allow easy, and exact length adjustment, in order to match specified femur and tibia dimensions.
- Reliable, repeatable and fast marker positioning with increased precision, for optical gait capture. Marker holders, fastened with screws/nuts on various positions, allow fast placement of alternate marker sets, to study e.g., different protocols, or the effect of marker color/size/texture on the recorded data accuracy.
- Reliable, fast and easy measurement/verification of joint angles.
- Ease of attaching force/acceleration/gyroscopic sensors as well as data loggers.
- Ease of attaching orthotics and/or exosuits, on and around the limbs.
- Ease of removing the limbs and replacing them e.g., with exoskeleton parts, under the same setup, maintaining stance and loads.

- Ease of switching knee joints (e.g., 4 to 6-bar mechanisms instead of simple rotational), and conducting experiments with off-the-shelf mechanisms for amputees, or biological knee specimens (ptomatic animal/human), or total knee arthroplastic surgery joints (TKA). Mounting plates, tibial and femoral clamps may be attached to facilitate the exchange of the various parts under study.

Note that if a different, more complex type of joint is used (e.g., 4, 5, or 6-bar mechanisms with springs, or structures including elastic materials/shock absorbers, or models emulating certain ailments), then joint play may not be zero. This results to a fuzziness, a multiplicity of potential knee locations per x,y coordinates, i.e., increased knee joint angle variance.

2.3 Stance and Gait Reproduction

The rig is agnostic to the type and number of the mounted limbs model(s), or to the motion under study. Its function is to drive the X1, Y1, X2, Y2 carriages (or a subset of them) to specific coordinates, at the required timing. 4 coordinates define a "stance". A series of stances with time stamps defines a gait, or a motion sequence in general, since the data may pertain e.g., to cycling or squats.

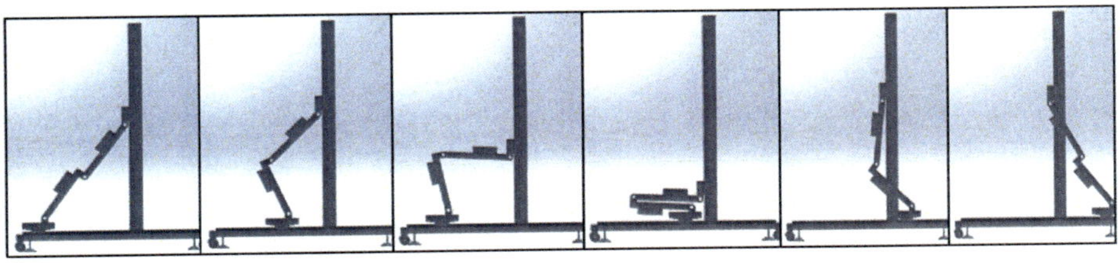

Fig. 4. A few stances of an exercise motion sequence. Note: last stance differs from the first.

If the motion sequence starts and ends with the same stance, then it forms a closed loop which can be repeated a specified number of times or indefinitely. Otherwise, the rig can fill the gap and automatically insert linear motion segments from the last to the first stance (see Fig. 4). This may be inappropriate for certain joints or arrangements as it may result in excessive stresses/attempts to drive a joint model beyond its nominal angle limits. If so, the experiment planner should insert a number of safe, intermediate stances. To protect sensitive exoskeletal limbs, the rig includes freely re-locatable terminal switches to interrupt motions beyond certain extents. Alternatively, the rig can back-trace a motion sequence when it ends. Thus, a non-looping sequence can be repeated with the limbs moving back and forth, without exceeding their limits.

2.4 Potential Applications and Fields of Usage

Once an experiment has been prepared, the rig offers repeatable performance and increased accuracy: the rig can repeat an exact stance or a particular portion of a specific gait cycle, unlike humans. Hence, the subject's fatigue, time constraints and the human Intra-Subject variability are excluded from gait studies (see paragraph 2.5). With movable/tilt-able vertical axes and easily replaceable/configurable leg models, the rig adjusts to various anthropometric factors. Reproducing test set-ups and gait cycles, or other motion sequences on the rig, allows manipulation of factors such as load, speed, direction and pausing/resuming/reversing. The rig facilitates measurements and motion capture from various angles, distances, and lighting conditions.

Repeatedly studying the gait cycle of a specific human, at different labs and countries, with different marker placement protocols, with varying personnel and equipment, at various time periods is almost impossible. But, if the gait data are downloaded to such a rig, the above scenarios become feasible and may assist in training the personnel, assessing the protocols, calibrating the equipment [13], etc.

In dual leg setups, the phase difference between the 2 models may be modified, producing a series of tests. E.g., the swing phases of both legs can be synchronized, vividly presenting left-right discrepancies. Alternatively, both models may act as right (or left) legs, one for a specific subject's gait, and the other for the mean (average) gait of the corresponding age group and gender/"ideal" healthy gait for this body type/gait of the same subject from a previous recording. To offer personalized services in arthroplasty, a patient with TKA surgery could be measured against pre-surgery gait data, obtaining comparable results against his own performance, and not against mean/standard healthy gait. An emerging era of customizable TKA parts, based on MRI measurement from the patients could benefit from such pre/post personalized TKA [14] gait comparisons. A dual setup, with one leg model wearing an orthotic device/exosuit can reveal the suitability and fitness of the orthotic, maturity for usage and if and what type of fine tuning would be necessary, thus, reducing the effort and the amount of time required for adjustments on the human subject.

Several active lower-body exoskeletons [15], as well as passive, soft, wearable exosuits [16] have been developed. Active exoskeleton mechanisms pose increased risks for injury, in case of failure or unexpected mishaps. Exoskeleton designs with incompatible joint motion can strain and even damage human joints [17]. As a potential countermeasure, ready-made robots, e.g., Nao, have been used for testing children's gaits compliance [18] before exoskeletons are fitted. Studies at the US National institute of standards and technology reveal that currently there are limited methods to measure exoskeleton fit to users [19]. The development of a quantifiable set of metrics to evaluate exoskeleton performance is still considered as a significant challenge [20, 21]. Thus, the rig can help to:

- fine-tune exoskeleton mechanisms, without involving humans,
- evaluate their compliance to the human motion,
- measure the percentage of aid provided (e.g., during walking or lifting),
- train and test exoskeletons by exposing them to various mal-gait cycles from different ailments/diseases/syndromes, exploring their ability to cope with diverse gaits, even when no such human volunteers are available,
- subject exosuits and exoskeletons to alternating gaits on real time sequences (e.g., from slow strides to fast walks, to jogging, to running, including tripping/falling), exploring all the transient phases which have been difficult to research,
- evaluate and benchmark procedures for ready-made exoskeletons,

- aid rehabilitation research, e.g., the development of low-cost gait trainers [22].

Prosthetics predate exoskeletons by far. Yet, they face common issues as achieving robustness, minimizing wear and optimally fitting to the motorial behavior of the complex knee joint under various speeds, gaits, forces and weights, e.g., by implementing polycentric mechanisms [23] or microprocessor-controlled prosthetics [24]. Benchtop characterization is applicable to prosthesis for animals, as well [25].

Research on anatomical protocols for gait capture and analysis often leads to variants of the widely used Conventional Gait Model (CGM) [26] or new protocols [27]. Some are poorly reported and others are not consistent with ISB recommendations. If an enhanced protocol appears, the main problem will be that gait databases should be rebuilt based on the new protocol. Otherwise: "lack of relevant reference data from patients completely restricts the use of these methods in clinical practice" [28]. The rig could address incompatibilities with the following set-up: a leg model with a set of targets according to the new protocol is driven with a patient's data recorded according to CGM. The gait is re-captured and thus, a new database can be built. Likewise:

- suggested protocols can be compared against CGM, or others,
- marker-less capture systems can be compared against conventional systems [28],
- motion capture equipment [29], algorithms/prediction models/inference engines, neural networks/AI can be developed, improved and meticulously/automatically tested.

2.5 Rapid Proofing

A fast method to verify the rig's precision and repeatability characteristics, as well as its suitability for gait cycle experiments, is to use an optical camera to capture a number of gait cycles and analyze them. For this purpose, we:

- downloaded the gait data of a unique gait cycle (duration about 2 s) to the rig,
- placed soft markers on knee, ankle and hip joints,
- instructed the rig to infinitely perform the same gait cycle (closed loop),
- recorded video of three consecutive gait cycles (frame rate around 30 fps),
- processed the video with the wide-spread, free Kinovea [29].

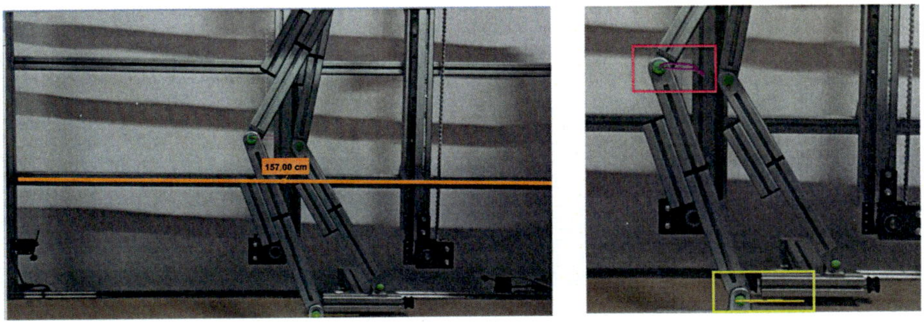

Fig. 5. Performing line calibration and path tracking in kinovea software.

Note that Kinovea supports marker-less coordinate capture. However, markers help the onscreen selection of the points of interest. We conducted line calibration in order to extract mm coordinates (see the orange line measuring 157.00 cm in Fig. 5 left), tracked the paths of joints (see purple and yellow paths in Fig. 5 right) and exported the x,y coordinates to spreadsheet. Ideally, the x,y time series of the 3 consecutive gait cycles should be same; and the superimposed plots (see Fig. 6 Top and Right with the knee's purple path) should be identical. Any difference may be attributed to the combination of the video camera in use, the optical-capture software and the hardware under study (rig). A discussion on the methodology to separate the error components can be found in [13]. Since the purpose of rapid proofing is to reveal the suitability of the combined software and hardware presented for analyzing gaits, we regard all errors as system-wide errors. For each joint, axis, and time instance, we calculate the maximum and minimum value of the available data triplet. Then error $e_{joint,axis,t} = |maximum - minimum|$ (absolute max minus min difference). The $maxError_{joint,axis}$ is the maximum of errors per joint and axis. Likewise, $meanError_{joint,axis}$ and $SD_{joint,axis}$ (mean and standard deviation) are calculated. E.g., for the knee X axis: $maxError_{knee,X} = 0.2$ mm, $meanError_{knee,X} = 0.063$ mm, and $SD_{knee,X} = 0.052$ mm. Accordingly, for Y axis: $maxError_{knee,Y} = 0.17$ mm, $meanError_{knee,Y} = 0.080$ mm, and $SD_{knee,Y} = 0.033$ mm.

The results are quite promising and verify the suitability of the presented rig for gait analysis purposes according to the set objectives. Based on the mean and SD, the rig offers repeatable performance.

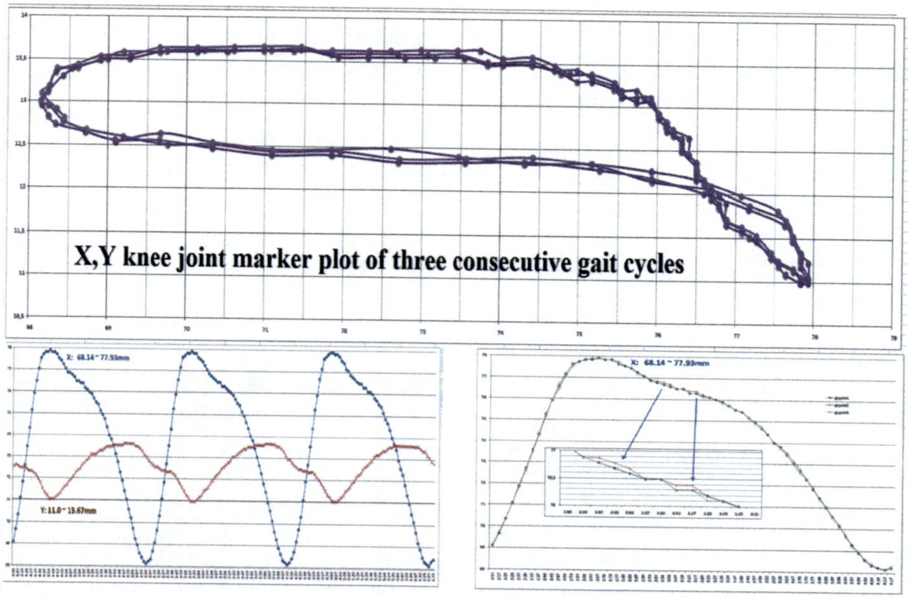

Fig. 6. Knee joint. Left: axis coordinates versus time (X in blue and Y in purple). Right: three overlapping gait cycles (x axis only). Inset detail showing the maximum error is 0.2mm.

3 Conclusions

This paper discussed the role of rigs in gait analysis systems, presented details of a configurable, active, 4 axes rig, focusing on the study of lower extremities, described an example setup with 2 passive leg models and presented an experimental technique to assess the rig's key characteristics illustrating its sub-millimeter accuracy. We commented on the potential applications including the rig's usefulness for assessing gait assisting/reforming systems such as exoskeletons and prosthetics. From the early design and testing phases of subassemblies, up to live tests in gait labs, the human gait cycle variability and constraints can be minimized by substituting the human subject with such a test rig on which a robotic device can easily be attached. Future work could include limbs' mechanisms with more DOF, e.g., rotational spherical joints in the coronal/transverse planes, and more driven rig axes, to extend research to ankle angles, and more complex experiments.

Acknowledgments. The authors would like to thank stakam S.A. for the support and funding. The registration fees were totally covered by the University of West Attica.

References

1. Dowson, D., et al.: A knee joint simulator. evaluation of artificial joints. Biol. Eng. Soc. **7**, 79–90 (1977)
2. Burgess, I.C., et al.: Development of a six station knee wear simulator and preliminary wear results. Proc. Inst. Mech. Eng. H. **211**, 37–47 (1997)
3. Walker, P.S., et al.: A knee simulating machine for performance evaluation of total knee replacements. J. Biomech. **30**, 83–89 (1997)
4. Maag, C., Metcalfe, A., Cracaoanu, I., Wise, C., Auger, D.: The development of simulator testing for total knee replacements. IET Biosurf. Biotribol. **7**, 70–82 (2021). https://doi.org/10.1049/bsb2.12001
5. Etoundi, A., Dobner, A., Agrawal, S., Semasinghe, C., Georgilas, I., Jafari, A.: A robotic test rig for performance assessment of prosthetic joints. Front. Robot. AI **8**, 613579 (2022). https://doi.org/10.3389/frobt.2021.613579
6. Richter, H., Simon, D., Smith, W., Samorezov, S.: Dynamic modeling, parameter estimation and control of a leg prosthesis test robot. App. Math. Modell. **39**, 559–573 (2014). https://doi.org/10.1016/j.apm.2014.06.006
7. Haider, S., Takhakh, A., Al-Waily, M., Saadi, Y.: Simulation of gait cycle in sagittal plane for above-knee prosthesis. AIP Conf. Proc. **2386**, 040002 (2022). https://doi.org/10.1063/5.0066819
8. Jung-Hoon, K., Jun-Ho, O.: Development of an above knee prosthesis using MR damper and leg simulator. In: IEEE International Conference on Robotics and Automation (2001). 10.000-7803-6475-9/01
9. Dockers, K., Stevens, V.: Onderzoek naar de krachtswerking van het patellofemoraal gewricht en ontwerp en constructie van een experimenteel belastingsmodel (2006). https://lib.ugent.be/catalog/rug01:001311795
10. Most, E.: Development of a 6-DOF Robotic Test System for Studying the Biomechanics of Total Knee Replacement (1998). http://hdl.handle.net/1721.1/9059

11. Ding, B., Stanley, R., Cazzolato, B., Costi., J.: Real-time FPGA Control of a Hexapod Robot for 6-DOF Biomechanical Testing (2011). https://doi.org/10.1109/IECON.2011.6119320
12. Abu-Faraj, Z., Harris, G., Smith, P., Hassani, S.: "Human gait and clinical movement analysis", wiley encyclopedia of electrical and electronics. Engineering (2015). https://doi.org/10.1002/047134608X.W6606.pub2
13. Kampouris, C., Azariadis, P., Moulianitis, V.: A methodology for assessing the impact of error components in gait analysis using closed-loop testing on a biomimetic rig. In: Frasson, C., Kabassi, K., Voulodimos, A. (eds.) Novelties in Intelligent Digital Systems: Proceedings of the 1st International Conference (NIDS 2021), Athens, Greece, September 30 – October 1, 2021. IOS Press (2021). https://doi.org/10.3233/FAIA210071
14. Koh, Y., Son, J., Kwon, O., Kwon, S., Kang, K.: Effect of post-cam design for normal knee joint kinematic, ligament, and quadriceps force in patient-specific posterior-stabilized total knee arthroplasty by using finite element analysis. BioMed Res. Int. **2018**, 1–11 (2018). https://doi.org/10.1155/2018/2438980
15. Sun, M., Ouyang, X., Mattila, J., Yang, H., Hou, G.: One novel hydraulic actuating system for the lower-body exoskeleton. Chin. J. Mech. Eng. **34**, 1–10 (2021). https://doi.org/10.1186/s10033-021-00535-w
16. Di Natali, C., et al.: Design and Evaluation of a Soft Assistive Lower Limb Exoskeleton. Cambridge Univeristy Press (2019). https://doi.org/10.1017/S0263574719000067
17. Wang, D., Guo, J., Lee, K.: An adaptive knee joint exoskeleton based on biological geometries. In: IEEE Conference Robotics and Automation (2011). 978-1-61284-380-3/11/2011
18. Wang, Z., Peyrodie, L., Cao, H., Agnani, O., Watelain, E., Wang, H.: Slow walking model for children with multiple disabilities via an application of humanoid robot. Mech. Syst. Signal Process. **68–69**, 608–619 (2015)
19. Bostelman, R., Li-Baboud, Y., Wyk, K., Shah, M.: Development of a Kinematic Measurement Method for Knee Exoskeleton Fit to a User. National Institute of Standards and Technology, Technical Note 2107 (2020). https://doi.org/10.6028/NIST.TN.2107
20. Koelewijn, A., Selinger, J.: Predictive simulations to replicate human gait adaptations and energetics with exoskeletons. IEEE Trans. Neural Syst. Rehabil. Eng. **30**, 1931–1940 (2022). https://doi.org/10.1109/TNSRE.2022.3189038
21. Ott, O., Ralfs, L., Weidner, R.: Framework for qualifying exoskeletons as adaptive support technology. Front. Robot. AI **9**, 951382 (2022). https://doi.org/10.3389/frobt.2022.951382
22. Shin, S., Deshpande, A., Sulzer, J.: Design of a single degree-of-freedom, adaptable electromechanical gait trainer for people with neurological injury. J. Mech. Robot. **10**, 044503 (2018). https://doi.org/10.1115/1.4039973
23. Al-Maliky, F., Chiad, J.: Study and evaluation of four bar polycentric knee used in the prosthetic limb for transfemoral amputee during the gait cycle. Mater. Today Proc. **42**, 2706–2712 (2020). https://doi.org/10.1016/j.matpr.2020.12.709
24. Cao, W., Yu, H., Chen, W., Meng, Q., Chen, C.: Design and evaluation of a novel microprocessor-controlled prosthetic knee. IEEE Access **7**, 178553–178562 (2019). https://doi.org/10.1109/access.2019.2957823
25. Park, H., Islam, M., Grover, M., Klishko, A., Prilutsky, B., DeWeerth, S.: A prototype of a neural, powered, transtibial prosthesis for the cat: benchtop characterization. Front. Neurosci. **12**, 471 (2018). https://doi.org/10.3389/fnins.2018.00471
26. Davis, R.B., Õunpuu, S., Tyburski, D., Gage, J.R.: A gait analysis data collection and reduction technique. Hum. Mov. Sci. **10**, 575–587 (1991). https://doi.org/10.1016/0167-9457(91)90046-Z
27. Zuk, M., Trzeciak, M.: Anatomical protocol for gait analysis: joint kinematics measurement and its repeatability. J. Theor. Appl. Mech. **55**, 3–15 (2017). https://doi.org/10.15632/jtam-pl.55.1.369

28. Tang, H., Pan, J., Munkasy, B., Duffy, K., Li, L.: Comparison of lower extremity joint moment and power estimated by markerless and marker-based systems during treadmill running. Bioengineering **9**, 574 (2022). https://doi.org/10.3390/bioengineering9100574
29. Yusof, Y., et al.: Walking gait analysis: kinovea versus motion capture system. In: IEEE, 13th ICSGRC (2022). https://doi.org/10.1109/ICSGRC55096.2022.9845736

Exploring Smart City Analytical Framework: Evidence from Select Case Studies

Apple Rose Alce, Jerina Jean Ecleo, and Adrian Galido(✉)

MSU-Iligan Institute of Technology, 9200 Iligan City, Philippines
adrian.galido@g.msuiit.edu.ph

Abstract. Smart city research, including its definition and implementation, has progressed with several locales employing smart city initiatives. Case studies present various contexts on how to achieve the cities' 'smartness'. This study aims to broaden smart city research through adapting an analytical framework — high-level objectives, action fields, and enabling factors — and content analysis on select case studies. The framework incorporates the smart city's desired outcomes (high-level objectives), "cross-cutting entry points for digital transformation" (enabling factors), and where solutions can be applied (action fields). Content analysis involved three stages – identification, selection of the literature, and literature grouping. Results indicate that apart from economic growth, sustainability, and quality of life, outcomes including administrative effectiveness and efficiency and knowledge-based development emerge. The uptake of action fields varies even in cities from the same country. The top-down approach to smart city development appeared to be more common from the select case studies. Open data is emerging in smart city development. The heterogeneity of results provides policymakers a staging point to further explore practices from different settings which could aid in tailoring accordingly to their cities' context. Further research could focus on updating the action fields and exploring the innovation process or partnership models of smart cities.

Keywords: smart cities · case study · analytical framework

1 Introduction

There is no apparent one-size-fits-all approach on the smartness of cities [1, 2]. While the smart city concept, implementation, and research have progressed, smart city continues to be an elusive concept with scholars spending considerable effort to clarify what it really is [3]. Several concepts including smart city definition, enabler, dimension, technology, indicator, and outcome will, continue to, benefit from further attention [3–7]. The opportunity to learn from different contexts becomes apparent.

The importance of and the value derived from contextual and data richness, along with a clear chain of evidence, cannot be discounted. Case study research, while scoring low on the generalizability of its results, shows promise for exploratory, theory building research and is "well-suited to ensure relevant IS research" as cited in [9]. The key elements of the case study evident from its structure are of note. Some of these elements, as illustrated

K. Kabassi et al. (Eds.): NiDS 2023, LNNS 783, pp. 164–175, 2023.
https://doi.org/10.1007/978-3-031-44097-7_17

in [10] include themes, research thrust, and sample selection, units of analysis, data collection method, and level of description about units.

The relation between smart cities and the discipline of information systems stems from employing information systems to the design and development of smart cities [11]. The current paper expands smart city research exploring the utility of analytical framework adapted from [12] to map and analyze the initiatives embedded in select case studies.

2 Prior Work

2.1 Current Direction of Smart Cities Research

The literature reviews demonstrated various perspectives to examine the smart city concept. Elements commonly observed include smart city definition, components, and enablers. For instance, [4] focused on the definitions, key components and dimensions, measures of performance, and initiatives of smart cities. A study [1] highlighted contextual conditions, governance models, and assessment of public value as relevant to further understand smart cities. As cited in [3], the widely cited smart city research of [13] highlighted six components including smart economy, smart people, smart governance, smart mobility, smart environment, and smart living.

Table 1. Concepts analyzed from a sample of recent smart city research.

Article*	Number of areas analyzed	Focus areas
[14]	4	Smart cities dimensions, smart applications associated with each dimension, risks associated with smart applications, and risk assessment tools and techniques
[15]	5	Smart city objectives, challenges, main components (strategic requirements, dimensions, risks, implementation policy, architecture), stakeholders, and resources
[16]	8	Research methodology, research purpose, data collection method, use of research results, research topics and keywords, analysis unit, general classification criteria, and security keywords
[8]	4	Smart cities dimensions, digital transformation, sustainability, and resilience

* Note: Sample of recent studies, i.e., 2022, retrieved from Google Scholar using the search phrase ("smart city" OR "smart cities") AND ("systematic review" OR "systematic literature review" OR "systematic map" OR "systematic mapping" OR "mapping study" OR "scoping review" OR "meta-analysis")

Table 1 presents related studies discussing different focus areas in reviewing smart city concepts. [14] revealed that not all smart cities adapt all of the smart city dimensions. [15] highlights relevant approaches such as strategic vision definition, action plan development, and management strategy definition to aid the ambiguity of smart cities strategies. In [16], the literature was observed to conduct surveys and interviews as research methods for identifying future directions of smart city research. [8], in tackling smart cities, conclude that research on smart cities can be performed as multidisciplinary studies.

Studies explore different focus areas as smart city implementation involves complex and multifaceted systems. As observed, prior research has primarily looked into smart city definition or smart city components such as that emanating from [13]. A holistic perspective apparent from such frameworks as [12], applied to case studies, have emerged and could contribute to advancing the field and practice. Further, as in [11] and [17], the planning process of smart cities is also inadequately addressed.

2.2 Advance of Smart Cities in Different Regions

The ASEAN Smart Cities Network, or ASCN, launched in 2018, could potentially advance collaboration and growth of the region through a pioneer batch of 26 cities [18]. The ASEAN Smart Cities Framework articulated key features as (i) strategic outcomes (high quality of life, competitive economy, sustainable environment); (ii) urban systems (integrated master planning and development, dynamic and adaptive urban governance); (iii) focus areas (civic and social, health and well-being, safety and security, quality environment, built infrastructure, industry and innovation); and (iv) enablers (digital infrastructure and applications, partnership and funding). In the Philippines' context, smart city initiatives could benefit from and to policy interventions such as the RA 11032 or the ease of doing business [19]. Fast forward, smart cities in the region could benefit from the experiences of their neighbors in the Asia-Pacific or their counterparts in/or Europe.

Smart city research has also been evident for North America and Europe. Countries in Europe have exhibited smart city initiatives as early as 2010 [20]. Spain and Italy topped the list in terms of the number of smart city initiatives. Further, of the 22 smart city categories identified, the community, inclusion, and participation was exhibited the most. North American smart cities were observed to be more bottom-up oriented compared to their counterparts in Europe [21].

3 Methods

Case studies are considered as vehicles for knowledge sharing and potential collaboration. Content analysis, adapting [14] with three main stages – identification, selection of the literature, and literature grouping – was employed. As a caveat, biases could manifest when employing the coding scheme of content analysis. For the identification stage, research questions based on the model of [12] were drawn. The model incorporated focus areas including (1) high-level objectives, (2) action fields, and (3) enabling factors. A fourth focus area aptly noted as the approach to smart city development encompassing the planning process and strategic planning tools and techniques from [11] and [17] was incorporated as well. For the selection of literature stage, two criteria were set. This paper attempts to address the questions including: (1) What were the objectives for transforming into smart cities? (2) What action fields were undertaken? (3) What factors enabled cities to be 'smart'? (4) What were the planning process, tools, and techniques employed to develop into a smart city?

The sample of cities to be included were identified and retrieved from Google Scholar using the search phrase ("smart city" OR "smart cities") AND (Asia or Europe) AND ("case study" OR "case studies"). Case studies published in peer-reviewed articles and available in the English language were considered. Lastly, for the literature analysis and grouping stage, the case studies' contents relevant for the focus areas were identified and subsequently summarized.

Table 2. Select case studies from different countries

Author*	Year	Local of interest (of current study)		Article*	Year	Local of interest (of current study)	
[28] Stratigea	2021	Greece	Trikalla	[33] Desouza et al	2020	South Korea	Songdo
[29] Supangkat et al	2018	Indonesia	Multiple, 31 cities	[31] Angelidou	2017	Spain	Barcelona
[30] Fietkiewicz et al	2015	Japan	Kyoto, Osaka, Tokyo, Yokohama	[34] Bakıcı et al	2013	Spain	Barcelona
[26] Yau et al	2016	Malaysia	Greater Kuala Lumpur	[23] Bibri et al	2020	Spain	Barcelona
[31] Angelidou	2017	Malaysia	Cyberjaya	[35] Orejon-Sanchez et al	2022	Spain	Multiple, 61 cities
[32] Lim et al	2021	Malaysia	Kuala Lumpur, Putrajaya	[25] Leu et al	2021	Taiwan	Multiple, 10 cities
[27] Yang et al	2020	South Korea	Daegu City, Siheung City	[36] Sontiwanich et al	2022	Thailand	Phuket
[16] Myeong et al	2020	South Korea	Gimpo, Namyangju	[37] Hoang Viet Bach et al	2019	Vietnam	Dalat City

* Note: Sorted based on country names

Table 3. Summary view of high-level objectives

Author	Local of interest (of current study)		Administrative effectiveness and/or efficiency	Knowledge-based development	Economic growth	Sustainability	Quality of life
[28]	Greece	Trikalla	/		/		
[29]	Indonesia	Multiple, 31 cities				/	
[30]	Japan	Kyoto, Osaka, Tokyo, Yokohama				/	
[26]	Malaysia	Greater Kuala Lumpur					/
[31]	Malaysia	Cyberjaya		/			
[32]	Malaysia	Kuala Lumpur, Putrajaya			/	/	/
[19]	Cauayan, Davao, Mandaue, San Fernando, Tagum	Philippines			/	/	/
[27]	South Korea	Daegu City, Siheung City			/		
[16]	South Korea	Gimpo, Namyangju	/				/
[33]	South Korea	Songdo			/		
[34]	Spain	Barcelona	/				/
[23]	Spain	Barcelona					/
[35]	Spain	Multiple, 61 cities			/		
[25]	Taiwan	Multiple, 10 cities	/		/		/
[36]	Thailand	Phuket		/			
[37]	Vietnam	Dalat City	/			/	/
		Count	**5**	**2**	**7**	**5**	**8**

4 Results and Discussion

The sample of case studies on cities (see Table 2) retrieved resulted in a collection including Barcelona for Europe, cities in South Korea, Japan, and Taiwan for Asia-Pacific, and cities in Indonesia, Kuala Lumpur, and Phuket for the ASEAN context among others. Four areas – high-level objectives, action fields, and enabling factors of [12] and the approach to smart city development of [11] and [17].

Table 4. Summary view of action fields

Author	Local of interest (of current study)		ICT infrastructure	Urban and land use	Mobility and transport	Energy	Environment	Water	Climate resilience and disasters risk management	Food and agriculture	Public management and services	Economy and jobs	Education	Health	Society and culture	Safety and security	Count
[28]	Greece	Trikalla		/	/						/			/		/	**5**
[29]	Indonesia	Multiple, 31 cities			/	/	/	/			/	/	/	/	/	/	**10**
[30]	Japan	Kyoto	/		/												**2**
[30]	Japan	Osaka	/				/										**2**
[30]	Japan	Tokyo	/			/											**2**
[30]	Japan	Yokohama	/			/	/	/							/		**5**
[26]	Malaysia	Greater Kuala Lumpur	/	/	/	/	/				/	/	/		/		**9**
[31]	Malaysia	Cyberjaya	/								/		/	/			**4**
[16]	Malaysia	Kuala Lumpur	/		/		/				/	/	/		/		**7**
[27]	South Korea	Daegu City	/	/	/				/		/	/				/	**7**
[27]	South Korea	Siheung City	/	/	/	/	/					/		/	/		**8**
[16]	South Korea	Gimpo	/						/								**2**
[16]	South Korea	Namyangju	/	/	/			/		/	/			/	/	/	**9**
[33]	South Korea	Songdo		/	/	/											**3**
[31]	Spain	Barcelona	/			/	/				/		/	/	/		**7**
[34]	Spain	Barcelona	/	/	/	/	/				/	/	/		/	/	**10**

(continued)

Table 4. (*continued*)

Author	Local of interest (of current study)		ICT infrastructure	Urban and land use	Mobility and transport	Energy	Environment	Water	Climate resilience and disasters risk management	Food and agriculture	Public management and services	Economy and jobs	Education	Health	Society and culture	Safety and security	Count
[23]	Spain	Barcelona			/	/	/				/						**4**
[35]	Spain	Multiple, 61 cities	/	/	/		/				/	/			/		**7**
[25]	Taiwan	Multiple, 10 cities	/		/		/			/		/	/	/			**7**
[36]	Thailand	Phuket			/		/				/	/	/	/	/		**7**
[37]	Vietnam	Dalat City		/	/		/			/	/	/	/	/			**8**
		Count	**15**	**9**	**15**	**9**	**13**	**3**	**2**	**3**	**13**	**10**	**9**	**9**	**10**	**5**	

Table 5. Summary view of enablers

Author	Local of interest (of current study)		Policy and institutions	Technology and innovation	Business and finance	Digital skills and capacity	Planning and coordination
[28]	Greece	Trikalla	/	/			/
[29]	Indonesia	Multiple, 31 cities	/	/		/	/
[30]	Japan	Kyoto	/				/
[30]	Japan	Tokyo	/				
[26]	Malaysia	Greater Kuala Lumpur	/	/	/	/	
[31]	Malaysia	Cyberjaya	/			/	
[16]	Malaysia	Kuala Lumpur	/	/		/	/
[27]	South Korea	Daegu City, Siheung City	/	/	/	/	/
[16]	South Korea	Gimpo	/	/	/		/
[16]	South Korea	Namyangju	/			/	/
[33]	South Korea	Songdo	/	/			
[31]	Spain	Barcelona					/
[34]	Spain	Barcelona	/	/		/	/
[23]	Spain	Barcelona	/	/		/	
[35]	Spain	Multiple, 61 cities	/	/	/	/	/
[25]	Taiwan	Multiple, 10 cities	/	/	/		/
[36]	Thailand	Phuket	/	/	/	/	
[37]	Vietnam	Dalat City	/	/	/	/	/
		Count	**17**	**13**	**7**	**11**	**12**

High-Level Objectives. The cities included in the sample demonstrated differing aspirations for becoming a smart city (see Table 3). The case studies have also shown that smart cities have exceeded the strategic outcomes – high quality of life, competitive economy, and sustainable environment – of the ASCN framework. Quality of life and economic growth have clear uptake as outcomes and appear more prevalent compared to sustainability. As indicated in the results, the connection between the smart and sustainability agenda is lagging. This observation is consistent with the literature such as [22] and [23]. In [24], smart city policies demonstrate techno-centricity. While there are existing measurement frameworks that incorporate sustainability indicators when assessing smart city performance, smart city definition, policies, and subsequent initiatives on sustainability need to be strengthened. Interestingly, the additional set of outcomes obtained from the case studies included administrative effectiveness and/or efficiency and knowledge-based development.

Administrative effectiveness and efficiency. This could be characterized as enabling better design and delivery of services from improved interaction with citizens. In Trikalla [28], the push for public participation in the decision-making process – "creation of new participatory culture" – enabling better identification of the needs and expectations of the community was evident. For Namyangju [16], smart city platforms were framed to improve the city's services to citizens. The smart city in Barcelona [34] through the Open Data project from Smart Governance intended to improve the efficiency of public services and increase social value. In terms of implementation, government information – territory, population, management and procedure indicators, urban environment and documental data – would be made available for public access.

Knowledge-based development. In [31], for Cyberjaya, knowledge-based development translates to enabling academia and industry to "collaborate and/or engage in innovative activities.“ Phuket's [36] push for smart city and a knowledge-based economy uncover the value of "a shift in the mindset of policymakers and city managers… Allow(ing) for more citizen-centric participatory processes in urban planning and management" (somehow relating to administrative effectiveness). More importantly, "smart city education initiatives must go beyond technological and academic skills."

Action Fields. The uptake of the action fields for the cities included in the sample are varied even when they are placed in the same country (see Table 4). There are a number of use cases for each of the action fields available from different countries. ICT infrastructure is, evidently, a foremost consideration for smart city development. The action fields relating to climate resilience, disaster risk management, and food and security could benefit from further advance in the context of the cities included in the sample.

Enabling Factors. Enabling factors contribute to the development and success of smart cities. The data suggest policy and institutions, planning and coordination, and technology and innovation are more commonly indicated and recognizable from the narratives of cities included in the sample (see Table 5). The availability of multiple enabling factors denotes their interdependence to advance smart city initiatives.

Approach to Smart City Development. The top-down approach appeared to be more common among the cities in the current study sample. The dual development model, as cited, in [25] combine top-down/bottom-up approaches. The dual development model could be a suitable alternative to the top-down approach.

A Closer Look at Technology. The technology advance of cities focused on data management and communication. Some of the technologies evident in the sample of case studies was also disparate. Big data, geographic information systems, and internet of things found utility in Namyangju [16]. Open data was of interest to Greater Kuala Lumpur [26] while the data hub was for Siheung and Daegu [27].

The heterogeneity of results provides policymakers a staging point to further explore practices from different settings which could aid in tailoring accordingly to their cities' context.

5 Conclusion

This paper intends to expand smart city research by utilizing the analytical framework adapted from [12]. The analytical framework of [12] provides a holistic and convenient approach to draw perspectives from smart city initiatives embedded in case studies. Case study research could effectively contribute to the different approach of moving to be a smart city. Further research initiatives are directed to review the existing areas of [12] particularly the action fields. Other categories such as in [20] could well be integrated in the analytical framework. Exploring the innovation process and partnership models of smart cities could benefit from further research.

Acknowledgements. This work was supported with the internally-funded research grant of MSU – Iligan Institute of Technology.

References

1. Meijer, A.J., Gil-Garcia, J.R., Bolívar, M.P.R.: Smart city research: contextual conditions, governance models, and public value assessment. Soc. Sci. Comput. Rev. **34**(6), 647–656 (2016)
2. Zhao, F., Fashola, O.I., Olarewaju, T.I., Onwumere, I.: Smart city research: a holistic and state-of-the-art literature review. Cities **119**, 103406 (2021)
3. Yu, W., Xu, C.: Developing smart cities in China: an empirical analysis. Int. J. Public Adm. Digit. Age (IJPADA) **5**(3), 76–91 (2018)
4. Albino, V., Berardi, U., Dangelico, R.M.: Smart cities: definitions, dimensions, performance, and initiatives. J. Urban Technol. **22**(1), 3–21 (2015)
5. Anthopoulos, L.G.: Understanding the smart city domain: a literature review. Transforming city governments for successful smart cities, 9–21 (2015)
6. Perboli, G., Rosano, M.: A taxonomic analysis of smart city projects in North America and Europe. Sustainability **12**(18), 7813 (2020)
7. Ramos, T.P., Lorenzo, P.J.M., Ancheta, J.A., Ballesteros, M.M.: How far have Philippine LGUs come in developing into smart cities? (No. 2022–07). PIDS Policy Notes (2022)
8. Reis, J., Marques, P.A., Marques, P.C.: Where are smart cities heading? A meta-review and guidelines for future research. Appl. Sci. **12**(16), 8328 (2022)
9. Akkermans, H., van Helden, K.: Vicious and virtuous cycles in ERP implementation: a case study of interrelations between critical success factors. Eur. J. Inf. Syst. **11**(1), 35–46 (2002)
10. Benbasat, I., Goldstein, D.K., Mead, M.: The case research strategy in studies of information systems. MIS Quart. **11**, 369–386 (1987)
11. Ismagilova, E., Hughes, L., Dwivedi, Y.K., Raman, K.R.: Smart cities: advances in research—an information systems perspective. Int. J. Inf. Manage. **47**, 88–100 (2019)

12. Yoon, S.Y., Hong, S.L., Zelt, T.: Smart city pathways for developing Asia: an analytical framework and guidance (2020)
13. Giffinger, R., Fertner, C., Kramar, H., Kalasek, R., Pichler-Milanovic, N., Meijers, E.J.: Smart cities. ranking of European medium-sized cities, Vienna University of Technology, Wien, Austria (2007)
14. Al Sharif, R., Pokharel, S.: Smart city dimensions and associated risks: review of literature. Sustain. Cities Soc. **77**, 103542 (2022)
15. Korachi, Z., Bounabat, B.: Towards a smart city approach: a comparative study. In: Ahmed, M.B., Teodorescu, H.-N.L., Mazri, T., Subashini, P., Boudhir, A.A. (eds.) Networking, Intelligent Systems and Security. SIST, vol. 237, pp. 619–633. Springer, Singapore (2022). https://doi.org/10.1007/978-981-16-3637-0_44
16. Myeong, S., Kim, Y., Ahn, M.J.: Smart city strategies—technology push or culture pull? a case study exploration of Gimpo and Namyangju South Korea. Smart Cities **4**(1), 41–53 (2020)
17. Mora, L., Deakin, M., Reid, A.: Strategic principles for smart city development: a multiple case study analysis of European best practices. Technol. Forecast. Soc. Chang. **142**, 70–97 (2019)
18. Center for Liveable Cities. ASEAN Smart Cities Network. CLC Publications: Singapore. Retrieved from https://www.clc.gov.sg/docs/default-source/books/book-asean-smart-cities-network.pdf (2018)
19. Ramos, T.P., Lorenzo, P.J.M., Ancheta, J.A., Magno-Ballesteros, M.: Readiness of Philippine cities to smart city development (No. 2021–33). PIDS Discussion Paper Series (2021)
20. Correia, D., Marques, J.L., Teixeira, L.: The state-of-the-art of smart cities in the European union. Smart Cities **5**(4), 1776–1810 (2022)
21. Kubina, M., Šulyová, D., Vodák, J.: Comparison of smart city standards, implementation and cluster models of cities in North America and Europe. Sustainability **13**(6), 3120 (2021)
22. Ahvenniemi, H., Huovila, A., Pinto-Seppä, I., Airaksinen, M.: What are the differences between sustainable and smart cities? Cities **60**, 234–245 (2017)
23. Bibri, S.E., Krogstie, J.: Smart sustainable cities of the future: an extensive interdisciplinary literature review. Sustain. Cities Soc. **31**, 183–212 (2017)
24. Yigitcanlar, T., Kamruzzaman, M., Foth, M., Sabatini-Marques, J., da Costa, E., Ioppolo, G.: Can cities become smart without being sustainable? A systematic review of the literature. Sustain. Cities Soc. **45**, 348–365 (2019)
25. Leu, J.H., Lin, B.C., Liao, Y.Y., Gan, D.Y.: Smart city development in Taiwan. IET Smart Cities **3**(3), 125–141 (2021)
26. Yau, K.L.A., Lau, S.L., Chua, H.N., Ling, M.H., Iranmanesh, V., Kwan, S.C.C.: Greater Kuala Lumpur as a smart city: a case study on technology opportunities. In: 2016 8th International Conference on Knowledge and Smart Technology (KST), pp. 96–101. IEEE (2016)
27. Yang, J., Kwon, Y., Kim, D.: Regional smart city development focus: the South Korean national strategic smart city program. IEEE Access **9**, 7193–7210 (2020)
28. Stratigea, A.: The concept of ‘smart cities’. Towards community development. Netcom Réseaux, communication et territoires **26–3/4**, 375–388 (2012)
29. Supangkat, S.H., Arman, A.A., Nugraha, R.A., Fatimah, Y.A.: The implementation of Garuda smart city framework for smart city readiness mapping in Indonesia. J. Asia-Pac. Stud. **32**, 169–176 (2018)
30. Fietkiewicz, K.J., Stock, W.G.: How “Smart” are Japanese cities? An empirical investigation of infrastructures and governmental programs in Tokyo, Yokohama, Osaka, and Kyoto. In: 2015 48th Hawaii International Conference on System Sciences, pp. 2345–2354. IEEE (2015)
31. Angelidou, M.: The role of smart city characteristics in the plans of fifteen cities. J. Urban Technol. **24**(4), 3–28 (2017)

32. Lim, S.B., Malek, J.A., Yussoff, M.F.Y.M., Yigitcanlar, T.: Understanding and acceptance of smart city policies: practitioners' perspectives on the Malaysian smart city framework. Sustainability **13**(17), 9559 (2021)
33. Desouza, K.C., Hunter, M., Jacob, B., Yigitcanlar, T.: Pathways to the making of prosperous smart cities: an exploratory study on the best practice. J. Urban Technol. **27**(3), 3–32 (2020)
34. Bakıcı, T., Almirall, E., Wareham, J.: A smart city initiative: the case of Barcelona. J. Knowl. Econ. **4**, 135–148 (2013)
35. Orejon-Sanchez, R.D., Crespo-Garcia, D., Andres-Diaz, J.R., Gago-Calderon, A.: Smart cities' development in Spain: a comparison of technical and social indicators with reference to European cities. Sustain. Cities Soc. **81**, 103828 (2022)
36. Sontiwanich, P., Boonchai, C., Beeton, R.J.: An unsustainable smart city: lessons from uneven citizen education and engagement in Thailand. Sustainability **14**(20), 13315 (2022)
37. Bach, K.H.V., Kim, S.-K.: Towards evaluation the cornerstone of smart city development: case study in Dalat City, Vietnam. Smart Cities **3**(1), 1–16 (2019)

Augmented Intelligence Assisted Deep Learning Approach for Multi-Class Skin Cancer Classification

Amreen Batool and Yung-Cheol Byun(✉)

Department of Computer Engineering, Major of Electronic Engineering, Institute of Information Science Technology, Jeju National University, Jeju 63243, South Korea
ycb@jejunu.ac.kr

Abstract. Skin cancer is the most common type of cancer in the world. An important aspect of skin cancer diagnosis is detecting lesions as early as possible. Over the past five years, deep learning-based techniques have performed better on image classification tasks. There still needs to be a compromise in analysis accuracy due to image occlusions and imbalanced datasets. The main drawbacks of image occlusion and imbalanced datasets are to need for help to produce generalized models, increased computational overhead, loss of information of occluded regions, etc. Therefore, a robust approach is required to tackle the aforementioned challenges to produce a generalized model for lesion cancer cell classification. This study proposes an augmented-assisted deep learning approach based on DenseNet for multi-class lesion cancer cell classification and detection. In this study, an open-source dataset of lesion cancer cell images is used. Furthermore, data augmentation generates synthetic images to balance the class distribution to avoid over-fitting and poor generalization-related issues. In addition, a robust multi-class classifier based on a DenseNet121 architecture is developed to classify seven skin lesion classes that resize all images to the same size. According to the experimental results, our proposed multi-class classifier achieved an accuracy of 93% and 90.17% using training and validation images set. Similarly, our proposed classifier achieved an accuracy of 90% using a testing (unseen) image set. In the case of skin lesion identification, this research has a positive impact on supporting dramatists in the clinic with more accurate decisions.

Keywords: Multi-class skin cancer · Data augmentation · Deep learning · Dense-net121

1 Introduction

The most prevalent type of cancer in people is skin cancer, which develops due to abnormal cell growth on the top layer of the skin. The development of this type of cancer is usually related to exposure to ultraviolet rays. It may occur in such parts of the skin if there is an abnormality in skin cells' deoxyribonucleic acid (DNA) [1]. Cancer grows because of the uncontrolled growth of cells based on mutations. The sun's UV rays

K. Kabassi et al. (Eds.): NiDS 2023, LNNS 783, pp. 176–186, 2023.
https://doi.org/10.1007/978-3-031-44097-7_19

and tanning beds are the most common causes of skin cancer, exposure to toxic chemicals or illnesses that impair the immune system, an unhealthy lifestyle, and genetics [2]. It is predicted that 96,530 new cases of melanoma will be detected in the USA in 2023, with 57,110 male cases and 40,370 female cases. The number of deaths from melanoma will also reach 7,990, with 5,420 men and 2,570 women affected [3]. Melanoma is increasingly prevalent over time, indicating a growing number of cases. Based on recent statistics, more skin cancer cases have been reported since the 1990s. Increasing use of solariums and solariums and a deteriorating ozone layer may explain current statistics about UV radiation. In [4]5, Lesions caused by melanoma can be pink, red, black, or brown in color, and they can treat according to serve and sickness. A dermatologist should examine a naevus bigger than 6 mm in diameter and has unusual colors for potential melanoma infection. Because there are so many different skin textures and blemishes, screening for skin cancer is a stren- uous task. Dermatologists first carefully assess this behavior by analyzing the ensuing lesion's width, shape, and odd skin color [6]. In [7] Author classified skin lesions using single model of CNN with input of only pixels and disease labels images and identified the common cancer and deadliest cancer. Dermatologists now iden- tify skin cancer using microscopic imaging and dermoscopy [8]. A mobile camera can be used to see low-resolution images of tiny objects. This unique imaging technique interferes with the skin's high reflectivity and resolution, allowing der- matologists to detect deeper features. CAD techniques are constantly required in the field of medical imaging to obtain a second opinion, particularly for a der- matologist to maintain their decision and diagnose skin lesions more rapidly and accurately [9, 10]. Building blocks for identifying skin lesions in CAD systems include image acquisition, feature extraction, segmentation, selection and fusion of skin image datasets, and classification of different skin lesions [11]. Get deep features from the CNN model's fully connected layers, which are subsequently utilized for classification [12]. Instead of traditional methods like color, texture, and shape, in-depth features include worldwide and local information about an image. The convolutional layer provided image input information, while the 1D layer provided global pooling and fully connected information. The localization of the wound area and the borders of the infected area are thinner in the first segmentation process, whereas the classes related to the restricted lesion are broader in the second classification stage. The contributions of this study are as follows:

- Develop a robust dense net-based multi-class classifier to predict and identify skin lesions.
- Develop an augmentation model to balance class distribution by generating synthetic samples to enhance results quality.
- Provide a comparative analysis to evaluate the multi-class classifier's perfor- mance in terms of accuracy, precision, F1 score, and recall performance.

2 Materials and Method

In this section, explain and present all material and methods we used to build and classify the proposed model, including pre-trained model data preparation and training. we explain the proposed methodology in Fig. 1.

2.1 Dataset

In the images, all significant diagnostic criteria for pigmented lesions are shown, including melanocytic nevi (nv), Melanoma (mel), Benign keratosis-like lesions (bkl) basal cell carcinoma (bcc), benign keratosis-like lesions (solar lentigines, seborrheic keratoses, and lichen-planus-like lesions, bkl), dermatofibroma (df), and vascular lesions are all examples of actinic keratoses and intraepithelial carci- noma/(angiomas, angiokeratomas, granulomas, pyogenic granulomas, vasc) The details of the dataset of skin cancer images are shown in Table 1. The majority of lesions are confirmed by histopathology (histo), with follow-up experiments, ex- pert consensus, and in-vivo confocal microscopy providing the remaining ground truth. Figure 2 shows the random samples of 7 classes of skin cancer images. In addition, it can be seen that the distribution of images is not balanced; therefore, it leads to cause several issues, such as over-fitting, increased computational overhead, and produce biased results.

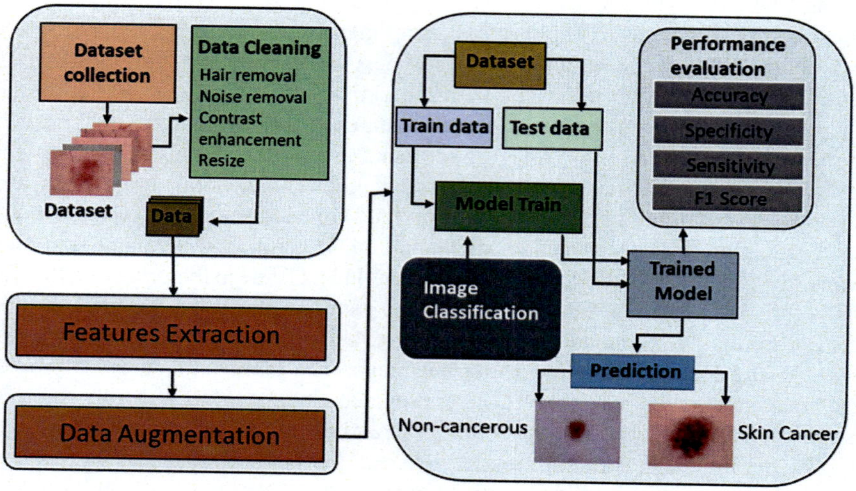

Fig. 1. Proposed Methodology

3 Model Training

3.1 Loading/Resizing Images

Image loading and resizing are necessary preprocessing steps in skin cancer clas- sification tasks. Skin cancer classification tasks generally involve training a ma- chine learning algorithm to predict whether a skin lesion is benign or malignant. The first step in this process is to load the skin lesion images from a dataset.

This involves reading the image files and converting them into a format that can be processed by the machine-learning model. 10015 dermatoscopic images were resized from 450×600 to 224×224, where 3 signifies the RGB color channel.

Table 1. Seven Classes of Skin Cancer Image Dataset

Classes	Images
Melanocytic nevi (nv)	6705
Melanoma (mel)	1113
Benign keratosis-like lesions (bkl)	1099
Basal cell carcinoma (bcc)	514
Actinic keratoses (akiec)	327
Vascular lesions (va)	142
Dermatofibroma (df)	115

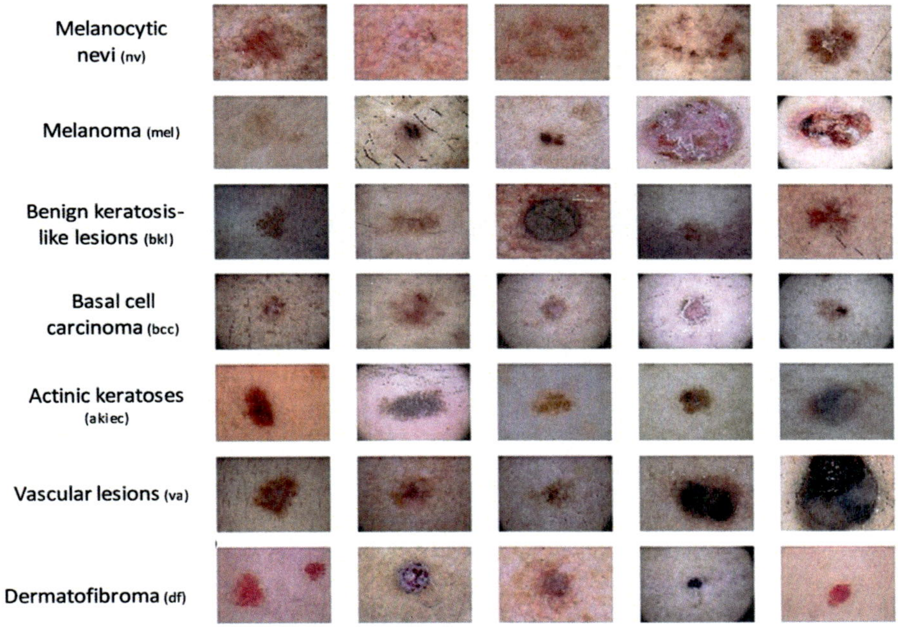

Fig. 2. Random sample images from HAM10000 dataset.

3.2 Train and Test

The training set is used to train a model, while the testing set is used to evaluate the model's performance.

The data is randomly split into training and testing sets with an 80:20 ratio. The model is trained on the training set and evaluated on the testing set. Only 7470 of the 10015 lesion IDs are distinct. The implication is that some lesion IDs are represented by several images. The training set has 9077 photos, whereas the validation set has 938 images.

3.3 Normalization

We applied the Min-Max normalization approach This normalization method scales the pixel values to be between 0 and 1.

Mathematically, it is defined as:

$$Pixel_{normalized} = \frac{pixel - Min_{pixel\_value}}{max_{pixel\_value} - min_{pixel\_value}} \quad (1)$$

An image's minimum and maximum pixel values are specified by the min and max pixel values.

3.4 Data Augmentation

Data augmentation is a technique used to generate additional training data by applying various transformations to the existing images in a dataset [13]. We used ImageDataGenrator.An imbalanced class in the HAM10000 dataset is one of its key issues. Melanocytic nevi (nv) contains 6705 pictures, far more than any other class. We will employ the data augmentation approach to solve this problem. Also, it will aid in reducing overfitting. Several data augmentation techniques include picture rotation and random nonlinear transformation. Data augmentation often seems to be a very effective way to increase accuracy. Param- eters such as h and v flip, zoom range, rotation range, ws range, and hs range are applied for data augmentation. For the training set, data augmentation was used, except for the melanocytic nevi(nv) class, because it is already well-represented. Each class generated a total of 43918 images. The further details are shown in Table 2.

Table 2. Description of image training and validation before and after augmentation

Classes	Before Augmentation		After Augmentation		Original Images
	Training	Validation	Training	Validation	
Melanocytic nevi	5954	751	5954	751	6705
Melanoma	1074	39	6894	39	1113
Benign keratosis-like lesions	1024	75	6894	75	1099
Basal cell carcinoma	484	30	6826	30	514
Actinic keratoses	301	26	6070	26	327
Vascular lesions	131	11	6157	26	327
Dermatofibroma	109	6	5123	6	115
Total	9077	938	43918	938	10015

4 Deep Learning Technique

Object detection, image segmentation, and medical imaging are among the deep learning applications. As a result of the deep learning algorithm's end-to-end training, the algorithm can determine whether an image is benign or malignant directly from the raw image pixels and the image label. Typically, deep learning models are trained on large datasets of images of skin lesions and then applied to new images.

4.1 DenseNet

DenseNet can be encapsulated as a feed-forward network that uses a dense trans- form to connect all network layers. DenseNet was developed by [14]. It needs fewer parameters than conventional networks since the feature maps are not re- quired to be relearned. DenseNet can be encapsulated as a feed-forward network that uses a dense transform to connect all network layers. In DenseNet, a limited number of feature maps are added to the network's general information while keeping the other aps' integrity. The DenseNet layers are thin; for example, each layer contains 12 feature maps. According to all feature mappings on the network, the final classifier determines the input's class. DenseNet [15] provides advanced information and gradient fluxes, simplifying training throughout the network. Input signal gradients can be directly accessed by each layer based on the loss function. As a result, complex network designs can be trained more easily. Moreover, too strict compliance with activities requiring smaller training sets is easily undermined by intense connections [15]. DenseNet is divided into Dense- Blocks. Each block has the same dimensions but has a different number of filters. As part of CNN, batch normalization is accomplished through downsampling by the transition layer. The architecture of DenseNet is shown in Fig. 3. This model has been enhanced with dense blocks. Each layer employs the "ReLu" activation function, and there is a pool of convolution layers with 3×3 filters to 1×1 filters, followed by batch normalization. Transition blocks combine the out- put and input of these dense blocks with those of earlier blocks. A convolution and pooling layer with 2×2 to 1×1 filters and dropout layers is present in each transition block. This concertation of blocks preserves the feature dissemination character.

5 Implementation and Results

We used the pre-trained DenseNet-121 model to train the dataset. GlobalAverage-Pooling2D, Dropout, and an activation function called "softmax" were all part of our sequential model. Neural networks that use the dropout approach are prevented from overfitting.

The model has 7044679 parameters, of which 6961031 are trainable, and 83648 are not. The model was created using TensorFlow's Keras API [16]. The ideal weights were determined by using a dropout rate of 0.4 and a learning rate of 0.001. Batch sizes were set at ten for validation and training using categorical cross-entropy as the loss function.

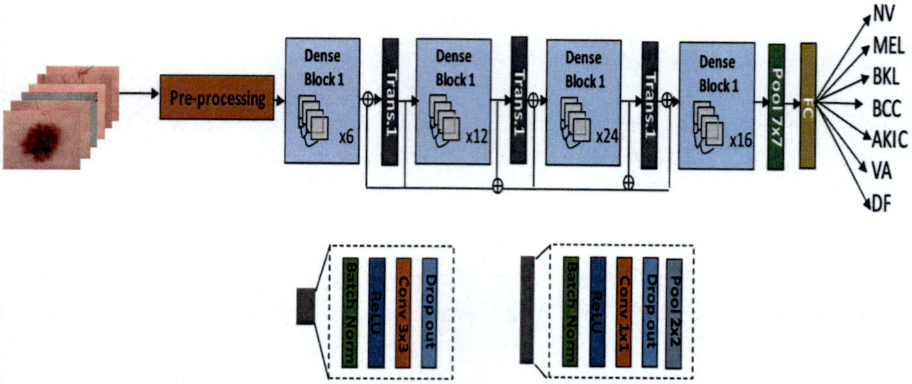

Fig. 3. Architecture of DenseNet.

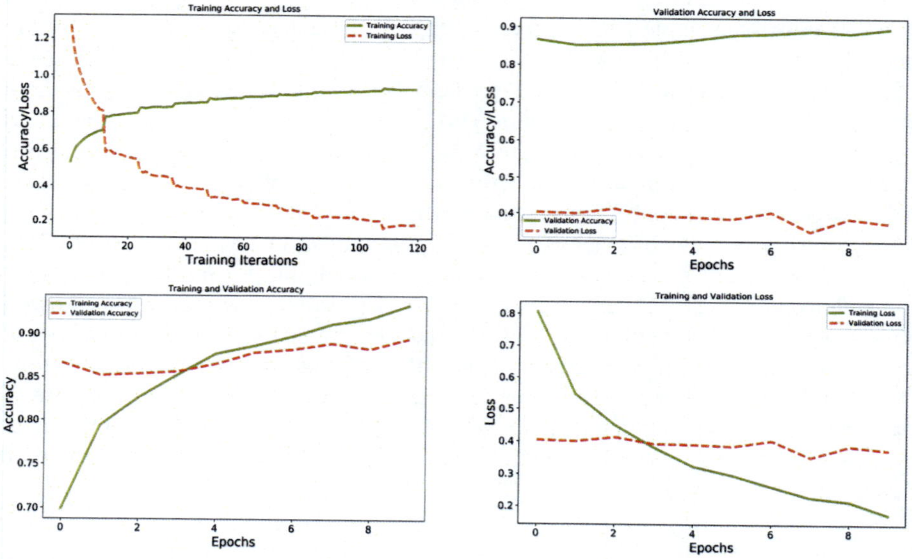

Fig. 4. Accuracy and loss performance curves of the proposed classification model for both validation and training on HAM10000.

Several metrics commonly used in classification problems were compared with the trained model's results. Based on these metrics, we can summarize them as follows:

$$Accuracy = TP + TN/N$$
$$Recall = TP/TP + FN$$
$$Precision = TP/TP + FN$$
$$F1\,score = 2_{*}(Precision_{*}\ Recall)/(Precision = Recall)$$

TP and TN represent true positives/negatives, and FP and FN represent false positives/negatives, respectively [17, 18]. We calculated all metrics for each fold and averaged ten protection folds per test. Performance results for 7 skin cancer classes are shown in Fig. 4.

Table 3. Observation of Images facts from the mmatrix

Classes	Total Images	Correctly Predicted
Melanocytic nevi (nv)	751	731
Melanoma (mel)	39	16
Benign keratosis-like lesions (bkl)	75	53
Basal cell carcinoma (bcc)	30	25
Actinic keratoses (akiec)	26	17
Vascular lesions (va)	11	11
Dermatofibroma (df)	6	4

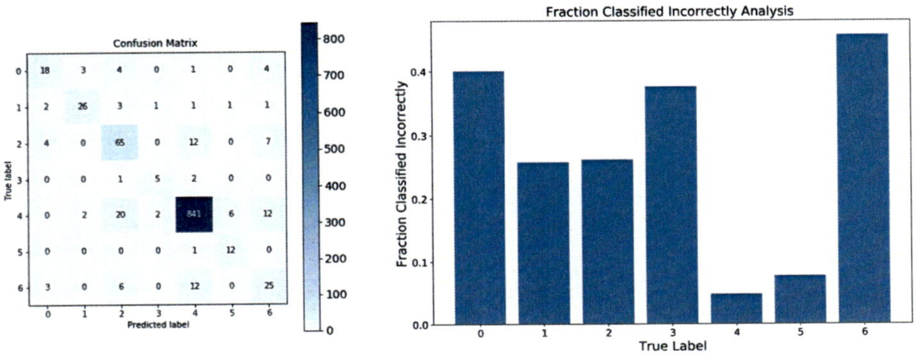

Fig. 5. (a) Confusion matrix (b) DenseNet121 fraction classified incorrectly.

The research findings and tests compare the effectiveness of the categorization model on the seven classes NV, MEL, BKL, BCC, AKIEC, VA, and DF. Based on dataset, we evaluated the performance of the confusion matrix, the wrongly classified fraction, the image classification output, and the classification reports.

Using the confusion matrix, we can assess performance across all classes more effectively and provide the results of seven-class forecasts. Confusion matrixes are tables demonstrating how accurately a calculation was made and how much it can be understood. Figure 5 illustrates the confusion matrix that explains the performance analysis results. Following the confusion matrix analysis, 751 NV images were used for the experiment; 731 NV images were classified correctly as NV with 96% accuracy using the confusion matrix table. 39 mel images were also used for the experiment, and 16 mel images were correctly predicted, and they were all correctly identified as such with an accuracy of 47%. Furthermore, 53 bkl images were correctly predicted with a classification accuracy of 69%. A total of 25 images were correctly classified out of 30, with an accuracy of 82%. Additionally, 6 akiec images were predicted incorrectly with 90% accuracy, and 2 DF images were incorrectly predicted with 90% accuracy. The prediction of images is shown in Table 3, and the classification accuracy of 7 classes is shown in Table 4 . This study benchmarked the entire 7-class dataset with the state-of-the- art to evaluate the performance of the proposed classification approach. Overall, the suggested technique performs better than the state-of-the-art. We compared our proposed model with the previous study on the same dataset performed with a different model. The compersion models are shown in Table 5

Table 4. DenseNete-121 Classification results

Class	Precision	Recall	f1-score	Support
nv	0.98	0.93	0.96	883
mel	0.41	0.57	0.47	46
bkl	0.62	0.78	0.69	88
bcc	0.76	0.89	0.82	35
akiec	0.68	0.50	0.58	30
va	0.71	0.77	0.74	13
df	0.60	0.75	0.67	8
micro-avg	0.89	0.89	0.89	1103
macro-avg	0.68	0.74	0.70	1103
weighted-avg	0.91	0.89	0.90	1103

Table 5. Compariosn of some previous related work with current study

Source	Dataset	Model	Classes	Accuracy (%)
[19]	HAM10000	VGGNet	7	85
[20]	HAM10000	MobileNet	7	83
[21]	HAM10000	MobileNet	7	82
Our Model	HAM10000	DenseNet121	7	**90**

6 Conclusion

Nowadays, the most common type of cancer is skin cancer. Early diagnosis is imperative to ensure the patient's complete recovery. We used a publicly avail- able dataset consisting of 10015 images, which were divided into 80:20 test and training sets. The train set was further divided into 10% validation. As a re- sult of training our DenseNet121 model for ten iterations, we achieved around 93% accuracy during training, 90.17 % on the validation set, and approximately 90% accuracy on the test set. To improve the quality of the results, we gener- ate synthetic samples to balance class distribution. Using comparative analysis, we evaluate the performance of multi-class skin cancer diagnostics in terms of accuracy, precision, and recall.

However, one of the main limitations of the proposed study is to utilize a hold-out approach to estimate the model performance on unseen samples, which is prone to high variance due to a single split of train validation. In addition, the DenseNet121 classifier was trained with a default set of hyperparameters which cause to produce low performance in terms of accuracy. Furthermore, our proposed classifier is evaluated on a single dataset; therefore, its generalizability is still limited and will be ensured to consider multiple datasets.

In the future, we plan to combine our model for detecting skin cancer with more advanced deep-learning models. In addition, our proposed classifier can be enhanced by tuning the hyperparameters to determine the optimal set of hyperparameters to enhance the performance for multi-class skin classification. Furthermore, to ensure the reliability and generalization of the proposed classifier, experiments will be conducted on different benchmark datasets.

References

1. Didona, D., Paolino, G., Bottoni, U., Cantisani, C.: Non-melanoma skin cancer pathogenesis overview. Biomedicines **6**(1), 6 (2018)
2. Parker, E.R.: The influence of climate change on skin cancer incidence–a review of the evidence. Int. J. Women Dermatol. **7**(1), 17–27 (2021)
3. American Cancer Society: Key statistics for melanoma skin cancer (2023). https://www.cancer.org/cancer/melanoma-skin-cancer/about/key-statistics.html
4. Posada, E.F., et al.: Alphafold2 workflow optimization for high throughput predictions in hpc environment. In: Practice and Experience in Advanced Research Computing, pp. 1–5 (2022)
5. Zhang, H., Wang, D.: Neural cascade architecture for multi-channel acoustic echo suppression. IEEE/ACM Trans. Audio Speech Lang. Process. **30**, 2326–2336 (2022)
6. Nachbar, F., et al.:. The abcd rule of dermatoscopy: high prospective value in the diagnosis of doubt- ful melanocytic skin lesions. J. Am. Acad. Dermatol. **30**(4), 551–559 (1994)
7. Esteva, A., et al.: Dermatologist-level classification of skin cancer with deep neural networks. Nature **542**(7639), 115–118 (2017)
8. Zexian, F., An, J., Yang, Q., Yuan, H., Sun, Y., Ebrahimian, H.: Skin cancer detection using kernel fuzzy c-means and developed red fox optimization algorithm. Biomed. Signal Process. Control **71**, 103160 (2022)
9. Dissanayake, T., Fernando, T., Denman, S., Sridharan, S., Ghaemmaghami, H., Fookes, C.: A robust interpretable deep learn- ing classifier for heart anomaly detection without segmentation. IEEE J. Biomed. Health Inform. **25**(6), 2162–2171 (2020)

10. Stacke, K., Eilertsen, G., Unger, J., Lundström, C.: Measuring domain shift for deep learning in histopathology. IEEE J. Biomed. Health Inform. **25**(2), 325–336 (2020)
11. Qu, J., Shi, Y., Xie, W., Li, Y., Wu, X., Du, Q.: Mssl: Hyperspectral and panchromatic images fusion via multiresolution spatial–spectral feature learning networks. IEEE Trans. Geosci. Remote Sens. **60**, 1–13 (2021)
12. Dey, S., Roychoudhury, R., Malakar, S., Sarkar, R.: An optimized fuzzy ensemble of convolutional neural networks for detecting tuberculosis from chest x-ray images. Appl. Soft Comput. **114**, 108094 (2022)
13. Batool, A., Byun, Y.C.: Lightweight EfficientNetB3 model based on depthwise separable convolutions for enhancing classification of leukemia white blood cell images. IEEE Access (2023)
14. Gazioğlu, B.S.A., Kamaşak, M.E.:. Effects of objects and image quality on melanoma classification using deep neural networks. Biomed. Signal Process. Control **67**, 102530 (2021)
15. Gao, H., Liu, Z., Van Der Maaten, L., Weinberger, K.Q.: Densely connected convolutional networks. In: Proceedings of the IEEE Conference on Computer Vision and Pattern Recognition, pp. 4700–4708 (2017)
16. Zhang, T., et al.: Cervical precancerous lesions classification us- ing pre-trained densely connected convolutional networks with colposcopy images. Biomed. Signal Process. Control **55**, 101566 (2020)
17. Tensorflow: Key statistics for melanoma skin cancer (2023). https://www.tensorflow.org/
18. Batool, A., Byun, Y.-C.: An ensemble architecture based on deep learning model for click fraud detection in pay-per-click advertisement campaign. IEEE Access **10**, 113410–113426 (2022)
19. Çevik, E., Zengin, K.: Classification of skin lesions in dermatoscopic images with deep convolution network. Avrupa Bilim ve Teknoloji Dergisi, pp. 309–318 (2019)
20. Sae-Lim, W., Wettayaprasit, W., Aiyarak, P.: Convolutional neural networks using mobilenet for skin lesion classification. In: 2019 16th International Joint Conference on Computer Science and Software Engineering (JCSSE), pp. 242–247. IEEE (2019)
21. Chaturvedi, S.S., Gupta, K., Prasad, P.S.: Skin lesion analyser: an efficient seven-way multi-class skin cancer classification using mobilenet. In: Hassanien, A.E., Bhatnagar, R., Darwish, As. (eds.) AMLTA 2020. AISC, vol. 1141, pp. 165–176. Springer, Singapore (2021). https://doi.org/10.1007/978-981-15-3383-9_15

Using Recorded Lectures in Teaching Higher Education in an Online Remote Learning Context

Kenneth Louis Cavanlit(✉), Ericka Mae Encabo, and Aurelio Vilbar

University of the Philippines Cebu, Gorordo Avenue, Lahug, Cebu City, Philippines
kmcavanlit@up.edu.ph

Abstract. The COVID-19 pandemic caused academic institutions to shift to online remote teaching. In the Philippines, slow internet connectivity has been a concern since it can disrupt the flow of online classes. To address the issue, recorded lectures, which can be viewed with or without an internet connection, were used in teaching an online communicative grammar course. This study shows that the students used different viewing strategies such as pausing, filtering and scanning text, and scrolling and rewatching portions of the recorded lectures. Moreover, the performance rating of the students indicated that they were closely approximating mastery in relation to grammatical accuracy. The findings revealed that using recorded lectures in teaching higher education in an online remote learning context impacted convenience, topic understanding, time management, and interaction.

Keywords: Recorded Lecture · Viewing Strategy · Online Learning

1 Rationale

Due to the COVID-19 pandemic, academic institutions have shifted to remote teaching and learning. Emergency Remote Teaching (ERT) temporarily sets aside the physical classroom as a traditional venue for learning [1]. Hence, teachers and students attend virtual classes through various technology platforms. Given such a setup, internet connectivity, without a doubt, has become a necessity in pursuing remote education. However, in the Philippines, a slow internet connection hinders participation in online activities [2]. With low or no connectivity, using live lectures or activities that require a stable internet connection is not a good option [3].

From my experience in teaching content courses in the Master of Education program, students experienced internet connectivity issues that hampered the flow of live lectures and presentations. Some students requested the lessons and tasks to be recorded to address the challenge. [4] pointed out that recorded lectures enable students to watch and review the content at their own pace. Such a claim implies that recorded lectures do not affect the quality of learning, for such lessons can be an alternative to physical attendance when it comes to class attendance.

K. Kabassi et al. (Eds.): NiDS 2023, LNNS 783, pp. 187–194, 2023.
https://doi.org/10.1007/978-3-031-44097-7_20

Though a few of my students found recorded lectures to be an excellent substitute for live classes, some perceived lecture recordings as ineffective as a means of learning. According to [5], students only access recorded lectures at the beginning of the semester and tend to lose interest in the entry of the recordings as the semester progresses. In addition, the study of [6] implies that recorded lectures do not allow students to interact with the teacher and the learning material.

Recorded lectures enabled students to access recordings despite common connectivity issues without limiting their capacity to learn and comprehend the learning materials. Creating recorded lectures and posting the lecture recordings on a virtual platform would enable the students to download and make the materials available offline, allowing them to view and review the learning materials. For online teaching and learning to be effective, [7] state that formative feedback in online education about the use of recorded lectures should be relayed to the student's learning environment.

This study aimed to examine the experiences of graduate students in using recorded lectures in an online remote learning context. Specifically, the study sought to determine the viewing strategies used by the graduate students, the performance rating of the graduate students on the performance tasks after viewing the recorded lectures, and the graduate students' feedback on using recorded lectures in online remote learning.

2 Literature Review

A recorded lecture is an instructional format that involves recording live classes and disseminating these recordings to students through various technologies as substitutes for classroom instruction [8] or as supplementary to online learning [9]. Using recorded lectures in teaching depends on the learning context, comfort with technology, and budget [10].

Several studies show that recorded lectures promote learning. In the survey of [11], 66.8% of students believed that recorded lectures helped improve their results, and 79.9% of students felt that recorded lectures made learning more accessible. Also, [12] discovered that 69% of the students surveyed in their study claimed that recorded lectures "reduced their anxiety" about the course. In addition, [13] found that 99% considered lecture capture as 'essential' or 'very useful' to their studies.

However, recordings alone do not create a successful online remote teaching-learning experience. [14], who developed a blended learning strategy with recorded lectures and suggested creating means to improve student interaction after watching these videos. In their design, students asynchronously watched recorded videos on YouTube to study at their own pace and then synchronously discussed them using conferencing platforms. Their findings positively impacted student learning and suggested developing more strategies to sustain student interest and active engagement during online classes. The suggestions complemented the results of [15] that interactive meetings with students must supplement recorded lectures in a one-way communication format. They stated that the interactive discussions could promote feedback, provide emotional and psychological support, and clarify lecture information.

Recorded lectures have educational benefits but need further research. [16] discovered that students had a positive attitude towards using recorded lectures but challenged further research on the issue of having one-size-fits-all recording styles for students with varied learning styles. [17] claimed the learning benefits of recordings but discovered that students could be too reliant.

3 Methods

3.1 Research Participants

This study involved graduate students with bachelor's degree in secondary education majoring in English. These students were enrolled in the Master of Education – English as a Second Language degree program in the graduate course, Communicative Grammar.

The request for the consent of the students to be part of the study was made to meet the ethical requirements for research. The performance task outputs were part of the course requirement, while the answers to the survey questionnaire, the reflection, and the focus group discussion were not graded. It was emphasized that the students could withdraw their participation anytime if the reflection and the focus-group discussion would affect their priorities. Pseudonyms were used in this study to protect the privacy of the students.

3.2 Data Collection and Analysis

Based on the Philippines' Commission on Higher Education's Memorandum Order No. 75, Series of 2017, which details the policies, standards, and guidelines for Bachelor of Secondary Education (BSEd), it is reflected that the BSEd English program covers nine units of linguistics courses and twenty-four units of education courses. This memo implies that a graduate of such a program has more exposure to pedagogy than syntax. Considering that the research participants of this study were graduates of secondary education majoring in English and enrolled in a communicative grammar course in the Master of Education program, the course instructor created a Google Form survey to determine which grammar topics the graduate students found challenging. It was identified that the students found verb tenses and prepositions as the two most challenging grammar topics. Hence, recorded lectures on verb tenses and prepositions were used in teaching the two topics. The design of the recorded lectures was based on the instructional model of [18]. It follows the phases of activation, demonstration, application, and integration. Regarding the duration of the recorded lectures, a minimum of 10 min and a maximum of 20 min were observed. According to [19], lectures should be limited to 10–20 min in length as research shows that a duration of 20 min can already disrupt a learner's focus on one piece of information.

To meet the first objective of this research, a survey questionnaire based on the study of [20], was utilized to identify the viewing strategies used by the students in viewing the recorded lectures. For the second objective, which was to determine the performance rating of the students on the performance tasks after viewing the recorded lectures, the students had to come up with a biographical essay using verb tenses and a descriptive-narrative story using prepositions. The performance tasks were adapted from the book of [21], Understanding and Using English Grammar, Third Edition. According to [22], writing tasks as performance tasks enable students to demonstrate their knowledge and proficiency of the target content. To avoid biases, three external language professionals rated the students' writing outputs based on the mean percentage scoring in assessing grammatical competence as adapted from the study of [23]. Finally, a reflection question was given to the students to determine the students' feedback on the use of recorded lectures in online learning, and a focus group discussion was conducted.

4 Results and Discussion

4.1 Viewing Strategies Used by the Students in Viewing the Recorded Lectures

Based on the survey questionnaire, pausing was one of the students' viewing strategies in viewing the recorded lectures. Student A wrote, "I sometimes pause the video to memorize some new concepts." Student B mentioned, "In viewing the recorded lectures, I kept pausing to clarify the information presented." Student C added, "When I paused, I internalized and tried to understand the information." The students used pausing to clarify, understand, and memorize the concepts presented in the recorded lectures. According to [24], pausing can be a strategy to gain meaningful concepts and enhance engagement.

Aside from pausing, the students used filtering and scanning text as a viewing strategy. Student D noted, "Sometimes I feel that the content of the video lecture is not that detailed due to time constraints, so I look for references that will give me more facts." Student E wrote, "To validate my understanding of the recorded lecture, I scanned some specific video content." The students applied filtering and scanning text to find specific information needed to make better sense of content. [25] mentioned that a viewing strategy could employ a successful learning strategy beneficial to comprehension.

In addition, the students used scrolling and rewatching portions of the video lecture as a viewing strategy. Student F specified, "I scrolled back to rewatch a portion of the video to locate significant examples." Student G added, "It made me realize that I need to get back to this info or move forward if it's something I know." The students scrolled back and rewatched portions of the video lecture to recall important concepts. [26] consider re-watching as a strategy for students to reiterate points and stay focused on the target content.

4.2 Performance Task Ratings of the Students After Viewing the Recorded Lectures

The students did two writing tasks as performance tasks. Three external language professionals rated the students' writing outputs on verb tenses and prepositions to avoid biases.

For the first task, to write a biographical essay using verb tenses, the students had an average rating of 91.3% with the qualitative description, "Closely approximating mastery." The rating suggests that the students could still understand the content of the recorded lecture, which dealt with some confusing rules on verb tenses. On the other hand, in the second task, which was to come up with a descriptive-narrative story using prepositions, the students obtained an average rating of 97.3% with the qualitative description "Mastered." This rating implies that the students could master the confusing rules on prepositions as discussed in the recorded lecture.

Overall, the students had a final average rating of 94.3% for the two writing tasks as performance tasks, with "Closely approximating mastery" as its qualitative description. Though the students could not reach the "mastered" definition, this suggests that the students still achieved comprehension of the contents of the recorded lectures with the aid of the viewing strategies.

4.3 Students' Feedback on Using Recorded Lectures in Online Learning

A reflection question was given to the students, and a focus group discussion was conducted to determine the students' feedback on using recorded lectures. The results show that recorded lectures impacted convenience, topic understanding, time management, and interaction.

The students found the use of recorded lectures to be convenient. Student H wrote, "It gave me control over the pacing of the discussions." Student I added, "It allows me to repeatedly review and replay the video. Also, I can view it anytime." Moreover, the students claimed that recorded lectures facilitated topic understanding. Student J mentioned, "I can revisit forgotten concepts, and it allows mastery of lessons." Student K said, "I can pause the video to process my thoughts." Furthermore, the use of recorded lectures had significance to the use of time. Student L noted, "Missing the class is not much of a hassle, and I can save time." Student M wrote, "I can control the time for viewing." Finally, the use of recorded lectures had an impact on interaction. Student N wrote, "I cannot address follow-up questions. It doesn't provide room for clarifications." Student O added, "It cannot provide further discussions."

The students responded positively to recorded lectures because they promoted accessibility, topic familiarity, and time management. The students' feedback shows that lecture recordings should have much learner control. These recorded lectures are designed for adult learners, such as graduate students, who prefer to settle for self- directed learning approaches [27]. In addition, accessing recorded lectures at the learners' own pace and time has been equally valued for empowering the learner with control of the course and convenience [28].

On the contrary, some students conveyed that those recorded lectures in remote learning didn't provide much interaction. This perception could be attributed to the contents of the recorded lectures, which were fixed and limited. [14] suggest that means should be implemented to promote student interaction, especially after watching the videos. The suggestion is related to the idea of [15], which states that recorded lectures only establish one-way communication and that interactive meetings with students should be conducted to provide feedback and clarify information.

5 Conclusion

The implementation of Emergency Remote Teaching (ERT) due to the COVID-19 pandemic paved the way for this study. Since ERT involves internet connectivity which can pose low to no connectivity issues, recorded lectures that can be downloaded and viewed online or offline can be effective learning materials. With Direct Instruction, recorded lectures in online remote learning substitute the traditional teaching and learning venue in the physical classroom.

The use of viewing strategies enabled the graduate students to understand the target contents presented in the recorded lectures. The students paused, filtered, and scanned text and re-watched portions of the video lectures. The graduate students gained topic understanding through viewing strategies based on the "closely approximating mastery" rating they obtained from the performance tasks. Though the use of recorded lectures provided limited opportunities for real-time interaction, it nevertheless allowed the graduate students to access and review target contents at their own pace and time with convenience, enabling them to understand the course content and the completion of the performance tasks.

In the context of this study, using recorded lectures in teaching communicative grammar was considered to be effective in enabling graduate students understand the course content. Hence, recorded lectures, as learning materials, may also be used in teaching a wide array of topics to different levels and types of learners.

Since this research only focused on a limited number of participants, it is recommended that future studies on the use of recorded lectures with a larger population can be undertaken. Moreover, related studies to teaching and learning in an online remote learning context with research variables such as giving tasks before, during, and after viewing recorded lectures may be conducted. Increasing the population size and adding more variables to this kind of study could further test the reliability of recorded lectures as learning materials.

With Research Ethics Clearance: Yes.

References

1. Hodges, C., Moore, S., Lockee, B., Trust, T., Bond, A.: The difference between emergency remote teaching and online learning. er.educause.edu (2020). https://er.educause.edu/articles/2020/3/the-difference-between-emergency-remote-teaching-and-online-learning
2. Fabito, B.S., Trillanes, A.O., Sarmiento, J.: Barriers and Challenges of Computing Students in an Online Learning Environment: Insights from One Private University in the Philippines. Int. J. Compu. Sci. Res. **5**(1), 441–458 (Sep. 2020). Available: https://www.stepacademic.net/ijcsr/article/view/179
3. Teräs, M., Suoranta, J., Teräs, H., Curcher, M.: Post-Covid-19 education and education technology 'solutionism': A seller's market. Postdigital Science and Education **2**(2) (Jul. 2020). https://doi.org/10.1007/s42438-020-00164-x
4. Screen Literacy for the Next Generation: How the Arts and Humanities Matter. HuffPost (Jul. 04, 2014). https://www.huffpost.com/entry/millennials-and-technology-screen-literacy_b_5556389, accessed 30 Mar. 2023
5. Karnad, A.: STUDENT USE OF RECORDED LECTURES A report reviewing recent research into the use of lecture capture technology in higher education, and its impact on teaching methods and attendance. Available: https://eprints.lse.ac.uk/50929/1/Karnad_Student_use_recorded_2013_author.pdf
6. Brockfeld, T., Müller, B., de Laffolie, J.: Video versus live lecture courses: a comparative evaluation of lecture types and results. Med. Educ. Online **23**(1), 1555434 (2018). https://doi.org/10.1080/10872981.2018.1555434. Jan.

7. McCarthy, J.: Enhancing feedback in higher education: students' attitudes towards online and in-class formative assessment feedback models. Act. Learn. High. Educ. **18**(2), 127–141 (2017). https://doi.org/10.1177/1469787417707615. Jun.
8. Cilesiz, S.: Undergraduate students' experiences with recorded lectures: towards a theory of acculturation. High. Educ. **69**(3), 471–493 (2014). https://doi.org/10.1007/s10734-014-9786-1
9. Khalil, R., et al.: The sudden transition to synchronized online learning during the COVID-19 pandemic in Saudi Arabia: a qualitative study exploring medical students' perspectives. BMC Med Educ, 285 (2020). Accessed: 31 Mar. 2023. [Online]. Available: https://search.bvsalud.org/global-literature-on-novel-coronavirus-2019-ncov/resource/en/covidwho-733044
10. Newton, G., Tucker, T., Dawson, J., Currie, E.: Use of Lecture Capture in Higher Education - Lessons from the Trenches. TechTrends **58**(2), 32–45 (2014). https://doi.org/10.1007/s11528-014-0735-8. Feb.
11. Gosper, M., McNeill, M., Phillips, R., Preston, G., Woo, K., Green, D.: Web- based lecture technologies and learning and teaching: a study of change in four Australian universities. ALT-J **18**(3), 251–263 (2010). https://doi.org/10.1080/09687769.2010.529111. Nov.
12. Traphagan, T., Kucsera, J.V., Kishi, K.: Impact of class lecture webcasting on attendance and learning. Edu. Technol. Res. Develop. **58**(1), 19–37 (Jun. 2009). https://doi.org/10.1007/s11423-009-9128-7
13. Williams, J., Fardon, M.: Perpetual connectivity: lecture recordings and portable media players (2007). www.learntechlib.org, https://www.learntechlib.org/p/46189/, accessed 31 Mar. 2023
14. Elsevier Enhanced Reader: reader.elsevier.com. https://reader.elsevier.com/reader/sd/pii/S1749772821000129?token=4F6E0A5F22EC6FE1C0E470E7E6573A9A39B02AF3E636535767FF7E8B07155A5553C60F42C84E3AA867A46357DA9FCBC6&originRegion=eu-west-1&originCreation=20230331095037, accessed 31 Mar. 2023
15. Choi, E.P.H., Ho, M., Smith, R.: What can we do for our part-time nursing students during COVID-19 pandemic? Medical Education (Apr. 2020). https://doi.org/10.1111/medu.14196
16. View of Lecture-recording technology in higher education: Exploring lecturer and student views across the disciplines. ajet.org.au. https://ajet.org.au/index.php/AJET/article/view/3068/1422, accessed 31 Mar. 2023
17. MacKay, J.R.D.: Show and 'tool': how lecture recording transforms staff and student perspectives on lectures in higher education. Computers & Education **140**, 103593 (Oct. 2019). https://doi.org/10.1016/j.compedu.2019.05.019
18. Ou, C., Joyner, D.A., Goel, A.K.: Designing and developing videos for online learning: a seven-principle model. Online Learning **23**(2) (Jun. 2019). https://doi.org/10.24059/olj.v23i2.1449
19. Baylor University: Effective Lectures. Academy for Teaching and Learning Baylor University. https://www.baylor.edu/atl/index.php?id=965135
20. Costley, J., Fanguy, M., Lange, C., Baldwin, M.: The effects of video lecture viewing strategies on cognitive load. J. Comp. High. Edu. (Apr. 2020). https://doi.org/10.1007/s12528-020-09254-y
21. Betty Azar - Understanding and Using English Grammar by Betty Schrampfer Azar - PDF Drive. www.pdfdrive.com. https://www.pdfdrive.com/betty-azar-understanding-and-using-english-grammar-d183993622.html, accessed 31 Mar. 2023
22. Designing Authentic Performance Tasks and Projects: Tools for Meaningful Learning and Assessment. ASCD. https://www.ascd.org/books/designing-authentic-performance-tasks-and-projects?chapter=implementing-performance-tasks-and-projects-what-and-why, accessed 31 Mar. 2023
23. Sioco, E., De Vera, P.V.: Grammatical competence of junior high school students. TESOL International Journal **13**(1) (2018). Available: https://files.eric.ed.gov/fulltext/EJ1247221.pdf

24. Stop and Pause for Engagement in Online Video Lectures. Teaching Online Pedagogical Repository. https://topr.online.ucf.edu/stop-and-pause-for-engagement-in-online-video-lectures/
25. L2 students' perceptions and practices of both giving and receiving ... [Online]. Available: https://www.researchgate.net/profile/Daniel-Bailey-12/publication/332171132_L2_Students'_Perceptions_and_Practices_of_both_Giving_and_Receiving_Online_Peer-Feedback/links/5ca455ab299bf1b86d60fdaa/L2-Students-Perceptions-and-Practices-of-both-Giving-and-Receiving-Online-Peer-Feedback.pdf. Accessed: 31 Mar 2023
26. Bruff, D., Fisher, D., Mcewen, K., Smith, B.: Wrapping a MOOC: student perceptions of an experiment in blended learning. MERLOT J. Online Learn. Teach. **9**(2) (2013). Available: https://jolt.merlot.org/vol9no2/bruff_0613.pdf
27. Knowles, M.: THE MODERN PRACTICE OF ADULT EDUCATION From Pedagogy to Andragogy REVISED AND UPDATED 4 What Is Andragogy? In the Beginning Was Pedagogy. Available: https://colllearning.info/wp-content/uploads/2019/03/The-Modern-Practice-of-Adult-Education.pdf
28. Simpson, N.: Asynchronous access to conventional course delivery: a pilot project. Br. J. Edu. Technol. **37**(4), 527–537 (2006). https://doi.org/10.1111/j.1467-8535.2006.00534.x. Jul.

Mental Confusion Prediction in E-Learning Contexts with EEG and Machine Learning

Maria Trigka[1](✉), Elias Dritsas[2], and Phivos Mylonas[1]

[1] Department of Informatics and Computer Engineering, University of West Attica, Athens, Greece
{mtrigka,mylonasf}@uniwa.gr

[2] Department of Electrical and Computer Engineering, University of Patras, Patras, Greece
dritsase@ceid.upatras.gr

Abstract. In cognitive science, the term confusion is used to capture the decline in learners' cognitive ability, which affects their ability to think, solve a problem, learn and understand. Unlike classroom education, in online e-learning contexts (e.g., in Massive Open Online Courses (MOOC)), confusion hinders the smooth evolution of the learning process from the learners' side, as the educator can't immediately interact with the students to restore cognitive equilibrium. This paper presents a Machine Learning (ML) based approach by comparing several classifiers that were trained and tested exploiting Electroencephalogram (EEG) data (namely, band power, attention and mediation features) acquired by the MindSet device to efficiently distinguish "Confused" from "Not-Confused" subjects. In particular, J48 was the dominant model reaching an optimal performance with accuracy, precision and recall equal to 99.9% and an Area Under the Curve (AUC) of 100%.

Keywords: E-learning · MOOC · Confusion Detection · EEG · Machine-Learning

1 Introduction

During the COVID-19 pandemic, classroom education was shifted to online education establishing the latter as an alternative learning mode. The rapid advances in information and communication technology have facilitated the migration of the majority of learning activities to the online mode of education but without deficiencies. Students' mental confusion while watching MOOC videos is among the drawbacks that should be properly handled [6]. Using EEG to quantify the confusion that occurs in the learning process as well as intervening has gained great interest from researchers [4,8,15]. EEG is a physiological signal that records brain activity in different areas (called lobes) through the scalp. EEG is generally the most effective non-invasive method for assessing a subject's cognitive functions[16]. Wearable EEG-based devices and Artificial Intelligence

K. Kabassi et al. (Eds.): NiDS 2023, LNNS 783, pp. 195–200, 2023.
https://doi.org/10.1007/978-3-031-44097-7_21

have come to support this trend facilitating the collection of enough data for the development of efficient prediction models. Apart from online education, its recognition can be beneficial in various domains such as healthcare, cognitive psychology, virtual online games, etc [8]. Since confusion is a dynamic process, an EEG-based recognition system can help educators quantify and monitor the students' cognitive state (which spans into attention, meditation, concentration, frustration and boredom, level of stress, anxiety, etc.), early identify if students feel confused (due to difficulties in solving a problem or understanding the conveying knowledge), and adapt the teaching plan without the students' feedback or intervention. Recently, various works [9,14,17] have experimented with EEG-based features that are fed as input to ML or Deep Learning (DL) models aiming to reason the mental state of participants (confused or not-confused) in online education platforms.

In this study, we used an EEG-based publicly available dataset whose records are represented by a feature set related to the mean value of raw data, the mean power of five EEG frequency bands at the frontal channel Fp1 and two additional features that capture attention and meditation. The confusion detection was treated as a binary classification problem and solved by investigating three types of classifiers: distance-based, probabilistic and tree-based. We aim to find a model with high sensitivity and separation ability of the mental states (confused, not-confused). The main suggestion of this paper is from an ML perspective and, especially, a decision tree-based model for confusion detection. Using 10-fold cross-validation (cv), J48 was the best-performing in all performance metrics achieving accuracy, precision, recall of 99.9% and AUC of 100%.

The remainder of the paper is organized as follows. In Sect. 2, a description of the dataset and adopted methodology is outlined. Furthermore, in Sect. 3, we discuss the ML models' performance outcomes. Finally, conclusions and future directions are presented in Sect. 4.

2 Methodology

In this section, the dataset and its characteristics are illustrated, the adopted methodology is noted, the ML models have been described, as well as the evaluation metrics with which the experimental evaluation was carried out.

2.1 Presentation of Data Collection Process

The dataset was derived from Kaggle [1]. For the data collection, ten students were arranged and assigned to watch ten MOOC video clips dedicating 2 min to each one. During this process, the students wore a single-channel wearable MindSet device that measured and recorded the brain's spontaneous electrical activity (over the frontal parietal lobe - channel Fp1) for a specific time period. Raw data recorded from the Fp1 channel were used to quantify the mental state of 10 students and recognize the occurrence of confusion. The target class was confirmed and self-labelled by the students as confused or not. This process was repeated in all sessions of watching online videos. Initially, the subjects rated their confusion level on a scale of 1–7 from low to high and then quantized into binary, confused or not confused [13]. By using NeuroSky's API, the raw EEG signals were sampled at 512 Hz. Also, MindSet's proprietary "attention" and "meditation" signals measured (at 1 Hz) mental focus and calmness. Moreover, the MindSet device measures and outputs the average power at five frequency bands, i.e., delta (1–3 Hz), theta (4–7 Hz), alpha (8–11 Hz), beta (12–29 Hz), and gamma (30–100 Hz) [8].

2.2 Data Understanding

To characterize the attention and mediation level of learners, we advised the manual of the MindSet device[1]. More specifically, the eSense, NeuroSky's proprietary algorithm is used for characterizing mental states. First, the NeuroSky ThinkGear technology processes the raw brainwave signal in order to remove the ambient noise and muscle movement. Then, the eSense algorithm interprets eSense meter values, which describe ranges of activity. The meter value is reported on a relative eSense scale of 1 to 100, where the different categories are: i) 1–20: Strongly Reduced, ii) 20–40: Reduced, iii) 40–60: Neutral, iv) 60–80: Slightly Elevated and v) 80–100: Elevated. Based on the aforementioned scaling, we obtained Fig. 1. Interpreting these outcomes, it is observed that attention and mediation levels don't considerably differ among confused and not-confused subjects verifying the complex nature of the specific mental state and the need for diverse features not only from EEG but also from eye tracking [11]. The age of 10 subjects varied between 24 and 31, with a mean age of 25.54 and a standard deviation of 2.27 years. Further statistical details of the EEG features

Table 1. Statistical description of the Dataset.

Features	Mean ± std	Min	Max
Attention	46.4 7± 19.06	1	100
Meditation	53.08 ± 16.26	1	100
Raw	34.18 ± 131.55	−2048	1440
delta	588937.42 ± 634988.04	440	3960000
theta	159174.09 ± 236645.72	17	2570000
low-alpha	38520.89 ± 67646.02	2	1370000
high-alpha	28216.66 ± 48541.36	2	1020000
low-beta	20330.76 ± 29239.49	3	841000
high-beta	24283.97 ± 36256.25	2	1080000
low-gamma	16961.54 ± 25819.75	1	658000
mid-gamma	8291.37 ± 11585.23	2	284000

[1] https://developer.neurosky.com/docs/doku.php?id=mindset_instruction_manual.

in the final dataset are captured in Table 1. It should be noted that, from this dataset, we excluded subjects with zero attention and meditation. As a result, the final dataset consisted of 5463 Not-confused and 5925 Confused samples.

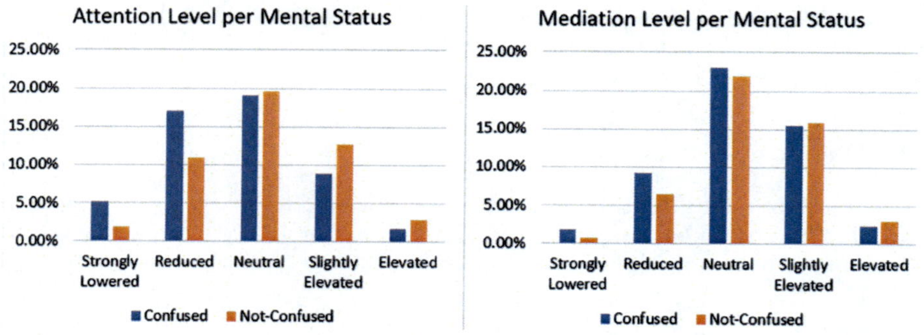

Fig. 1. Attention and Mediation Level as measured by MindSet device.

2.3 Machine Learning Models and Evaluation Metrics

The evaluation of our ML models was carried out with a widely known free software, namely WEKA [2], which contains tools for data pre-processing, classification, regression, clustering, association rules, and visualization. As for the ML methodology, we selected Logistic Regression-LR [5] and Naive Bayes-NB [7], which are probabilistic models. Also, a distance-based classifier and especially k-NN [18] was evaluated. Moreover, 3 tree-based models were assessed, namely, Random Tree-RT, Random Forest-RF (an ensemble of decision trees where the final prediction is based on voting) [3] and Decision Tree-DT (J48) [12]. To evaluate the ML models, we relied on metrics [10] commonly used in the ML field, namely accuracy, precision, recall, and AUC. Note that the final score in each metric is derived by averaging the scores from all folds. The definition of these metrics is based on the confusion matrix consisting of the elements true positive (Tp), true negative (Tn), false positive (Fp) and false-negative (Fn). Hence, the aforementioned metrics are defined as follows: $\mathrm{Acc} = \frac{\mathrm{Tn+Tp}}{\mathrm{Tn+Fn+Tp+Fp}}$, $\mathrm{Prec} = \frac{\mathrm{Tp}}{\mathrm{Tp+Fp}}$, $\mathrm{Recall} = \frac{\mathrm{Tp}}{\mathrm{Tp+Fn}}$. To assess the separability of a model, the AUC $\in [0, 1]$ metric is used.

3 Results and Discussion

In this section, the performance behaviour of the selected ML models is presented and analyzed. As a first approach, the models were trained and tested with all available features (apart from demographics) in the dataset. In Table 2, we demonstrate the average values of performance metrics, which were acquired assuming 10-fold cv. Note that for the k-NN classifier, we experimented with the parameter k and verified that for k = 1 the highest performance was achieved. Comparing the assessed models, we see that the 1-NN classifier outperformed the RT and RF in terms of accuracy, recall and precision, while RF indicated higher AUC. The higher AUC of RF reveals that for the specific model, there is a 98.6% chance that the model will be able to distinguish between the confused class and the not-confused class. Also, LR and NB were the least efficient models in identifying confused subjects. For the specific dataset, J48 was the most efficient model for confused/non-confused students' prediction as this specific classifier managed to keep the Fp, Fn at the lowest level (just a few subjects were misclassified with Fp = 3 and Fn = 7).

Table 2. Performance Results of ML Models.

	Accuracy%	Recall%	Precision%	AUC%
Naive Bayes	54.76	54.8	59.5	65.0
Logistic Regression	61.47	61.5	61.4	65.4
Random Tree	83.39	83.4	83.4	83.3
Random Forest	93.79	93.8	94	98.6
1-NN	95.64	95.6	95.6	95.6
J48	99.90	99.9	99.9	100

4 Conclusions

To sum up, this work relied on an EEG-based dataset which helped us identify a powerful ML model for confused subjects detection. A limitation of the specific dataset is that we didn't have access to the whole time series data from the Fp1 channel to extract several other EEG features in the time, frequency or time-frequency domain. However, it is adequate to train efficient models for the prediction of the human mental state of subjects who attended lectures in an online education environment. From our analysis, J48 was the prevailing model with accuracy, precision and recall equal to 99.9% and AUC of 100%. In future work, we anticipate applying feature selection techniques to understand the significance and correlation of the features to the specific task and reevaluate the ML models' performance. Also, we will focus on the design of personalized confusion detection models which will be compared to the current global models. Finally, our research will be directed to eye-tracking-based data [11].

Acknowledgements. This research was funded by the European Union and Greece (Partnership Agreement for the Development Framework 2014–2020) under the Regional Operational Programme Ionian Islands 2014–2020, project title: "Indirect costs for project "Smart digital applications and tools for the effective promotion and enhancement of the Ionian Islands bio-diversity"", project number: 5034557.

References

1. Confused student EEG brainwave data. https://www.kaggle.com/datasets/wanghaohan/confused-eeg?select=EEG_data.csv. Accessed 12 May 2023
2. Weka. https://www.weka.io/. Accessed 12 May 2023
3. Alexiou, S., Dritsas, E., Kocsis, O., Moustakas, K., Fakotakis, N.: An approach for personalized continuous glucose prediction with regression trees. In: 2021 6th South-East Europe Design Automation, Computer Engineering, Computer Networks and Social Media Conference (SEEDA-CECNSM), pp. 1–6. IEEE (2021)
4. Balamurugan, B., Mullai, M., Soundararajan, S., Selvakanmani, S., Arun, D.: Brain-computer interface for assessment of mental efforts in e-learning using the nonmarkovian queueing model. Comput. Appl. Eng. Educ. **29**(2), 394–410 (2021)
5. Boateng, E.Y., Abaye, D.A.: A review of the logistic regression model with emphasis on medical research. J. Data Anal. Inf. Process. **7**(4), 190–207 (2019)
6. Chakraborty, P., Mittal, P., Gupta, M.S., Yadav, S., Arora, A.: Opinion of students on online education during the COVID-19 pandemic. Hum. Behav. Emerg. Technol. **3**(3), 357–365 (2021)
7. Chen, S., Webb, G.I., Liu, L., Ma, X.: A novel selective naïve bayes algorithm. Knowl.-Based Syst. **192**, 105361 (2020)
8. Daghriri, T., Rustam, F., Aljedaani, W., Bashiri, A.H., Ashraf, I.: Electroencephalogram signals for detecting confused students in online education platforms with probability-based features. Electronics **11**(18), 2855 (2022)
9. He, S., Xu, Y., Zhong, L.: EEG-based confusion recognition using different machine learning methods. In: 2021 2nd International Conference on Artificial Intelligence and Computer Engineering (ICAICE), pp. 826–831. IEEE (2021)
10. Hossin, M., Sulaiman, M.N.: A review on evaluation metrics for data classification evaluations. Int. J. Data Min. Knowl. Manage. Process **5**(2), 1 (2015)
11. Khan, A.R., Khosravi, S., Hussain, S., Ghannam, R., Zoha, A., Imran, M.A.: EXECUTE: exploring eye tracking to support E-learning. In: 2022 IEEE Global Engineering Education Conference (EDUCON), pp. 670–676. IEEE (2022)
12. Maulana, M.F., Defriani, M.: Logistic model tree and decision tree J48 algorithms for predicting the length of study period. PIKSEL: Penelitian Ilmu Komputer Sistem Embed. Log. **8**(1), 39–48 (2020)
13. Ni, Z., Yuksel, A.C., Ni, X., Mandel, M.I., Xie, L.: Confused or not confused? Disentangling brain activity from EEG data using bidirectional LSTM recurrent neural networks. In: Proceedings of the 8th ACM International Conference on Bioinformatics, Computational Biology, and Health Informatics, pp. 241–246 (2017)
14. Reñosa, C.R.M., Bandala, A.A., Vicerra, R.R.P.: Classification of confusion level using EEG data and artificial neural networks. In: 2019 IEEE 11th International Conference on Humanoid, Nanotechnology, Information Technology, Communication and Control, Environment, and Management (HNICEM), pp. 1–6. IEEE (2019)
15. Tahmassebi, A., Gandomi, A.H., Meyer-Baese, A.: An evolutionary online framework for MOOC performance using EEG data. In: 2018 IEEE Congress on Evolutionary Computation (CEC), pp. 1–8. IEEE (2018)
16. Trigka, M., Dritsas, E., Fidas, C.: A survey on signal processing methods for EEG-based brain computer interface systems. In: Proceedings of the 26th Pan-Hellenic Conference on Informatics, pp. 213–218 (2022)
17. Xu, T., Wang, J., Zhang, G., Zhang, L., Zhou, Y.: Confused or not: decoding brain activity and recognizing confusion in reasoning learning using EEG. J. Neural Eng. **20**(2), 026018 (2023)
18. Zhang, Z.: Introduction to machine learning: k-nearest neighbors. Ann. Transl. Med. **4**(11) (2016)

Exploring the Use of Augmented Reality in Teaching History to Students with Attention-Deficit Hyperactivity Disorder

Chrysa Fraggista, Akrivi Krouska(✉), Christos Troussas, and Cleo Sgouropoulou

Department of Informatics and Computer Engineering, University of West Attica, Egaleo, Greece
akrouska@uniwa.gr

Abstract. Augmented reality is an emerging technology that aims to enrich the elements of the real world with virtual information through the use of appropriate devices. These new technological devices, namely smart mobile devices such as smartphones and tablets, can serve as a link between augmented reality and the user. Augmented reality technology has not left the education industry unscathed, attracting the interest of many teachers and researchers, since it is seen as an important, cutting-edge learning and educational tool. Augmented reality provides students with unique learning experiences by supplementing the sensory stimuli of sight and hearing with three-dimensional images, sounds, videos, etc. The purpose of this study is to investigate the effectiveness of augmented reality in high school history classrooms for the teaching of students with attention-deficit hyperactivity disorder (ADHD). The evaluation was conducted with semi-structured interviews with students who have been diagnosed with ADHD, using a questionnaire at the end of the procedure. Qualitative analysis was mainly used to conclude. The method used to carry out this work is the case study. Research findings show that augmented reality technology can make a positive contribution to the learning process. Through the answers of most students, it seemed that they accepted augmented reality and considered it essential for learning better. Furthermore, student behaviors such as attention, engagement and learning motivation showed an improvement. Finally, the educational objectives of this didactic intervention were achieved to a satisfactory degree according to the results of the test.

Keywords: ADHD · Augmented Reality · Educational tool · E-learning · Special Education

1 Introduction

In today's context of digital transformation, new learning and communication networks, students are challenged to develop new skills, search for information, think critically and cooperate with others [1, 2]. This fact requires changes in the traditional school learning environment and expects teachers to be trained in new theories of teaching and learning [3–5]. The teacher is no longer the center of the classroom, but has the role of coordinator and assistant for the children [6, 7].

K. Kabassi et al. (Eds.): NiDS 2023, LNNS 783, pp. 201–214, 2023.
https://doi.org/10.1007/978-3-031-44097-7_22

The digital world has become part of everyone's life. Many surveys emphasize the fundamental importance of introducing new digital technologies in education [8, 9]. The groundbreaking digital tools strengthen the role of the teacher, putting the student at the center of the learning process [10].

A wide range of educational software has been developed in recent years, such as simulation and modeling systems, supernatural intelligent systems for teaching, robotic systems, multimedia e-books, and real-world software that incorporates advanced technologies such as multimedia, virtual reality and remote communication via the Internet [11, 12]. The issue is the teacher's ability to deal with these systems in the classroom and what appropriate and innovative teaching strategies are suitable to achieve the expected learning outcomes [13].

As in general educations, the learning activities in special education are developed with the help of educational technology [14]. Many new technological media are successfully used in the education of children with learning difficulties, such as difficulty on focusing on learning, need more time for learning, split attention and need for constant reinforcement and feedback [15]. New technologies offer the opportunity to create a digital environment that encourages the use of the student's senses, which is not always easy in the traditional classroom [16].

One new technology that has gained prominence in recent years is augmented reality, which has attracted the interest of the scientific community, since it is a technology that has been developed in the last few decades and can be applied in many areas. One of these areas is education [15, 17]. The educational benefits of augmented reality inevitably make it a great educational tool and one of the most emerging technologies in education [18].

The present work investigates the importance of technological tools and, in particular, augmented reality in the education of students with learning disabilities and attention deficit disorders. Through the simple and straightforward use of technological tools and modern mobile devices such as portable computers, smartphones and tablets, children with learning disabilities gain motivation to learn and show more interest in educational activities. The aim of this research is to develop an augmented reality application for high school students who have been diagnosed with attention deficit disorder, and therefore have difficulties in learning. Thus, the goal is to explore and evaluate augmented reality applications to enhance and sustain the attention of students with attention deficits in history classes.

In particular, the research questions are as follows:

- How can students with ADHD interact with augmented reality in the learning process?
- Can augmented reality technology improve the attention of students with attention deficits in the learning process?
- Does the use of instructional materials and applications of augmented reality in the learning process increase the motivation for active participation of students with attention deficits?

2 Background

This section reviews the use of augmented reality for teaching students with learning disabilities or special learning needs. The limitations and benefits of using AR in the educational process are presented based on research conducted in recent years.

2.1 Limitations AR

After a thorough review of the literature, the limitations in supporting the education of students with learning disabilities are the following:

1. The technical problems in the use of AR applications are an important limitation when researches refer to students with disabilities, whether physical or mental, because the levels of frustration increase in case technical problems and difficulties appear to students while handling such applications [19]. Thus, the proper preparation and previous experience with AR by the teachers is essential for the successful completion of the learning process.
2. The lack of research on the use of AR applications in the special education. Students with ADHD have significant learning difficulties, needing special treatment to help them in the learning process. AR technology can significantly help improve the learning outcomes of these students; as such, it is fertile ground for research [20].
3. Teachers fall behind in digital literacy, making difficult the adoption of new technologies in their instruction [18]. As AR is more complex technology in use comparing to traditional educational software, it is important the teachers to be trained in this technology.
4. It is not possible to replicate or repeat the AR learning scenario in other educational settings because they are oriented to the special needs of students with learning difficulties [15].
5. Researchers report difficulty in finding students with learning disabilities. The overall process of evaluating systems with these students requires teachers to obtain the necessary consent forms to allow them to participate in the studies [16]. In the case of conducting the survey in a school setting, the research should be conducted under safe conditions and with respect for the participants to avoid marginalizing a group of students, not only students with learning disabilities but also other students at risk of marginalization.
6. It is important to teach students new technologies. In many cases, students with learning disabilities are not familiar with the use of the necessary technology that supports augmented reality applications. Therefore, it is suggested that they be trained by specialized teachers [18]. This can also help them avoid distractions in their attention during the learning process.

2.2 Benefits AR

Using an AR system can support real-time feedback of information. It can also improve concentration and promote motivation to learn [18, 21]. Moreover, the "curiosity" about such digital applications, can improve student interaction as they can develop more interaction opportunities, which could be of great benefit to traditional learning methods. Finally, these learning environments can improve the frequency of work and enable the learning process through game models, which increases learning motivation [19, 22]. For teachers, the use of augmented reality is a promising choice for teaching students with learning difficulties, even though it is challenging to use. Teachers can collaboratively create instructional materials based on augmented reality and share them via cloud applications. Augmented reality in special education supports intuitive and interesting learning processes for children with learning needs by connecting the real and potential worlds [18]. However, no teaching method or approach is universally good. The benefits of using this technology best accrue to students who have difficulty watching or maintaining attention and who need more support [15, 19]. Teachers can organize an entire lesson with activities that use an augmented reality system. Through these teaching materials, students can repeat tasks independently, reducing dependence on teachers and enabling teachers to develop multi-level teaching strategies to help children adapt to independent learning [22]. In addition, teachers can collaborate with researchers in this field to design appropriate individualized educational programmes. On the other hand, the design of educational materials using AR technology requires the acquisition of specific computer science skills that need to be taught to these teachers.

In all studies where augmented reality has been used, positive results have been reported in both content learning and motivational enhancement [14, 15]. There are studies that have been conducted in real-world conditions and show detailed results of their application. In [23], the authors examined the outcomes of teaching with augmented technology reality science terminology to four students with intellectual disabilities. Students were given a printed booklet containing the terms and a tablet with an application installed. Students were instructed to scan the words to activate the digital content, which included a video with the 78 definition of the word and a visual example of its use. The result was that students understood the words and were able to respond correctly to the tasks included in the application. In [24], the authors present an educational application of a magic book that runs on mobile devices and allows young children to handle two- and three-dimensional images of plants in a simple and intuitive way.

3 Research Methodology

The scope of this paper is to develop an original augmented reality application to be integrated into the context of teaching a historical learning object to students with ADHD. The focus is to create a pedagogical teaching scenario that includes activities based on the use of augmented reality software that can be easily used by any teacher who wishes to integrate this intervention into his or her classroom. The ultimate goal is to use the new technology to enrich instruction so that it is better understood by students with ADHD. The first step is to select the appropriate AR tool. The chosen app is Blippar, an augmented reality app for smartphones and tablets that can transform any static image (entry, printed, packaging, television, outdoor, etc.) into something alive.

The implementation of the teaching intervention was based on three basic stages. First, we had to investigate what students know about AR technology and help them get familiar with it. The second part is the creation of the AR application that will enrich the existing educational material and its implementation in the classroom. The third part is the evaluation of the developed application through a survey. In the first step, the selection of the book is important to define the type and amount of information on which the learning scenario will be built in combination with the use of augmented reality. After analyzing all the information that the book offers about the Ancient History of the First High School class (pictures, diagrams, written examples), several points were selected to be enhanced by the innovative way of viewing augmented reality. In particular, five themes were created: Arthur Evans, Map of Minoan Crete, Knossos, Phaistos' Disc and Bull. The textbook images corresponding to the above themes were enlarged. In order to make sure that the students are familiar with the use of smartphones and the Blippar application, a one-week trial application was conducted to identify possible difficulties and malfunctions.

The course of history was chosen since it is an information-rich course and this kind of course material can hardly keep students with ADHD focused in the learning process. In this research, the chapter of the book that refers to the Minoan culture was chosen. The main reason the Blippar app was chosen is because it is easy to use for students but also for teachers. It is age-appropriate to perceptions, abilities and potential of the students who use it. Information can be shared freely regardless of time and geographical area. It is a free of charge tool and the applications developed via it can be shared with students or other users. The students can very easily start learning after selecting the printed images and focus with the mobile device where the application is installed.

3.1 Educational Scenario

The design of the activities is based on images from the high school's Ancient History book, in which an augmentation was made using the Blippar application. The students, together with the teacher, scan the images of the book during the learning process and see on the screen of their mobile device the virtual element that appears. After that, the students and the teacher have a conversation about the element that appeared and everyone expresses their opinion about it. The goal of this scenario is to make history lessons more interesting and exciting for students with ADHD and to motivate them to participate more in the learning process. Below is the analysis of one of the AR activities included in the scenario, with topic title "Arthur Evans".

The general goal of this activity is to introduce the students to the archaeologist, named Arthur Evans, who discovered the largest palace of the Minoan civilization. It is therefore important that students know who this great archaeologist was, to whom we owe the discovery of all these great finds for the ancient Greek civilization, but also in which year it was discovered. As such, an augmentation point was made where the name of the archaeologist is in the textbook, assigned more information about him. When students scan with their cell phone camera this augmentation point, they show the related material (Figs. 1 and 2).

The activity was done at the beginning of the lesson and lasted about 10 min. Students seemed very excited and showed great attention during this first augmented reality activity. They seemed to understand who was behind the discovery of Knossos and what this man did. During the activity, the evaluators observed the maintenance of interest of the students, their progressive development, the careful observation of what they say and what they do, the production of spontaneous dialogue between them, but also with the teacher, their expressions and in general their reactions during their work with the tablet and application.

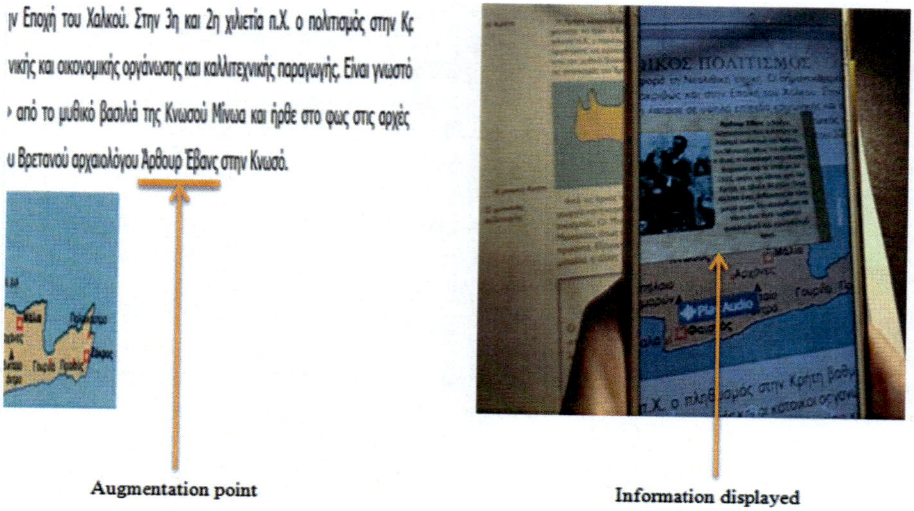

Fig. 1. 1st AR example.

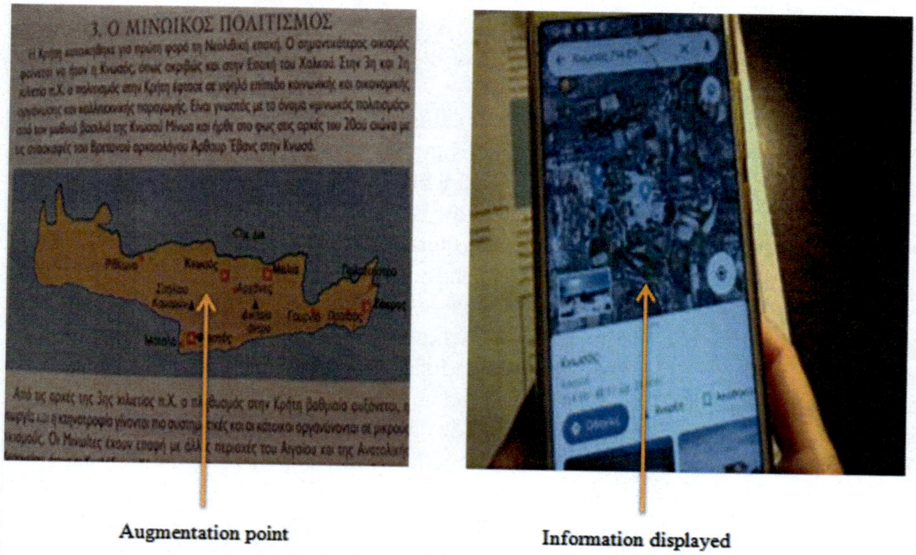

Fig. 2. 2nd AR example.

4 Evaluation

This work is an experimental study conducted on 10 students with attention-deficit hyperactivity disorder (ADHD). To reinforce and evaluate the maintenance of their attention in the learning process, the tool Blippar is used. This is a new technology that combines the user's environment with virtual objects such as images, sound and videos. The system recognizes the augmentation points and then enhances the object with the virtual or sound elements that have been matched. The method used to evaluate the present work is the interview to understand the children's attitudes, thoughts and views about AR technology.

The sampling strategy chosen is that of availability sampling, in which the researcher seeks and collects individuals who are available for research and share common characteristics in order to study and better understand this population. The site chosen for the sample was at a secondary school in the western suburbs of Athens, which has special sections for students with learning difficulties. The duration of the experiment was one month, during the history lessons. The augmentation was done through tablets provided by the school lab. The study involved 10 first grade students of a high school, 8 boys and 2 girls, who have difficulties in learning because they suffer from attention deficit disorder. All students have Greek as their mother tongue and have an assessment from relevant public and private organizations (KEDDY, Mental Health Center).

The most appropriate data collection method for this research was the quality approach. In particular, the research process follows the examples of developmental research and case study in conjunction with observation. The case study contributes to research and clinical practice. In this case, the purpose of the case study is to determine if an AR tool (Blippar) can help students with attention deficits maintain their concentration to the learning process and achieve their learning goals. As part of the case study on the use of AR applications, an experiment was conducted to determine whether AR can significantly contribute to improving the learning experience and strengthening and maintaining students' attention. To validate the research cases, it was necessary to create a semi-structured interview to capture students' thoughts, attitudes, and opinions while using Blippar and to engage them with the application of Blippar.

Respondents' answers were recorded by the evaluators' mobile tape recorder and then transcribed. Then, they were processed and the details of the objectives and questions of the research were selected. Important in the analysis of interviews is the analysis of the content of what was said by the interviewees. Content analysis primarily refers to evidence that comes from written or oral communication and aims to systematically and quantitatively describe the content of written or oral communication with the goal of interpreting that content [25]. Thematic analysis chosen by the evaluators aims to analyze their meaning of the groups of words said about interests, opinions, emotional state of the interviewer e.g. joy, fear, anxiety. In general, thematic analysis consists of systematic recognition, ordering and understanding of recurring patterns of meaning within a data set. In the first phase, it is important to ask the right questions to the participants in order to conduct the interview correctly. Therefore, some topics should first be elicited into axes, each of which belongs to both the corresponding and the topic of the axis questions. The following table shows the questions asked to the students according to the thematic axis they belong to, while the remaining four tables below reflect some of the students' answers according to the questions of each thematic axis (Table 1).

Table 1. Questionnaire.

Usefulness	Suggestions improvements	Distraction	Acceptance/Participation
U1. Have you ever use something similar in history lesson?	SI1. What would you change from teaching?	D1. Were you thinking of something else when we implemented the activities	AP1. With augmented reality in class do you felt that you wanted to participate more from other times?
U2. How would you feel if the other lessons were also done in this way?	SI2. Did you understand the instructions given to you at the beginning of the course?	D1. Is it easy for you to follow the instructions?	AP2. Can you tell something about the experience you had with this alternative course?
U3. Did you learn the historical information more slowly or faster than at other times?	SI3. Did you feel stressed while using the AR application?	D1. Was there anything else that caught your attention during the lesson?	AP3. What did you like most or least about this experience?
U4. Did the AR help you learn more easily or not the lesson that was taught?	SI4. Was it easy for you to use the Blippar app?	D1. Did you always feel ready to participate in the discussions that took place after the activities?	AP4. Did you find the AR interesting for the History lesson?
U5. What were the differences between this lesson and the lesson at school?	SI5. Did you have to do the activity several times to understand something?		AP5. Can you tell me some information you remember from today's lesson?
	SI6. What would you add, what would you omit?		AP6. Would you like to learn other lessons this way?

Regarding acceptance/participation axis (Table 2), the majority of the answers were positive for their experience with AR, as well as for their involvement in the lesson. There were also a few students who did not see the whole process positively and did not show great interest in AR technology.

Concerning AR application's usability (Table 3), the majority of students considers that AR is a helpful and useful tool in learning process. Moreover, they stated that they would like to use it in other courses as well. However, a few students did not believe

Table 2. Evaluating student acceptance/participation.

Question	Sample of answers	Statistics
AP1	S2: "I definitely participated a lot more than other times."	Positive answers 7 Negative answers 3
AP2	S7: "I liked that we used our mobiles telephones during class time, which showed us various information about the course."	Positive answers 8 Negative answers 2
AP3	S1: "I liked that we did something different in history lesson, I usually feel bored."	Positive answers 8 Negative answers 2
AP4	S6: "Every time we scanned book images I couldn't wait to see what would appear, it was amazing."	Positive answers 7 Negative answers 3
AP5	S1: "I remember using the augmented reality app with our smart phone scanning the book and going straight to the map of Crete appeared and we saw where Knossos is."	Positive answers 10
AP6	S6: "Why not, the lesson was more fun than other times."	Positive answers 7 Negative answers 3

that AR is useful in learning and continued to believe that the history lesson is "boring" even with the exploitation of this technology.

It is important for a new teaching intervention applied in the context of a lesson using a technological tool, the students to give their feedback and to suggest improvements or changes, in order the whole approach to be improved. This fact was the subject of the 3rd axis of the interview questions (Table 4). The students suggested some improvements to be added in the process, such as gaming and videos as augmented information. They also expressed their opinion about the instructions given to them about the whole process, and according to their answers most of them got tired of them and became stressed. However, this was expected as children with ADHD get tired and get distracted when they hear a lot of information. In addition, some students were also stressed handling the Blippar application during the lesson, which was probably due to the fact that they had not paid close attention to the instructions.

Regarding the distraction axis (Table 5), the questions that were asked to students concerned their attention to the lesson. The students said that comparing to other times they paid much more attention to the lesson and that the time passed quickly and without realizing it, indicating that there was a flow in their concentration. Of course, for students with ADHD, it is difficult not to think about something else during the learning process, but according to their answers, it appeared that in relation to other times, their frequency of distraction had decreased.

5 Discussion

According to a variety of research findings, AR can be used in many cognitive domains for the education of all students. However, the question in this research is whether it can be integrated into the learning process of students with ADHD. In addition to the easy

Table 3. Evaluating application's usability.

Question	Sample of answers	Statistics
U1	S2: "No, it's the first time for me to use application of augmented reality, I have never seen before something similar."	They have used it 3 They haven't used it 7
U2	S4: "It would be very nice, maybe we were not so bored."	Positive answers 7 Negative answers 3
U3	S10: "Maybe I was more concentrated and I was able to learn faster."	Positive answers 6 Negative answers 4
U4	S3: "Compared to other times it was easier for me to understand some important information."	Positive answers 6 Negative answers 4
U5	S6: "We don't have ever used our smart phones during the lesson at school."	Positive answers 10

Table 4. Suggestions for improvement

Question	Sample of answers	Statistics
SI1	S9: "There was a lot of information and I worried if I would I remember everything during the learning process"	Positive answers 4 Negative answers 6
SI2	S3: "Yes, I was a little nervous, I didn't want to do it some mistake, I was afraid that I wouldn't see what was showing up on my classmates' screens"	Positive answers 5 Negative answers 5
SI3	S9: "At first I was worried, if I would use it properly"	Positive answers 6 Negative answers 4
SI4	S10: "Yes, especially in the points where the content was pictures with information"	Yes 6 No 4
SI5	S7: "I would add a game, I would I removed the images, the videos I liked them"	

and frequent distraction that students with ADHD face, they also have difficulty with memory and information recall. The goal is to use AR not only to teach them to pay attention, but also to recall what they have learned. In order to answer the first research question, which is "How do students experience the interaction of ADHD with AR in the learning process?", some semi-structured interviews were conducted to explore students' experiences and views on all procedures. The majority of the interviewees stated that the augmented reality application helped them to participate more in class. According to the students, the history course is a difficult lesson with a lot of information, some of them even described it as "boring". The students with ADHD get tired quickly every time due to the amount of information they receive in the history lesson, but it seems that the students gain interest in the lesson through the use of AR. For example, they

Table 5. Distraction.

Question	Sample of answers	Statistics
D1	S10: "No, compared to other times that I don't pay attention to the lesson at all, this time I felt that my attention was there"	Positive answers 6 Negative answers 4
D2	S1: "When I have many instructions which I should follow, usually I get tired, but this time I didn't feel tired"	Positive answers 5 Negative answers 5
D3	S1: "It is one of the few times, I wasn't thinking about anything else"	Yes 4 No 6
D4	S5: "Most of the time I wanted to to participate in the conversation to say my opinion on what I saw"	Yes 6 No 4

stated that "I wasn't bored with the lesson." or "I liked that we did something different in the history lesson, usually I was bored.".

Regarding the second research question, which is: "Can AR technology improve the attention of students with attention deficits in the learning process?", by using AR to teach history, the majority of the children had positive results. AR helps the student with ADHD to increase attention and gain confidence. These students have a hard time remembering bulky written texts or verbal words, while the digital material and the power of the image keep them focused. As such, the virtual environment created helped mobilize students' interests and reinforced their observation. Students were generally more focused and this was shown by their responses: "A lot. Only for a very short while I thought of other things. I did not understand this time how fast the time went.", "It's one of the few times that something else didn't catch my attention." and "Yes, compared to other times when I didn't care for history class at all.". Despite the enthusiasm and positive response of students to the whole process, some students showed ADHD symptoms during the course, as the following responses showed them: "Yes, I sometimes got tired and thought about other things, but that happens every time in history class." or "hmm... I will always think of something else that has nothing to do with the lesson.". These responses were expected, since the sample consisted of ten students with different personalities who deal with disorder in different ways, some of them have greater difficulty with concentration, while others less.

Finally, in the last research question, namely "Does the use of instructional materials and applications of AR in the learning process increase the incentives for active participation of students with attention deficit disorder?", the majority of students said they were more engaged in the learning process compared to other times. For example, they stated that "I was definitely much more involved than usual." or "I didn't feel bored, I was involved in the class.". In this study, the use of AR technology showed that students were helped become more engaged in the history lesson and have a positive impression of the whole process. AR activated students' motivation and interest, and contributed to students' creativity. Moreover, it allowed students to learn through experience in real-virtual world. The findings also revealed that learning through AR is more interesting because it enhances students' perception, allows them to understand the learning content better and to participate actively in the educational process. In general, the teaching through AR has the primary objective of improving students' motivation, their participation in activities that they would not find interesting otherwise [26]. Finally, augmented content helped students engage in history learning and have positive impressions about this innovative method.

6 Conclusion

The results of this research confirm and verify the findings of other researches on the reinforcing role and positive influence of AR on the reinforcement and maintenance of attention in students with ADHD. AR can be a useful tool for the teachers who want to help their students who have learning difficulties due to their attention deficit. Moreover, students appeared to enjoy the educational process more and to have a more active role in it. Consequently, student engagement is enhanced and the learning outcomes are improved by the use of AR.

Future work includes the extension of the experiment to other learning subjects in order to investigate the potential of AR in their instruction. It is also necessary to enrich the evaluation procedure, combining the qualitative approach used with a quantitative one, in order to achieve greater reliability of results and validity. To this direction, a larger sample is desirable to be used. Finally, the exploitation of AR in teaching students with other learning difficulties, such as dyslexia and malnutrition, remains an area to be explored in our future research.

References

1. Troussas, C., Krouska, A., Sgouropoulou, C.: Collaboration and fuzzy-modeled personalization for mobile game-based learning in higher education. Comput. Educ. **144**, 103698 (2020)
2. Krouska, A., Troussas, C., Virvou, M.: SN-Learning: an exploratory study beyond e-learning and evaluation of its applications using EV-SNL framework. J. Comput. Assist. Learn. **35**(2), 168–177 (2019)
3. Abuhassna, H., Al-Rahmi, W.M., Yahya, N., Zakaria, M.A.Z.M., Kosnin, A.B.M., Darwish, M.: Development of a new model on utilizing online learning platforms to improve students' academic achievements and satisfaction. Int. J. Educ. Technol. High. Educ. **17**, 1–23 (2020)
4. Troussas, C., Krouska, A., Sgouropoulou, C.: A novel teaching strategy through adaptive learning activities for computer programming. IEEE Trans. Educ. **64**(2), 103–109 (2020)
5. Troussas, C., Krouska, A., Sgouropoulou, C.: Impact of social networking for advancing learners' knowledge in E-learning environments. Educ. Inf. Technol. **26**, 4285–4305 (2021)
6. Krouska, A., Troussas, C., Sgouropoulou, C.: A cognitive diagnostic module based on the repair theory for a personalized user experience in E-learning software. Computers **10**(11), 140 (2021)
7. Hsu, H.C.K., Wang, C.V., Levesque-Bristol, C.: Reexamining the impact of self-determination theory on learning outcomes in the online learning environment. Educ. Inf. Technol. **24**, 2159–2174 (2019)
8. Krouska, A., Troussas, C., Voulodimos, A., Sgouropoulou, C.: A 2-tier fuzzy control system for grade adjustment based on students' social interactions. Expert Syst. Appl. **203**, 117503 (2022)

9. Valverde-Berrocoso, J., Garrido-Arroyo, M.D.C., Burgos-Videla, C., Morales-Cevallos, M.B.: Trends in educational research about e-learning: a systematic literature review (2009–2018). Sustainability **12**(12), 5153 (2020)
10. Krouska, A., Troussas, C., Virvou, M.: A literature review of Social Networking-based Learning Systems using a novel ISO-based framework. Intelligent Decision Technologies **13**(1), 23–39 (2019)
11. Marougkas, A., Troussas, C., Krouska, A., Sgouropoulou, C.: Virtual reality in education: a review of learning theories, approaches and methodologies for the last decade. Electronics **12**(13), 2832 (2023). https://doi.org/10.3390/electronics12132832
12. Papakostas, C., Troussas, C., Krouska, A., Sgouropoulou, C.: Exploring users' behavioral intention to adopt mobile augmented reality in education through an extended technology acceptance model. Int. J. Hum.-Comp. Intera. **39**(6), 1294–1302 (2023)
13. Börnert-Ringleb, M., Casale, G., Hillenbrand, C.: What predicts teachers' use of digital learning in Germany? Examining the obstacles and conditions of digital learning in special education. Eur. J. Spec. Needs Educ. **36**(1), 80–97 (2021)
14. Kapetanaki, A., Krouska, A., Troussas, C., Sgouropoulou, C.: Exploiting augmented reality technology in special education: a systematic review. Computers **11**(10), 143 (2022). https://doi.org/10.3390/computers11100143
15. Lian, X., Sunar, M.S.: Mobile augmented reality technologies for autism spectrum disorder interventions: A systematic literature review. Appl. Sci. **11**(10), 4550 (2021)
16. Kapetanaki, A., Krouska, A., Troussas, C., Sgouropoulou, C.: A Novel Framework Incorporating Augmented Reality and Pedagogy for Improving Reading Comprehension in Special Education. In: Novelties in Intelligent Digital Systems, pp. 105–110. IOS Press (2021)
17. Papakostas, C., Troussas, C., Krouska, A., Sgouropoulou, C.: Personalization of the learning path within an augmented reality spatial ability training application based on fuzzy weights. Sensors **22**(18), 7059 (2022)
18. Kljun, M., Geroimenko, V., Čopič Pucihar, K.: Augmented reality in education: Current status and advancement of the field. Augmented Reality in Education: A New Technology for Teaching and Learning, 3–21 (2020)
19. Sytwu, T.-A., Wang, C.-H.: An investigation of the effects of individual differences on mobile-based AR english vocabulary learning. Mobile Learning Design. In: Churchill, D., Lu, J., Chiu, T., Fox, B. (eds.) Lecture Notes in Educational Technology, pp. 387–410. Springer, Singapore (2016). https://doi.org/10.1007/978-981-10-0027-0_23
20. Sheehy, K., Ferguson, R., Clough, G.: Augmenting learners: educating the transhuman. In: Sheeh, K., Ferguson, R., Clough, G. (eds.) Augmented Education, pp. 137–158. Palgrave Macmillan, New York, NY (2014)
21. Papakostas, C., Troussas, C., Krouska, A., Sgouropoulou, C.: User acceptance of augmented reality welding simulator in engineering training. Educ. Inf. Technol. **27**(1), 791–817 (2022)
22. Liu, G., Wu, N., Chen, Y.: Identifying emerging trends for implementing learning technology in special education: A state-of- the-art review of selected articles published in 2008–2012. Research in Developmental Disabilities **34**, 3618–3628 (2013)
23. McMahon, D.D., Cihak, D.F., Wright, R.E., Bell, S.M.: Augmented reality for teaching science vocabulary to postsecondary education students with intellectual disabilities and autism. J. Res. Technol. Educ. **48**(1), 38–56 (2016)
24. Parton, B.S., Hancock, R.: Animating the Inanimate using Aurasma: Applications for Deaf students. Society for Information Technology & Teacher Education International Conference, pp. 3743–3745. Association for the Advancement of Computing in Education (AACE) (2012, March)

25. Kansteiner, K., König, S.: The role (s) of qualitative content analysis in mixed methods research designs. In: Forum Qualitative Sozialforschung/Forum: Qualitative Social Research, vol. 21, no. 1 (2020, January)
26. Kapetanaki, A., Krouska, A., Troussas, C., Sgouropoulou, C.: Analyzing the impact and application of Augmented Reality in Education: The case of students with special educational needs. In: Proceedings of the 26th Pan-Hellenic Conference on Informatics, pp. 240–245 (2022, November)

Development of a Module for Generating Function Header Tasks Through the Analysis of Textual Comments

Vladislav Sukhoverkhov and Anton Anikin(✉)

Volgograd State Technical University, Volgograd, Russia
anton@anikin.name

Abstract. The relevance of this research topic stems from the efficacy of utilizing simulators to create function headers during programming training, as well as the labor-intensive nature of generating tasks for these simulators when dealing with extensive volumes of available commented program code. Analyzing commented program code enables the extraction of valuable information and data, which serve as the foundation for automating the generation of such tasks. This necessitates the development of suitable approaches, methods, and algorithms for program code analysis and the subsequent generation of tasks based on the acquired data. The development of a generator of function headers creation tasks aims to address the challenges of complexity and labor intensity associated with manual task creation.

Keywords: Text analysis · Training assignment generation · Feedback · Doxygen-comments · Abstract syntax tree

1 Introduction

Information analysis has always played a crucial role in the adoption of new technologies and in efforts to optimize and expedite information processing. Analyzing textual comments not only enables the extraction of valuable information but also provides a foundation for generating function header composition tasks. Contemporary programming instruction employs various approaches for creating tasks for learners. Analyzing textual comments within program code allows for the extraction of useful information and data, which serve as the basis for automating the generation of such tasks. This necessitates the development of suitable approaches, methods, and algorithms for both program code analysis and task generation based on the acquired data.

Recently, active learning methods are replacing traditional methods, when teachers continue to use both. According to teachers, the use of active methods involves a number of problems. To qualitatively use the active learning method requires a high level of independent training, which can be difficult to achieve when students exhibit low levels of intellectual engagement.

Feedback is a crucial component of the teaching and learning process, as it enables students to identify their knowledge gaps and evaluate their progress in learning. Feedback becomes necessary when specific information regarding a learning task or process

K. Kabassi et al. (Eds.): NiDS 2023, LNNS 783, pp. 215–223, 2023.
https://doi.org/10.1007/978-3-031-44097-7_23

is required, bridging the gap between the desired and actual comprehension of content or skill development. Through feedback, students refine their knowledge and learning abilities that were previously challenging or insufficient. Numerous studies have demonstrated the positive impact of effective feedback on learning outcomes [2].

Considering the aforementioned factors, the research objective was established as follows: to develop a function header compilation task generator that automates feedback and addresses the challenges of complexity and labor intensity associated with manual task creation. The efficacy of header composition simulators informed the primary approach, which entails analyzing large volumes of program code's textual comments to ultimately provide tasks for trainees and facilitate ease of use.

2 Concepts and Approaches

The creation of learning tasks requires the best possible assessment of the skills and success of the learner, so in order to simplify their creation and at the same time speed up this process, we need a tool that can automatically generate tasks. One of the tasks involved in studying various programming languages is writing function headers that accept diverse input parameters and yield specific results. In practice, task generation is used in training simulators, where the execution of actions based on the teacher's assigned task can be visualized [4].

To ensure that learners grasp a particular programming language construct while studying its syntax, the level of detail provided must be sufficiently comprehensive. As part of this work, it is necessary to analyze text comments for information regarding the various keywords of the programming language and its constructs, accurately presenting the results to the learner.

Various approaches are employed in analyzing text; this paper is particularly interested in examining the following methods:

1. Deep methods of analysis are used when it is necessary first to structure a given text as a whole and then identify certain words or word combinations, bringing the text to its normal form.
2. Superficial methods are necessary when deep analysis of the text is not required, and it is possible to focus on the analysis of individual parts of the text, as well as statistical metrics.

For comprehensive language parsing, an in-depth text analysis is essential, necessitating the construction of a parse tree that encompasses all relationships between auxiliary words. A parser is a combination of a lexer, which groups source code symbols into meaningful sequences called tokens, and a parser that processes these tokens to derive their syntactic structure [6].

A higher-level text parsing is represented by an abstract syntax tree (AST), which eliminates all irrelevant characters, such as brackets and commas. However, there exist parsers in which the lexing and parsing stages are combined.

For parsing specially formatted comments, custom-trained parsers are developed to directly construct a tree structure and provide access to it through API declarations. Currently, there are limited parsers available for C/C++ parsing, with Doxygen and QDoc being the most commonly used. Doxygen is a first project capable of parsing comments from C/C++ code and generating HTML documentation that includes hyperlinks, inheritance graph images, calls, and more. Doxygen is considered the de facto standard for documenting C/C++ code.

3 Generation Module Models and Algorithms

3.1 Conceptual Model of the Generation Module

The generation module works according to the following principle. Initially, the module queries a file containing the executable code of a third-party program. This file parses the Doxygen comments, from which the tags associated with the description of method parameters are "pulled", as well as the linguistic text of the method description itself. Next, the generator receives received data in the structure suitable for the generator. The generator by input parameters defines method data elements and method of storing each element. The generator will also allow you to choose programming language data types, on the basis of which you will finally need to build a method prototype.

3.2 Algorithm for Obtaining Information

Before parsing files, they must be obtained from a remote repository; when searching within the current task, source files with comments presented in the Doxygen format are required. In addition to the required comment format, the parser will also consider comments belonging only to functions/methods and having tags to their parameters and a linguistic description. The algorithm for obtaining the information required to build the task generator is reduced to performing the actions described below.

Searching for Files in Remote Repositories. When searching for the required comments, we access the repositories we have prepared in advance for analysis. In each repository, we go through all the files and try to find method/function descriptions in Doxygen format, which in addition have parameter tags and a linguistic description.

Determining How to Store Data Items. First you need to generate a list of method parameters, and for each of them you first define its direction (it can be either input or output), then the method of storage. Before sending the data, they are checked, and if no errors are found, the algorithm ends.

Defining Data Elements. To define the data elements, you first need to get all the function parameters, then define the keywords in the function description. At the end, the obtained data are stored in the internal structure.

Defining Data Types. In the beginning, you need to generate a list of method parameters, and for each of them, first define the data component, then the data type. Before sending the data, they are checked, and if no errors are found, the algorithm ends.

Building a Function Prototype. To build a function prototype, you need to write its name, define its return type and all parameters (names and return types of each).

4 Comment Categories

The analysis uncovers a majority of non-linguistic comments, encompassing source code, comments not in English, and other elements that are not within the scope of this paper. Additionally, the comments include copyrights, licenses, encoding declarations in file headers, and other service-related remarks [5].

Code/Math. Comments frequently contain source code or mathematical equations alongside linguistic remarks. The source code embedded in the comments is typically retained to preserve previously written code when there is no immediate plan to remove it, which may later be inadvertently left in the repository. While mathematical equations do not contribute significant value to the linguistic model analysis, they facilitate easier human comprehension of such comments.

Non-English. For the purposes of the current work, only English is considered, so all non-English words and symbols must be discarded.

Copyright. Some of the comments contain information about the copyright or license of the file. Since these comments do not actually describe the source code, they are not considered.

Non-linguistic. Comments can also consist only of punctuation marks (e.g. delimiters) or contain ASCII characters.

Duplicates. Comments can be duplicated, so this fact must also be taken into account.

5 Analysis of Doxygen Comments

Doxygen is a tool capable of converting specific comments within a program into a description file. The most valuable comments typically pertain to descriptions of functions, categories, and so on. As a result, if annotations can be processed and transformed into a well-organized reference file that aligns with the program's structure, the tool can be beneficial in three scenarios:

1. The system can generate documentation both for on-line viewing (in HTML) and as a reference description (in) from a set of source files.
2. Doxygen can be configured to extract the code structure from undescribed source files. This is very useful for quickly navigating through large distributed sources.
3. Doxygen is designed for Linux and Mac OS X, but is portable. As a result, it runs on most other Unix flavors as well. Windows versions are also available.

5.1 Types of Comments

Doxygen handles single-line comments marked with three slashes, i.e. ///. The single-line comments in a row are combined into a common block-paragraph. Single-line comments separated by lines of program statements fall into different paragraphs.

Doxygen also handles multi-line comments like /* */, but you must have at least two asterisks at the beginning of the comment.

5.2 Options for Documenting with Comments

Doxygen analyzes the structure of the program text, identifying individual modules (files), global definitions (variables, macros, etc.), and finding functions and local elements for functions. Comments that are "inside" a function block of a program are grouped. That is, comments within a function will refer to the documentation for that function, while comments scattered "between" functions will be assembled into documentation for the entire file.

To document the entire file, it is best to use a multi-line comment located at the very beginning of the file and in which special Doxygen parameters must be used. A parameter is a specific keyword which serves to notify Doxygen to perform a special handling of the next (or subsequent) comment words.

6 Overview of Methods and Types of Task Generators

6.1 Types of Task Generation Tools

There are notable differences among available generators. Firstly, the format of student assignments varies. The output format can either be presented on a computer screen or on paper. In the former case, there is the advantage of receiving immediate feedback on the same computer, eliminating the need for manual assignment checking, which is a drawback in the latter case, where the teacher must invest time in assessing the student's skills. The second distinction among generators lies in the subjects they can address, specifically the capacity to handle tasks related to diverse areas of application, ranging from mathematical problems to humanities disciplines, in which students are presented with questions accompanied by multiple answer options [1].

To provide the generator with additional information, various parameters are utilized. For instance, in a matrix determinant search problem, the generation parameters may allow the user to select the size of the matrix and the number of zero elements. In a number addition problem, the generation parameter could determine the range of values for the generated numbers. These parameters define the mode of operation (exploration, evaluation, or both), the user's initial level of knowledge, the sequence and subset of operations visible in the progress bar, the expressions to be displayed in the expression selection menu, and a few other aspects.

6.2 Types of Task Generation Tools

Developing generators is a complex task that typically necessitates programming and implementation of algorithms. Each generator is founded on a particular method that defines the structure of its description and the algorithm for transforming this description into the generator itself. These methods inherently restrict the class of tasks that can be produced using them. Generally, the broader the class of assignments, the more challenging the method becomes in terms of usage and learning.

The fundamental concept revolves around the division of labor between programmers and teachers, where the former develop the generator editor while the latter utilize it to prepare generator descriptions. This approach is known as object-oriented programming, and the language employed to describe the generators is referred to as a domain-specific language (DSL).

A crucial objective in the field of generation is to devise methods that are not only straightforward to use but also capable of producing a diverse range of multi-variant tasks.

7 Approaches to Generating Learning Tasks

7.1 Template-Based Generation

The first method of generation is the use of templates. The tutorial assignment is presented as a text with variables in some places. An example of a condition and answer pattern: "How much is [A] + [B]? Answer: [A] + [B] = [C]."

Variables can be not only numbers, but also other mathematical objects: matrices, polynomials, and lines. Mathematical objects: matrices, polynomials, straight lines defined analytically, mathematical expressions (functions), etc.

A generalization of pattern-based generation involves generation with condition text output in a context-free grammar (CC-grammar). This generalization arises from the assumption that each variable in the template is itself generated by the template. In other words, variables are nonterminal grammar symbols, and the various choices of values for these variables correspond to different grammar rules for that particular nonterminal.

7.2 Generating Mathematical Expressions

Generating mathematical expressions enables the creation of more diverse mathematical tasks compared to solely generating numbers. This type of generation is essential in problems involving integration, calculating derivatives, solving equations and inequalities, and graphing.

One method for generating a mathematical expression is to define the KC-grammar that describes it and then employ the algorithm for generating random output in the KC-grammar. To control the size of the generated expression, it is necessary to specify a range for the desired length of the resulting output, which could be the total number of terminals, or the number of terminals of a specific type, such as variables.

Researchers at Microsoft Research propose that instructors write just one expression, allowing the program to automatically generalize from it and generate a set of similar assignments.

7.3 Generation Based on Knowledge Bases

Template generation is not well-suited for humanities or for assessing knowledge of definitions and properties of objects. The reason is that, in such tasks, the entire text is typically altered rather than just fragments of the question text.

The data for generating such tasks are derived from a knowledge base, with methods varying based on the nature of the base. Broadly speaking, these methods can be divided into those that use a structured base and those that employ an unstructured base.

The simplest form of knowledge representation is tabular. A table can store a set of terms and their definitions. This enables the generation of questions that require choosing the correct definition for a term from multiple options, selecting the correct term for a given definition, or comparing several terms and their definitions.

8 Review of Analogues of Training Task Generators

Some of the generators developed for creating learning tasks are made publicly available by the educators themselves. Among these generators are those designed for creating math problems. By generating text files containing numerous problems, their purpose is to facilitate training through the practice and resolution of a vast array of exercises [7]. Other generators provide a collection of task creators, implemented using Microsoft Word macros. Integrated within Word, these generators appear as task templates consisting of conditional text with a variable count. During the generation process, all variables are replaced with specific numbers or words. It is worth noting that among all the generation systems examined, only the template-based system in Word allows users to create their own generators. However, this comes at the expense of generating relatively simple tasks. In contrast, other systems produce tasks that are significantly more complex and varied.

Some generators are accessible online, enabling teachers to create assignments directly from their web browsers. These generators are considerably more user-friendly than the previously described ones, as there is no need to manage software installation and configuration on personal computers.

Among the online generators, there are paid options that generally offer more advanced functionality compared to their free counterparts. These tasks often incorporate illustrations and are neatly organized on the page. However, a potential drawback is the limited availability of specific tasks; if a required task is not available, it cannot be generated [3].

Wolfram|Alpha, developed by Wolfram, a leader in mathematical and scientific software, is an innovative product from the creators of Mathematica, their well-known computer algebra system. Wolfram|Alpha enables users to ask and answer questions in English on a wide range of topics, drawing from an extensive collection of databases spanning various fields of knowledge and leveraging Mathematica's computational engine. It is not possible to create own generators in Wolfram|Alpha, but it can be done with Mathematica, on which it is based.

9 Problem Statement

The generation of class assignments is popular among educational institutions, leading to a range of commercial offerings as well as numerous independent developments, many of which are made publicly available by the educators themselves. Generators can take the form of desktop applications or web applications. The primary drawbacks of such software include a limited set of generated tasks.

The challenge lies in developing a generator that balances ease of use, accessibility, and the diversity of multi-variant tasks. The proposed approach entails using the analysis of text comments as the foundation for creating generators, thereby making the generator more flexible and scalable, without relying on constraints associated with template-based task creation.

10 Performance Evaluation

For the effectiveness and usefulness of the current implementation, an experiment was conducted in the course "Fundamentals of Programming". Twenty first-year students participated in the experiment, who were given a theoretical framework to create function prototypes in advance. Each student was given 10 tasks, which were first done manually and later with the provided design. The result was that with manual completion, only 17.6 percent of the assignments were successfully completed, whereas with the implementation of the design, success was already determined by 91.8 percent. The success of using the development is determined, in particular, by the fact that each student was provided with hints compiled by the teachers and intelligently provided by the module itself. The survey of students was conducted using a Likert scale, resulting in the following statistics:

hints helped to solve the problem: 91.9%;
the tips helped to understand the reasons for mistakes and avoid similar mistakes in the future: 94.5%;
using a tutor takes a long time: 6.5%;
it is desirable to use a tutor to solve similar problems: 97,2%.

11 Conclusions

From all of this we can conclude that in order to achieve our goal, we need to automate the analysis of functions in Doxygen format, which must be "parse" and identify in the function description itself the objects that are input and output parameters. Thus, this paper considers the following:

1. Developing teacher-oriented tools: providing tools to teachers will improve their teaching practices and even involve them in improving feedback to students.
2. Analysis of text comments for generating function header tasks: automation of task compilation will solve the problem of complexity and laboriousness of their manual compilation.

References

1. Ali, H., Chali, Y., Hasan, S.A.: Automatic question generation from sentences. In: Actes de la 17e conférence sur le Traitement Automatique des Langues Naturelles. Articles courts 213–218 (2010)
2. Cavalcanti, A.P., et al.: Automatic feedback in online learning environments: a systematic literature review. Comp Edu. Artifi. Intell. **2**, 100027 (2021)

3. Leung, H., Li, F., Lau, R., Li, Q.: Advances in web based learning–icwl 2007 6th international conference edinburgh, uk, august 15–17, 2007 revised papers. In: Conference proceedings ICWL, p. 219. Springer (2007)
4. Litovkin, D., Anikin, A., Kulyukin, K., Sychev, O.: Intelligent tutor for designing function interface in a programming language. In: Intelligent Tutoring Systems, pp. 293–302. Springer International Publishing (2022). https://doi.org/10.1007/978-3-031-09680-8_27
5. Matskevich, S., Gordon, C.: Preprocessing source code comments for linguistic models. arXiv preprint arXiv:2208.11235 (2022)
6. Padioleau, Y.: Parsing c/c++ code without pre-processing. In: Lecture Notes in Computer Science, pp. 109–125. Springer Berlin Heidelberg (2009). https://doi.org/10.1007/978-3-642-00722-4_9
7. Tomás, A.P., Leal, J.P.: A clp-based tool for computer aided generation and solving of maths exercises. In: Practical Aspects of Declarative Languages: 5th International Symposium, PADL 2003 New Orleans, LA, USA, January 13–14, 2003 Proceedings, pp. 223–240. Springer (2002)

E-Health Cloud Based Systems: A Survey on Security Challenges and Solutions

Ismahene Marouf[1,2](✉), Asma Saighi[1,3](✉), Sofia Kouah[1,3](✉), and Zakaria Laboudi[1](✉)

[1] University of Oum El Bouaghi, Mathematics and Computer Science Department, Oum El Bouaghi, Algeria
Ismahane.marouf@univ-constantine2.dz, {saighi.asma,Sofia.kouah, laboudi.zakaria}@univ-oeb.dz
[2] Research Laboratory on Computer Science's Complex System, Oum El Bouaghi, Algeria
[3] LIAOA Laboratory, University of Oum El Bouaghi, Oum El Bouaghi, Algeria

Abstract. Cloud based healthcare systems have transformed the healthcare sector in many ways. It has several advantages like scalability of the required service, flexibility, and resource sharing. However, the centralization of data on the cloud increases many security and privacy issues. Several surveys exist in the field; however, they do not do not deal with decentralized and hybrid eHealth cloud. This paper presents survey in the field of cloud-based e-Health security systems, including a comparison between the various approches (i.e., centralized, decentralized and hybrid). For this purpose, we start by defining the main concepts and common security issues in and summarizing the existing surveys that deal with e-Health cloud-based security. Then, we provide a state-of-the-art toward secure and privacy-preserving for cloud-based e-Health systems. Finally, we highlight open research challenges and possible future research directions.

Keywords: Cloud computing · e-Health · Cloud security

1 Introduction

Cloud computing provides several facilities, for data health computing power and storage without direct management by users. For this purpose, a variety of e-Health based-cloud applications have been developed to monitor health status of patient. Patient's clinical data are important and sensitive. It should be secured to avoid any data tampering by any means [1]. Health data must be as safe and reliable as possible, because its disclosure can have serious consequences for patients. This means that, there is a need for a more secure health research platform and technology that ensures the use of electronic health data while focusing on its security. Many related research works was proposed to deal with the problem of security and privacy concerning e-Health based-cloud systems and several surveys focused on various security aspects. However, the existing surveys do not treat decentralized and hybrid approaches, despite they offer scalability, reliability, and security. In this work, we present a definition of the main concepts and common

K. Kabassi et al. (Eds.): NiDS 2023, LNNS 783, pp. 224–229, 2023.
https://doi.org/10.1007/978-3-031-44097-7_24

questions in the field with some examples and case studies. Then, we summarize some related surveys and related works in the literature. The comparison criterion we used is related to some security issues. The remaining of this paper is structured as follows. Section 2 presents a definition of each related key word. Section 3, summarizes some existing surveys and proposed solutions in the studied field. Section 4 presents comparison between approches. Section 5 provides new challenges in terms of security and privacy solutions in the field. Lastly, Sect. 6 concludes the paper.

2 Background

2.1 E-Health

In 2001,Gunther Eysenbach, reported on [10], "What is eHealth" series, "eHealth: refer to health services and information delivered or enhanced through the Internet and related technologies". This term has become more common in the last 5 years, especially during the COVID-19 pandemic, which has highlighted the need for remote patient monitoring and other e-health technologies [3–9].

2.2 Common eHealth Security Issues

eHealth system argue that security threats to the cloud data include spoofing identity via an attacker pretending to be a valid user [10, 11]. As a case study David Lake and others deal with eHealth Security frameworks, and smart systems to address this threats [12–14].

2.3 Cloud Computing and Security

It is a model for enabling ubiquitous, convenient, on-demand network access to a shared pool of configurable computing resources [6, 8]. Weak cloud security can expose users to all types of cyber security threats [15] like malware, and fishing.

3 Related Literature Surveys and Works

Research articles are evenly distributed and cover all aspects of e-Health service development and implementation. The focus in this paper is the security aspect of e-Health systems. In Table 1 Table 2, we summarize, respectively some related surveys and proposals for securing e-Health Cloud Based systems.

Several proposals were carried out in the field of e-Health cloud Security and some works can be summarized as follows: [18] Investigates the privacy of e-health systems, they identified components of the modern e-health systems, [20] used hash functions because this last is the one-way procedure that designed to protect the integrity of data [17]. 5G digital used to produce a variety of electronic health record.

Table 1. A comparison of some related surveys in the literature.

Suervey	Reference	Objectif	Security Privacy	Block -chain	Comments
Blockchain in healthcare and IoT: literature review	[6]	Analyzes trends blockchain deployment in IoT and healthcare	YES	YES	Patient-centric
Security and privacy issues in e-health cloud-based system: A comprehensive content analysis	[16]	Focused on mechanisms b used to handle ecurity and privacy in e-Health	YES	NO	Efficient
Application of 5G network combined with AI robots	[17]	Aims to focus the5G in healthcare	YES	YES	Invest, educate
A Comprehensive Survey on Security for ehealth data	[5]	Identified major components of the modern e-health systems	NO	NO	Futer-proof

Table 2. Summary of some related works for secured e-Health designs and cloud

Reference	Security	Main focus	Advantage	Disadvantage	Limitation
[21]	YES	Security challenges	Explore the security and privacy issues in eHealth	Devices are not usually incorporated in the plan	Significant amount of advanced technology involved
[22]	NO	User attitudes	Knowledge from social media	Troubling informations	Not developed
[23]	NO	Application of 5G technology	5G technology inwireless BodyAreaNetwork	It is still in the process of being developed	Cost of its Infrastructure is expensive
[2]	YES	Healthcare data privacy	Attacks performed on eHealth systems -IoT (internet of things) devices	The difficulty of achieving both cost-effectiveness	Data shared indefinitely for predetermine Period of time

(*continued*)

Table 2. (*continued*)

Reference	Security	Main focus	Advantage	Disadvantage	Limitation
[24]	YES	Use of intelligent techniques in health	Security and privacy of medical data in IoT	Blockchain mining process benefits wealthiest participants	Lack of clarity on the quantity of online health data

4 Comparaison and Discussion of Various Works

In eHealth cloud, three fundamental approaches exist: centralized, decentralized and hybrid. The choice of the most convenient approach depends on specific needs of the organization. For example, if reliability is the most important factor, then a decentralized approach may be the best choice. In another case where a combination of scalability, reliability, and security is needed, a hybrid approach may be the best choice. Table 3 summarizes the key differences between the studied approaches.

Table 3. Different approaches of eHealth cloud system.

Approaches	Advantages	Disadvantages
Centralized [5, 15, 23]	Scalable, secure	Single point of failure
Decentralized [8]	Scalable, reliable	Complex to manage
Hybrid [24]	Scalable, reliable, secure	Requires coordination between multiple clouds

5 Future Directions and Challenges

- The use of multidisciplinary teams and formative research is important.
- New architectures should be developed based on the use of better technologies: 5G [25], requirement for specialized security services and protocols.
- Highly distributed, component-based and self-organizing architecture should be the goal of future R&D activities.

6 Conclusion

In this paper, we presented a background concerning the security and e-health cloud-based systems. Then, we reviewed the related surveys and we surveyed the state-of-the-art of existing security solutions for e-health cloud systems. The existing surveys do not cover all security aspects in the field. We provided also an overview of the related works and their main objectives. Most of these solutions do not deal with all the competing security needs. There still exist several open challenging purposes, such as building new cyber security datasets for the comparison of intrusion detection solutions which should be further investigated. We plan to propose solution balancing all the questionable requirements.

References

1. Kumar, D., Mandal, N., Kumar, Y.: A novel approach for securing e-health application in a cloud environment. In: Machine Learning, Big Data, and IoT for Medical Informatics, pp. 349–364. Academic Press (2021)
2. Abbas, A., Khan, S.U., Member, S.: A review on the state-of-the-art privacy-preserving approaches in the e-health clouds
3. Chowdhary, S.K.; Yadav, A.; Garg, N.: Cloud computing: future prospect for e-health. In: Proceedings of the 2011 3rd (2011)
4. Randine, P., Sharma, A., Hartvigsen, G., Johansen, H.D., Årsand, E.: Information and communication technology-based interventions for chronic diseases consultation: scoping review. Int. J. Med. Inf. 104784 (2022.)
5. Oh, S.R., Seo, Y.D., Lee, E., Kim, Y.G.: A comprehensive survey on security for electronic health data. Int. J. Environ. Res Public Health **18**, 9668 (2021)
6. Adere, E.M.: Blockchain in healthcare and IoT: a systematic literature review. Array **14**, 100139 (2022)
7. Mustafa, M., Alshare, M., Bhargava, D., Neware, R., Singh, B., Ngulube, P.: Perceived security risk based on moderating factors for blockchain technology applications in cloud storage to achieve secure healthcare systems (2022)
8. Mell, P., Grance, T.: Ce NIST Definition of Cloud computing (2022)
9. Yazan Al-Issa, M.A.O.A.A.T.: eHealth cloud security challenges: a survey. Vols. **2019**, Article ID 7516035
10. Eysenbach, G.: What is e-health? J. Med. Internet Res. **3**(2), E20 (2001)
11. Rahmani, A.M., et al.: Exploiting smart e-Health gateways at the edge of healthcare Internet-of-Things: a fog computing approach. Fut. Gener. Comput. Syst. **78**, 641–658 (2018)
12. Varghese, B., Buyya, R.: Next generation cloud computing: new trends and research directions. Futur. Gener. Comput. Syst. **79**, 849–861 (2018)
13. Gai, K., Qiu, M., Zhao, H.: Security-aware efficient mass distributed storage approach for cloud systems in big data. In: 2016 IEEE 2nd International Conference on Big Data Security on Cloud (bigdata security), IEEE International Conference on High Performance and Smart Computing (HPSC), and IEEE International Conference on Intelligent Data and Security (IDS), pp. 140–145. IEEE (2016)
14. Onumanyi, A.J., Ohize, H.O., Alenoghena, C.O.: eHealth: a survey of architectures, developments in mHealth, security concerns and solutions (2022). 54 pages
15. Al-Issa, Y., Ottom, M.A., Tamrawi, A.: eHealth cloud security challenges: a survey. J. Healthcare **2019**, 1–15 (2019)
16. Rahman, M.M., Khatun, F., Sami, S.I., Uzzaman, A.: The evolving roles and impacts of 5G enabled technologies in healthcare: the world epidemic COVID-19 issues. Array **14**, 100178 (2022)
17. Hameed, R.T., Mohamad, O.A., Hamid, O.T., Tapus, N.: Patient monitoring system based on e-health sensors web services (2020)
18. Annane, B., Alti, A., Lakehal, A.: Blockchain based context-aware CP-ABE schema for Internet of Medical Things security. Array **14**, 100150 (2022)

19. Zvikhachevskaya, A., Markarian, G., Mihaylova, L.: Quality of Service consideration for the wireless (2019)
20. Maeen, S., Zykov, S.: Towards social network—integrated e-health: identify user attitudes. Procedia Comput. Sci. **55**, 1174–1182 (2015)
21. Okoh, E., Awad, A.I.: Biometrics applications in e-health security: a preliminary survey. IEEE (2021)
22. Kotsiuba, I., Velvkzhanin, A., Yanovich, Y., Bandurova, I.S., Dyachenko, Y., Zhygulin, V.: Decentralized e-Health Architecture for medical journal (2019)
23. Butpheng, C., Yeh, K.H.: Security and privacy in IoT-cloud-based e-health systems—a review. Symmetry **12**, 1191 (2020)
24. Rahm, E., Do, H.H.: Data cleaning: problems and current approaches. IEEE Data Eng. Bull. **23**(4), 3–13 (2022)
25. Ayub, M.F., Mahmood, K., Kumari, S., Sangaiah, A.K.: Lightweight authentication protocol for e-health clouds in IoT-based applications through 5G technology. Digit. Commun. Networks **7**(2), 235–244 (2021)

Crafting Immersive Experiences: A Multi-Layered Conceptual Framework for Personalized and Gamified Virtual Reality Applications in Education

Andreas Marougkas(✉), Christos Troussas, Akrivi Krouska, and Cleo Sgouropoulou

Department of Informatics and Computer Engineering, University of West Attica, Egaleo, Greece
{amarougkas,ctrouss,akrouska,csgouro}@uniwa.gr

Abstract. This paper presents a clear and detailed conceptual framework for creating Virtual Reality (VR) applications that incorporate personalization and gamification techniques. The framework comprises several layers designed to deliver a more immersive and engaging experience that is customized to the unique needs and preferences of each individual user. By integrating these techniques, the framework aims to enhance user satisfaction, motivation, and overall engagement in VR environments. The paper provides an in-depth examination of each layer of the framework and includes a visual diagram to demonstrate the connections between them. In the proposed framework, personalization techniques are used to customize the VR experience to each user's needs and preferences. This can include the adaptation of the application's content, design, and functionality to suit the user's skill level, interests, and goals. The framework aims to provide researchers and developers with a valuable guide for the development of personalized and gamified VR applications.

Keywords: Virtual Reality · Personalization · Education

1 Introduction

Virtual reality (VR) is a disruptive technology with the potential to transform fields such as entertainment [1], automotive industry [2], education [3], training [4], and healthcare [5]. It provides a visually and audibly genuine simulated experience, producing an immersive experience. Personalization and gamification aspects, on the other hand, are required to enhance user enjoyment and engagement in VR experiences. To address this, the paper delivers a conceptual framework for personalized and gamified VR applications that aim at improving user engagement and satisfaction. User Profiling, Content Customization, Interaction and Gameplay, Feedback and Progression, and Social Integration are the five layers that work together to create a personalized and immersive VR experience by gathering user information, customizing content, incorporating game-like elements, providing real-time feedback, and enabling social interaction. The proposed

K. Kabassi et al. (Eds.): NiDS 2023, LNNS 783, pp. 230–241, 2023.
https://doi.org/10.1007/978-3-031-44097-7_25

framework provides a methodical way to create virtual reality applications that adapt to unique user preferences while incorporating game-like components for an enjoyable experience. The framework emphasizes the incorporation of personalization [6] and gamification features to enhance immersion and enjoyment, while also providing chances for personalized learning, skill development, and social interaction. The relevance, important components, and implementation strategies of each layer will be explored further in the piece, along with an illustration indicating the relationship between them.

2 Relevant Literature

The speedy progress and the lowering of costs of Virtual Reality (VR) technology has opened the possibility of its use in areas beyond just gaming. It has now become increasingly feasible to integrate this technology into other areas like education and training, thanks to its availability and affordability. However, developing such applications can be time-consuming and require significant research efforts. To address this challenge, [7] proposed an immersive virtual reality framework that simplifies application development and allows researchers to focus on designing educational or training experiences, while the framework handles technical aspects of VR development. The framework, based on Unreal Engine and compatible with most VR helmets, includes tools for creating VR experiences, such as player movement, object interactions, and scene objectives. To demonstrate the framework's effectiveness, the paper presents two educational experiences developed in two cultural heritage environments.

Hansen et al. [8] have introduced a multiplayer framework called "Let's VR" that assists developers and researchers on developing immersive VR applications. This framework allows users to collaborate and communicate in real-time and is also compatible with a wide range of virtual reality devices and platforms. The paper also offers a detailed overview regarding the framework's architecture, components and layers, which can be beneficial for developers to integrate it into their applications. The "Let's VR" framework is an important contribution to the VR industry, as it offers developers a comprehensive tool for building interactive and socially-oriented virtual reality experiences. The framework provides multi-user collaboration and communication support, compatibility with various devices and platforms, and a detailed technical architecture.

This paper presents a unique framework that stands out by offering a thorough and all-encompassing guide for developers and researchers in creating personalized experiences with gamified elements in virtual reality (VR) applications. This particular framework addresses an area that has not been extensively explored or evaluated in existing literature.

By utilizing this framework, designers and developers can take into consideration various aspects such as users' profiling, customization of content, interactive gameplay, feedback mechanisms, and social integration. This comprehensive approach allows for the creation of VR applications that adapt to individual user preferences, employ game-like mechanics to enhance engagement, and foster social interaction among users.

The potential implications of implementing personalized and gamified VR applications are significant across diverse domains. They can enhance learning outcomes, increase motivation, and facilitate skill development. As VR technology continues to evolve and become more accessible, the proposed framework offers valuable insights and guidelines to ensure the design of impactful and enjoyable virtual experiences.

3 Layer 1: User Profiling

The central layer of the framework for developing customized and gamified virtual reality applications is the user profiling layer. It collects and evaluates user data to create unique profiles that encompass their inclinations, behavioral trends, and physiological reactions. With a thorough understanding of each user's distinct attributes, this layer enables the other layers to tailor the virtual reality experience to meet their specific requirements and preferences.

3.1 Data Collection Methods

To collect information about users, the user profiling layer employs a range of data gathering techniques, which may involve surveys, interviews, questionnaires, and observational studies, among others. These methods provide valuable insights into user interests, preferences, and motivations. Furthermore, the user profiling layer may use biometric sensors like heart rate monitors or eye-tracking devices to record physiological responses and emotional states, enabling a more comprehensive understanding of user experiences within the virtual environment.

3.2 User Preference Analysis

After collecting user data, advanced analytical techniques, such as machine learning algorithms and data mining strategies, can be utilized to detect patterns and preferences in the data. These analyses can aid in the discovery of individual inclinations for content genres, interaction methods, visual design, and other aspects that pertain to the VR experience. By comprehending user preferences, the following layers can tailor the VR content to suit their particular requirements.

3.3 Personalized User Profiles

By using the gathered data and conducting analysis, personalized profiles for each user are constructed. These profiles provide a complete snapshot of the user, encompassing their demographic data, preferences, behavior trends, and physiological responses. These profiles form the groundwork for customizing the virtual reality experience throughout the entire application.

3.4 Dynamic Profile Adaptation

User profiles are not fixed and unchanging; they can grow and adjust as time progresses. The user profiling layer should include mechanisms for updating and modifying user profiles dynamically based on fresh data and interactions within the virtual reality setting [9]. This flexible strategy guarantees that the VR application remains in sync with the evolving preferences and requirements of the users.

3.5 Privacy and Ethical Considerations

Maintaining user privacy and ethical considerations [10] are extremely critical in the user profiling layer. It is essential to address user consent, data security, and confidentiality carefully. Adhering to ethical guidelines and regulations is vital to protect user privacy and ensure that user data is collected and used responsibly.

The user profiling layer serves as the foundation for personalization within the framework's concept. By capturing user preferences, behavior, and physiological responses, this layer facilitates subsequent layers in customizing the virtual reality experience, thereby creating a highly personalized and immersive environment. The insights obtained from user profiling enhance user satisfaction, create a sense of ownership, and establish a deeper connection between the user and the VR application.

In the following section, we will delve into Layer 2: Content Customization, which relies on the user profiles developed in Layer 1 to tailor the VR content to individual users' inclinations and characteristics.

4 Layer 2: Content Customization

The content customization layer is a crucial component of the conceptual framework for personalized and gamified virtual reality applications. Building upon the user profiles created in Layer 1, this layer focuses on adapting the VR content to align with the preferences, interests, and characteristics of individual users. By tailoring the content to meet the unique needs of each user, the content customization layer enhances immersion, engagement, and overall user satisfaction within the virtual reality environment.

4.1 Dynamic Content Generation

The content customization layer involves generating dynamic content as a crucial aspect. This means that virtual elements are created or changed in real-time based on the user's profile and interactions. This can include modifying visual assets, audio cues, or narrative arcs to align with the user's preferences, which results in a more individualized and captivating experience.

4.2 Adaptive Narratives

Another crucial element of the content customization layer is the incorporation of adaptive narratives. By considering the user's profile and actions, the VR application can modify the narrative structure [11], character interactions, or branching paths dynamically. This results in a distinct storytelling experience for each user that aligns with their interests and preferences. Such an approach fosters emotional engagement and investment in the virtual environment.

4.3 Personalized Aesthetics

The content customization layer also focuses on tailoring the aesthetics [12] of the virtual environment to match the user's preferences. This can involve tweaking features like lighting, texture, color, and overall visual style. By aligning the visual aesthetics with the user's preferences, the VR experience becomes more captivating and engaging, creating a greater sense of immersion and pleasure.

4.4 Contextual Relevance

The customization of content layer also considers the contextual relevance of the VR experience. This involves taking into account factors such as the user's location, time of day, or specific interests to provide content that is relevant to the user's immediate environment and situation. This customization based on context helps the user to feel more connected and relevant to the virtual world, making the experience more engaging and meaningful.

4.5 User-Controlled Customization

The content customization layer uses user profiles to customize the VR experience, but it is also important to allow users to have some control over the customization process. Offering options for users to modify certain content elements, such as difficulty levels, character appearances, or thematic preferences, gives them more control over their experience. This user-controlled customization fosters a sense of empowerment and ownership, increasing satisfaction and enjoyment within the virtual reality application.

The content customization layer aims to provide personalized experiences within the virtual reality application by incorporating dynamic content creation, adaptive storytelling, personalized aesthetics, contextual relevance, and user-controlled customization. By utilizing these features, this layer creates immersive and engaging experiences that align with each user's unique preferences and characteristics, promoting a deeper sense of connection and enjoyment within the virtual environment.

In the following section, we will explore Layer 3: Interaction and Gameplay, which introduces gamification elements into the VR experience to further enhance user motivation and engagement.

5 Layer 3: Interaction and Gameplay

The third layer of the conceptual framework for creating personalized and gamified virtual reality applications concentrates on interaction design and gameplay mechanics. This layer endeavors to elevate user motivation, engagement, and overall enjoyment by introducing gamification components into the VR experience. By utilizing the personalized aspects of the previous layers, such as user profiles and customized content, this layer generates interactive and immersive gameplay experiences within the virtual environment.

5.1 Interaction Design

The interaction design plays a vital role in designing the virtual reality application for intuitive and engaging user experiences. It involves creating user interfaces, input mechanisms, and control schemes that are specifically tailored for VR interaction. The primary objective is to offer users smooth and natural ways to interact with the virtual environment, allowing them to explore, manipulate objects [13], and navigate the virtual space effortlessly.

5.2 Gamification Elements

The interaction and gameplay layer employs different gamification elements within the VR experience to boost user motivation and engagement [14]. These elements may consist of challenges, rewards, achievements, leaderboards, and progression systems. The challenges can be designed to match the user's preferences and skill levels, creating a sense of purpose and fulfillment. Rewards and achievements act as incentives, reinforcing positive behaviors, whereas leaderboards encourage competition and social interaction. Progression systems enable users to monitor their progress and unlock new content or features, providing a sense of growth and continuous engagement.

5.3 Personalized Gameplay Mechanics

In Layer 3 of the conceptual framework for personalized and gamified virtual reality applications, gamification elements are customized based on the user's profile and preferences. The difficulty level of challenges can be automatically adjusted to correspond with the user's skill level, preventing any possible frustration or monotony. Furthermore, the types of rewards and achievements can be tailored to align with the user's interests, resulting in a more significant and engaging gameplay experience.

5.4 Immersive Feedback

In the interaction and gameplay layer, providing feedback is essential to the user's experience. Virtual reality offers various ways to provide immersive and multi-modal feedback to users, such as visual cues, haptic feedback, spatial audio, or physiological responses through biofeedback devices. Delivering real-time feedback helps to increase user engagement and provides users with a sense of achievement by understanding their performance within the virtual environment, which guides their actions.

5.5 User-Centered Social Interactions

Layer 3 of the conceptual framework for personalized and gamified virtual reality applications prioritizes interaction design and gameplay mechanics. It aims to increase user motivation and engagement by incorporating various gamification elements into the VR experience, which are personalized based on the user's profile and preferences. These elements can include challenges, rewards, achievements, leaderboards, and progression systems.

The difficulty level of challenges can be adjusted to match the user's skill level, while the types of rewards and achievements can be tailored to resonate with the user's interests. Feedback mechanisms, such as visual cues, haptic feedback, spatial audio, and biofeedback devices, are also essential for enhancing user engagement and guiding their actions.

Additionally, social interactions within the virtual environment, such as multiplayer modes and collaborative challenges, can foster a sense of community and social connection. All these elements work together to create immersive, challenging, and socially interactive experiences within the virtual reality application. In the next section, Layer 4 will be explored, which focuses on providing real-time feedback to users and tracking their progress within the VR application.

6 Layer 4: Feedback and Progression

Layer 4 of the conceptual framework for personalized and gamified virtual reality applications aims to deliver real-time feedback to users and monitor their progress within the VR environment. This layer is vital for boosting user motivation, engagement, and contentment by offering valuable feedback, gauging performance, and fostering a sense of advancement and success.

6.1 Real-Time Feedback Mechanisms

The VR application's real-time feedback mechanisms offer users instant information regarding their actions, performance, and results. Such feedback can take various forms, such as visual cues, auditory prompts, haptic feedback, and even adaptive responses within the virtual environment. Such feedback helps users comprehend the effects of their decisions, reinforces favorable behavior, and guides their actions, leading to a more immersive and engaging experience.

6.2 Performance Metrics and Evaluation

The fourth layer of the conceptual framework for personalized and gamified virtual reality applications involves utilizing performance metrics and evaluation methods to measure user performance in the virtual environment. Depending on the nature of the application, these metrics may include completion time, accuracy, scores, or other relevant measures. By quantifying user performance, the virtual reality experience becomes more focused on achieving goals and competitive, encouraging users to enhance their skills and strive for better performance.

6.3 Adaptive Difficulty and Challenge

The feedback and progression layer implements adaptive difficulty and challenge mechanisms to maintain an ideal level of challenge and user engagement. These mechanisms can adjust the difficulty [15] of the VR experience according to the user's performance and skill level. By providing challenges that are neither too easy nor too difficult, users are motivated to reach their potential and feel a sense of achievement, keeping them engaged and motivated.

6.4 Progress Visualization

In Layer 4, it is crucial to visualize user progress. Progress visualization mechanisms are employed to represent user advancement, milestones, and achievements within the VR application. Progress bars, level indicators, or achievement badges can be utilized to provide a clear visual representation of user progress and encourage further engagement. Such progress visualization enhances the sense of achievement and purpose and helps users stay focused and motivated within the VR experience.

6.5 Goal Setting and Personalized Objectives

In Layer 4, users can establish personalized goals and objectives within the virtual reality application. These objectives can be customized based on the user's interests, preferences, or specific hurdles they aim to conquer. This capability allows users to tailor their VR experience and establish a more meaningful and engaging connection, instilling a greater sense of ownership and motivation.

6.6 Continuous Progression and Unlockable

A system of continual advancement is featured in the fourth layer of the conceptual framework for personalized and gamified VR applications. This method provides customers with a continual sensation of progress as they go through the VR experience. Users can unlock additional features, levels, or content when they reach milestones, giving them a sense of success and motivating them to continue using the service [16]. The adoption of such technology also improves the user's immersion and enjoyment of the VR experience.

Layer 4 of the conceptual framework for personalized and gamified VR applications enhances the user experience by including a variety of aspects such as adaptable difficulty, real-time feedback, progress visualization, goal setting, performance metrics, and ongoing progression. These methods assist in motivating users, establishing a sense of accomplishment, and encouraging ongoing engagement, hence increasing user immersion and involvement in the personalized and gamified VR experience.

The next section will examine Layer 5, which concentrates on integrating social interaction and community building features into the virtual reality application.

7 Layer 5: Social Integration

Layer 5 of the conceptual framework for personalized and gamified virtual reality applications focuses on social integration, which involves incorporating social interaction and community building features into the virtual reality application. The goal is to foster social connections, encourage collaboration, and create shared experiences that enhance immersion, engagement, and overall enjoyment for users in the virtual environment.

7.1 Multiplayer Modes

The incorporation of multiplayer modes is a crucial aspect of social integration within the virtual reality application, as emphasized in Layer 5 of the framework. Multiplayer modes (allow users to engage in real-time with other users within the virtual environment, either collaboratively or competitively [17]. Team challenges, collaborative gameplay, or multiplayer competitions are among the ways in which social interaction can be fostered, creating opportunities for shared experiences and building mutual support.

7.2 Shared Experiences

Layer 5 of the conceptual framework for personalized and gamified virtual reality applications promotes the development of collective experiences within the virtual environment. This can be accomplished through coordinated activities, like group missions or events, where multiple users can take part simultaneously. By enabling joint experiences, users can participate in shared activities, communicate with each other, and interact, promoting social bonding and creating a feeling of community.

7.3 Social Challenges and Leaderboards

In order to encourage social interaction and friendly competition, Layer 5 suggests integrating social challenges and leaderboards into the virtual reality application. Users can engage in challenges together, strive for high scores, or compete against one another. Leaderboards display users' achievements and rankings, promoting competition and motivating users to improve their performance while fostering social interaction and engagement [18].

7.4 Communication Features

Layer 5 of the conceptual framework for personalized and gamified virtual reality applications includes communication features that allow users to interact with each other within the virtual reality application. These features may include voice chat, text chat, or gesture-based communication systems. The ability to communicate in real-time enhances social connections, facilitates coordination in multiplayer activities, and allows users to share experiences and collaborate within the virtual environment.

7.5 Virtual Social Spaces

Layer 5 of the conceptual framework for personalized and gamified virtual reality applications also involves building virtual social spaces within the virtual reality application. These spaces serve as designated areas where users can meet, socialize, and participate in non-gaming activities [19]. Virtual social spaces can be customized to simulate real-world locations or provide unique gathering areas with virtual events and entertainment, encouraging users to connect with each other on a social level and fostering a sense of community within the VR environment.

7.6 User-Generated Content and Sharing

Layer 5 aims to improve social integration by introducing features that enable users to create and share their own virtual environments, challenges, or experiences. User-generated content fosters collaboration, creativity, and community engagement, as users can explore and interact with content created by other users. This promotes a sense of ownership and participation within the virtual reality application, thus enhancing the overall user experience.

Layer 5 of the conceptual framework emphasizes social integration in the virtual reality application through multiplayer modes, shared experiences, social challenges, communication features, virtual social spaces, and user-generated content. By promoting social connections, collaboration, and a sense of community, this layer enhances the user experience and increases immersion, engagement, and enjoyment (Fig. 1).

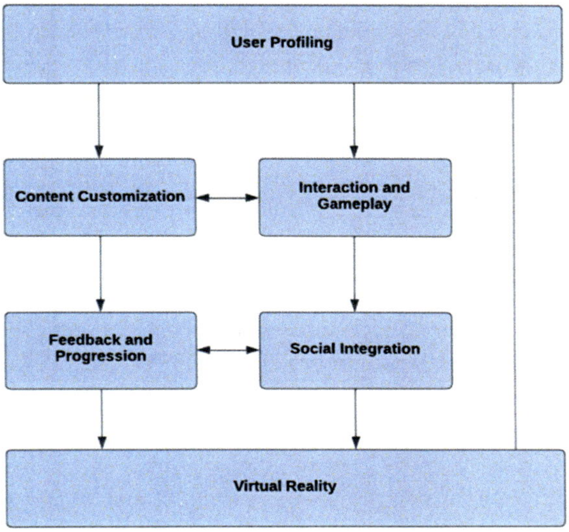

Fig. 1. Interconnected framework layers

This diagram portrays the layers as interconnected layers, highlighting their intricate relationships. The design elements and node arrangement enhance the diagram's visual appeal.

- User Profiling is the foundation, linking to both Content Customization and Interaction and Gameplay layers.
- Content Customization and Interaction and Gameplay layers are interdependent, indicating their role in developing personalized and engaging experiences.
- Feedback and Progression and Social Integration layers connect to both Interaction and Gameplay and the overarching Virtual Reality layer, indicating their impact on the user experience.
- The Virtual Reality layer represents the culmination of all interconnected layers, showcasing their collective influence on the immersive virtual reality experience.

Please note that this diagram's appearance and layout can still be tailored to suit your particular requirements and preferences.

8 Conclusion

This paper proposes a framework for designing personalized and gamified virtual reality (VR) applications that enhance user engagement and satisfaction. The framework consists of five layers: User Profiling, Content Customization, Interaction and Gameplay, Feedback and Progression, and Social Integration. Each layer plays a significant role in creating a personalized and engaging VR experience. By implementing this framework, designers and developers can create VR applications that adapt to user preferences [20], engage users through game-like mechanics, and promote social interaction. This can have significant implications across various domains. While the framework provides a comprehensive approach, its implementation may vary based on specific contexts and objectives. Personalized and gamified VR applications can have vast benefits, including enhancing learning outcomes, increasing motivation, and facilitating skill development. The proposed framework offers valuable insights and guidelines for designing impactful and enjoyable virtual experiences as VR technology continues to evolve and become more accessible [21].

References

1. Capasa, L., Zulauf, K., Wagner, R.: Virtual reality experience of mega sports events: a technology acceptance study. J. Theor. Appl. Electron. Commer. Res. **17**, 686–703 (2022). https://doi.org/10.3390/jtaer17020036
2. Winkler, I., Henriques, A.: The advancement of virtual reality in automotive market research: challenges and opportunities. Applied Sci. **11**(1) (2021). https://doi.org/10.3390/app112411610
3. Hu-Au, E., Lee, J.: Virtual reality in education: a tool for learning in the experience age. Int. J. Innov. Educ. **4**(4), 215226 (2017). https://doi.org/10.1504/IJIIE.2017.091481
4. Xie, B., et al.: A Review on Virtual Reality Skill Training Applications **2** (2021). https://doi.org/10.3389/frvir.2021.645153
5. Halbig, A., Babu, S.K., Gatter, S., Latoschik, M.E., Brukamp, K., von Mammen, S.: Opportunities and challenges of virtual reality in healthcare – a domain experts inquiry. Front. Virtual Real. **3**, 837616 (2022). https://doi.org/10.3389/frvir.2022.837616
6. Papakostas, C., Troussas, C., Krouska, A., Sgouropoulou, C.: Personalization of the learning path within an augmented reality spatial ability training application based on fuzzy weights. Sensors **22**(18), 7059 (2022). https://doi.org/10.3390/s22187059
7. Checa, D., Gatto, C., Cisternino, D., De Paolis, L., Bustillo, A.: A Framework for Educational and Training Immersive Virtual Reality Experiences (2020). https://doi.org/10.1007/978-3-030-58468-9_17
8. Hansen, A.S., Andersen, K., Sievert, B., Kiswani, J., Dascalu, S.M., Harris, F.C.. Let's VR: A Multiplayer Framework for Virtual Reality (2018)
9. Papakostas, C., Troussas, C., Krouska, A., Sgouropoulou, C.: Modeling the Knowledge of Users in an Augmented Reality-Based Learning Environment Using Fuzzy Logic (2022). https://doi.org/10.1007/978-3-031-17601-2_12

10. Steele, P., Burleigh, C., Kroposki, M., Magabo, M., Bailey, L.: Ethical considerations in designing virtual and augmented reality products—virtual and augmented reality design with students in mind: designers' perceptions. J. Educ. Technol. Syst. **49**, 004723952093385 (2020). https://doi.org/10.1177/0047239520933858
11. Vallance, M., Towndrow, P.A.: Perspective: narrative storyliving in virtual reality design. Front. Virtual Real. **3**, 779148 (2022). https://doi.org/10.3389/frvir.2022.779148
12. Gulhan, D., Durant, S., Zanker, J.M.: Aesthetic judgments of 3D Arts in Virtual Reality and Online Settings. Virtual Reality (2022). https://doi.org/10.1007/s10055-022-00671-1
13. Papakostas, C., Troussas, C., Krouska, A., Sgouropoulou, C.: User acceptance of augmented reality welding simulator in engineering training. Educ. Inf. Technol., pp.1–27 (2021). https://doi.org/10.1007/s10639-020-10418-7
14. Toda, A.M., Klock, A.C.T., Oliveira, W., et al.: Analysing gamification elements in educational environments using an existing Gamification taxonomy. Smart Learn. Environ. **6**, 16 (2019). https://doi.org/10.1186/s40561-019-0106-1
15. Lemmens, J., Münchhausen, C.: Let the beat flow: how game difficulty in virtual reality affects flow. Acta Psychologica **232**, 103812 (2023). https://doi.org/10.1016/j.actpsy.2022.103812
16. Neupane, A., Hansen, D., Fails, J.A., Sharma, A.: The role of steps and game elements in gamified fitness tracker apps: a systematic review. Multimodal Technologies and Interaction **5**(2), 5 (2021). https://doi.org/10.3390/mti5020005
17. Rychkova, A., Korotkikh, A., Mironov, A., Smolin, A., Maksimenko, N., Kurushkin, M.: Orbital battleship: a multiplayer guessing game in immersive virtual reality. J. Chem. Educ. **97**, 4184–4188 (2020). https://doi.org/10.1021/acs.jchemed.0c00866
18. Balci, S., Secaur, J.M., Morris, B.J.: Comparing the effectiveness of badges and leaderboards on academic performance and motivation of students in fully versus partially gamified online physics classes. Educ. Inf. Technol. **27**, 8669–8704 (2022). https://doi.org/10.1007/s10639-022-10983-z
19. Scavarelli, A., Arya, A., Teather, R.J.: Virtual reality and augmented reality in social learning spaces: a literature review. Virtual Reality **25**, 257–277 (2021). https://doi.org/10.1007/s10055-020-00444-8
20. Marougkas, A., Troussas, C., Krouska, A., Sgouropoulou, C.: A framework for personalized fully immersive virtual reality learning environments with gamified design in education. In: 1st International Conference on Novelties in Intelligent Digital Systems **338**, 95–104 (2021). https://doi.org/10.3233/FAIA210080
21. Marougkas, A., Troussas, C., Krouska, A., Sgouropoulou, C.: Virtual reality in education: reviewing different technological approaches and their implementations. In: Krouska, A., Troussas, C., Caro, J. (eds) Novel & Intelligent Digital Systems: Proceedings of the 2nd International Conference (NiDS 2022). NiDS 2022. Lecture Notes in Networks and Systems, **556** (2023). Springer, Cham. https://doi.org/10.1007/978-3-031-17601-2_8

Augmented Reality for Enhancing Linguistic Skills of International Students in Preschool Education

Andrianthi Kapetanaki, Akrivi Krouska(✉), Christos Troussas, Stavroula Drosou, and Cleo Sgouropoulou

Department of Informatics and Computer Engineering, University of West Attica, Egaleo, Greece
akrouska@uniwa.gr

Abstract. The utilization of new technologies in education is essential for providing an effective and of high quality learning environment. The preschool education is considered quite demanding, especially with the increasing number of multilingual and multi-ethnic families in European countries. This fact leads to constantly find new teaching methods and ways of learning that deviate from traditional ones, in order to promote the active learning and innovative teaching. As such, this study presents the development of an Augmented Reality (AR) application, using the Quiver tool, to enrich vocabulary acquisition of international students who attend Greek Kindergartens and are taught Greek as a second / foreign language. For the needs of this study, qualitative research method – and specifically, case study and observation – was used to investigate the impact of the developed AR application in oral communication of foreign preschool students. The results show that the use of AR has a positive impact in vocabulary and linguistic skills acquisition, fluency in oral communication, motivation, commitment to the learning process and satisfaction.

Keywords: Augmented Reality · Bilingual Education · Linguistic skills · Metalinguistic awareness · Preschool Education

1 Introduction

The use of Information and Communication Technologies (ICT) in teaching and learning has increased substantially over the last few years in most developed countries [1]. It is an indisputable fact that the incorporation of technological advances has altered the way with which learners gain knowledge, as digital learning environments can be better tailored to learners' personal and cognitive interests and preferences [2].

Early childhood is a critical period in a child's social, emotional, linguistic and cognitive development, and preschool education serves as the first transitional step from home to the wider social environment and socialization of the child [3]. The education of preschool children is considered quite demanding, since it requires constant alertness and ingenuity from each instructor. Nevertheless, as a result of global migration processes

K. Kabassi et al. (Eds.): NiDS 2023, LNNS 783, pp. 242–249, 2023.
https://doi.org/10.1007/978-3-031-44097-7_26

and the increasing number of multilingual and multi-ethnic families, many children in the twenty-first century are bilingual or even multilingual [3].

As technology has invaded our lives, its exploitation in preschool education is considered crucial for developing effective and meaningful learning environments. Alternative learning experiences focused on students who take an active role in learning process, impose the main challenge for instructors to adjust teaching to the demands of modern times. Augmented Reality (AR) is a rapidly emerging technology, which has been widely used in educational settings in recent years [4]. An AR system is a combination of real and virtual objects based in a real environment that interact and align with each other in real time [5]. Actually, several studies present that AR technology offers notable advantages when used in education [6–8].

To this direction, this paper investigates Augmented Reality (AR) use to international children who attend Greek Kindergartens and are taught Greek as a second / foreign language. For the needs of this study, Quiver, an Augmented Reality application, was used in order to explore its effectiveness in teaching vocabulary to preschool students who are taught Greek as a second language. By this research, the importance of using digital educational tools while teaching Greek as a second language was featured, as the utilization of Quiver application showed a positive effect on learning and on enhancing communicative and collaborating skills in a gamified educational context.

2 Related Work

Researchers have explored the positive effects of AR technology in education, as compared to traditional methods of teaching and learning [4]. However, only few studies of AR in early childhood have been published in the literature [9]. Studies focused on AR use in preschool education reveal that pupils obtained positive results in skills such as collaboration, communication, attention and active participation, as well as specific knowledge in particular subjects such as Mathematics, Arts, Sciences and Language Learning [10–14]. Moreover, both parents and teachers have evaluated AR use positively, as it enhances concentration and participation, providing adaptability to different leaning styles as well as students' satisfaction [10].

In [15], the authors, by using an AR picture book, explored how children and parents read the book through analysis of behavioural patterns and cognitive attainment. A total of 66 individuals (33 child–parent pairs) participated in this study via collaborative learning [15]. In [16], the authors developed educational magic toys (EMT) with AR technology. Virtual objects, such as story animations, 3D objects and flash animations, appeared on the toys. EMT included puzzles, flash cards and match cards to teach animals, fruits, vegetables, vehicles, objects, professions, colors, numbers and shapes for pupils in preschool education via collaborative and interactive learning. The sample consisted of 30 teachers and 33 children aged 5–6 years [16]. In [14], the authors designed an AR featured educational game addressed to 4–6 years old preschool children. The main objective of the game was to teach children about color mix, mathematics and 2D-3D geometrical shapes. In [13], the authors developed a didactic unit enriched with AR contents about the topic "animals" via constructive learning approach. The sample consisted of 36 children between 4–5 years old divided in 2 groups (experimental and control group) [13].

Furthermore, AR – as interactive material – can be a valid instrument for inclusive education. The development of educational technology combined with constructivism learning theory enables instructors to design learning environments adapted to meet the specific educational, social or cultural needs of the students in multicultural classes. Technology in the context of constructivism can offer to international students the opportunity to be active, autonomous and creative in a multicultural environment demanding cooperation between the participants [17].

3 System Overview

In this study, Quiver, an AR tool, was used for preschool vocabulary learning. Quiver application offers a variety of images, helping children paint and visualize their images in 3D. For this study, printed pages were distributed to the students and then they were asked to paint in the colors they wanted. These images were brought to life on a smartphone/tablet screen by looking at them through the app's viewer, enabling students to focus by arousing their interest and providing them an immediate feeling of satisfaction. Quiver application offers the opportunity to individuals to express their artistic creativity and interact with their own creation. The requirements for this AR system are an Internet connection, a computer and a smartphone/tablet, as well as the Quiver application accessible by https://quivervision.com/.

For the needs of this study, a learning scenario was designed to cultivate oral communication skills. Namely, the learning scenario included oral communication activities and was organized in the context of structural model according to which knowledge and language are gradually structured in a supportive learning environment. These activities concerned four subject areas, namely seasons, shapes, animals and colours, and were developed using digital tools, in accordance with curriculum and learning objectives of preschool education.

Initially, a testing implementation for a week took place in order to acquaint pupils with Quiver application and its use. During the learning scenario, the oral communication and AR activities were shaped according to specific learning techniques and articulate objectives in order its pedagogical impact to be assessed.

For the subject area of four seasons, the objectives of this activity were to understand circularity of time, to recognize the seasons according their characteristics, to verbally describe their characteristics and to enrich their vocabulary (Fig. 1).

For the subject area of shapes and solid figures, it was aimed to recognize and classify the basic shapes and solid figures, to verbally describe them, to construct them and to enrich their vocabulary (Fig. 2).

For the subject area of animals, the objectives were to recognize and name the animals, verbally describe them, widen their knowledge about organisms and enrich their vocabulary (Fig. 3).

For the subject area of colours, it was aimed to recognize colours and their blends, to classify objects according to their colour, to verbally describe colours and to enrich their vocabulary (Fig. 4).

During each activity, pupils either individually or in teams of two had to draw printed pages from Quiver application, and then they had to create a story inspired from the drawing and present it to their classmates.

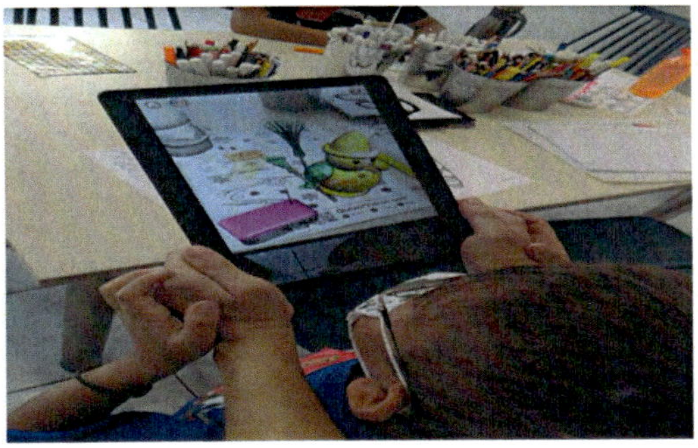

Fig. 1. AR Snowman

Fig. 2. AR Shape

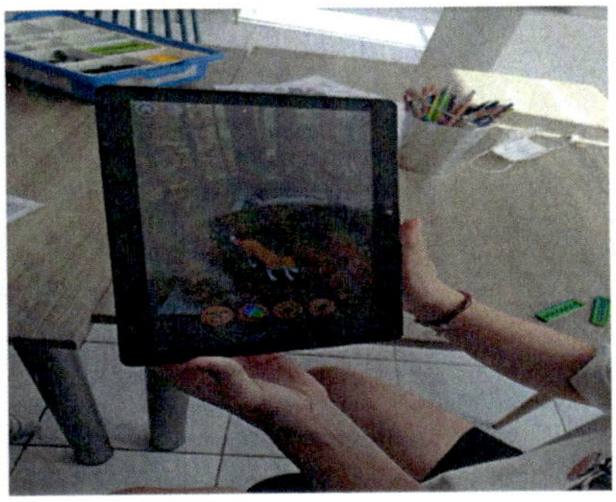

Fig. 3. AR Fox

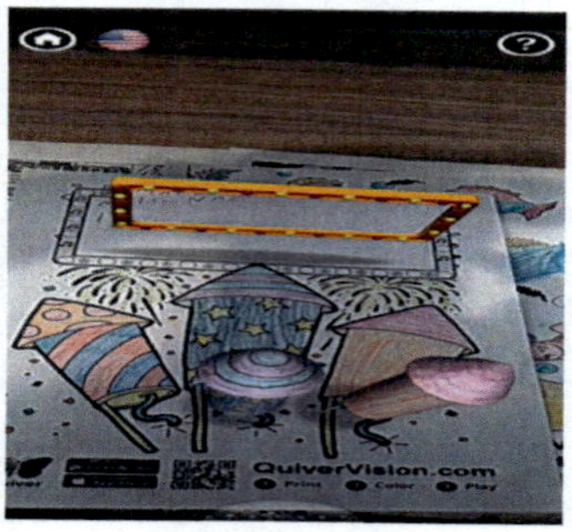

Fig. 4. AR Colours

4 Evaluation

The object of this study was to explore the impact of AR application, using Quiver tool, on vocabulary enhancement of preschool pupils who attend Greek Kindergartens and are taught Greek as a second/foreign language. In particular, the following research questions were addressed:

RQ1: Does the use of AR in teaching Greek vocabulary to bilingual preschool students promote vocabulary enhancement?
RQ2: Does the use of AR enhance motivation and active participation of bilingual preschool students?
RQ3: Does the use of AR improve metalinguistic awareness?

The sample group was selected with sampling method of non-probability. This sampling method could be characterized as convenience sampling, since the participants were selected only because they were conveniently available [18]. The research was conducted in the Athens (Greece), and namely in two private kindergartens. In total, 8 pupils, 5 boys and 3 girls, aged 5 years old, who were taught Greek as a second language, took part in the study.

For the needs of this study, qualitative research method – and specifically case study and observation – was used to investigate the impact of AR application in oral communication of international preschool pupils. All the process was recorded by video camera and students' interactions were observed and noted.

For the assessment of the learning scenario, Kirkpatrick Evaluation Model was applied. The Kirkpatrick Model is a widely recognised model for analysing and evaluating the results of training and educational programs. It takes into account any style of training, both informal and formal, to determine attitude based on four levels criteria: reaction, learning, behaviour, results [19]. Reaction criterion measures how participants react to the training. The objective of this level is to evaluate how individuals react to the training model by asking questions that establish the trainees' thoughts, so as to figure out if the participants enjoyed their experience and if they found the material in the program useful for their work. This evaluation criterion is applied orally by interviews and questionnaires [19]. Learning criterion analyses if they truly understood the training. In this level, individuals take the test or evaluation prior to the training (pre-test) and following training (post-test) to figure out the effect of the intervention on participant's learning [19]. Behaviour criterion examines the differences in the participant's behaviour after completing the program. For the assessment of this level, observation before and after learning process are necessary to evaluate significant change [19]. Results criterion determines the overall success of the training model. Actually, evaluation of the program takes place in terms of intended results such as eager response to activities, quicker task completion, attitude change towards the lesson, cooperation with peers, interest attraction and satisfaction boost.

In view of the above, four criteria were implemented for the assessment of the AR activities created for the learning scenario:

C1. Students enjoyed the content created by AR tool.
C2. Students knew and in the same time reproduced the material used via Augmented Reality.
C3. Students recognised and named the worksheet beyond the use of the tool.
C4. Students used the vocabulary obtained by these activities in spontaneous speech.

Table 1 illustrates the evaluation results for each learning scenario designed.

Table 1. Evaluation results.

Evaluation criteria	Learning Activities			
	Four seasons	Shapes	Animals	Colours
C1	Yes	Yes	Yes	Yes
C2	Yes	A few times	Yes	Yes
C3	Yes	A few times	Yes	Yes
C4	A few times	Yes	Yes	Yes

5 Discussion

The use of Information and Communication Technologies in preschool education is the main object of this study. In particular, the Quiver tool was chosen to be used to preschool education in order to examine whether it had a positive impact on international students' vocabulary acquisition.

According to the findings of this study, significant improvement of the Greek vocabulary in spontaneous oral communication of bilingual preschool students who participated in the study has been noticed. Concerning the research question about vocabulary enhancement, it was found that there had been a significant improvement of the participants' fluency. The positive response and impressions of the participants act upon the successful use of these tools in learning process. As for the research question of the motivation and active participation enhancement, it has been noticed that during the implementation of the scenario students enjoyed the activities demonstrating their interest, enthusiasm and commitment to the learning process. In respect of the research question of the improvement of the metalinguistic awareness, students used in oral communication bigger sentences with better articulation and more intense effort for memory recall of the appropriate vocabulary. Finally, none of the participants was confronted with technical or functional problems related to the tablet and the use of AR application which indicates students' familiarity with the tool as well as its suitability for the specific age group. The only disadvantage mentioned was tablet's weight as it was tiring for preschool students to hold it, so as to scan the cards.

To sum up, the exploitation of AR in Preschool Education can enhance learning process, providing an effective and motivating learning environment. The Quiver tool is a representative example of AR use and implementation in learning process. Preschool students can benefit from the use of this digital tool, as learning employs a playful and joyful way.

6 Conclusions and Future Work

The scope of this study was to examine the positive impact of AR in preschool education. It was concluded that using AR in learning activities of international preschool students improved participants' vocabulary, attracted their interest, enhanced motivation, enthusiasm and metalinguistic awareness.

On the other hand, few limitations in the use of AR in preschool education have been detected. This study is limited in a sample of international preschool students who attend Greek Kindergartens. The evaluation of skills acquired by the use of AR refers to the impact of the Quiver tool that used, in context of a well-structured learning scenario. Furthermore, the number of the participants in this study was quite small, and the implementation time of AR activities limited. Additionally, the fact that participants had the opportunity to be involved in the activities optionally in a daily basis for as long they were interested, in order to evaluate their motivation, does not allow us to generalize the results of this intervention program. Another limitation of this study is that the research method used was only qualitative, including case study and observation. Thus, a combination of qualitative and quantitative research method would be more efficient in order to achieve greater validity.

A major challenge for future work is the design and implementation of a long-term research, examining the same research questions in a bigger and more heterogeneous sample consisted of international preschool students, including and students with special educational needs. In that way, validity of the results will be more assured. Finally, another challenge for future work is to design research with more complex activities for phonological awareness, so as to investigate further students' metalinguistic skills.

References

1. Fernández-Gutiérrez, M., Gimenez, G., Calero, J.: Is the use of ICT in education leading to higher student outcomes? Analysis from the Spanish autonomous communities. Comput. Educ. **157**, 103969 (2020)

2. Troussas, C., Krouska, A., Sgouropoulou, C.: Improving learner-computer interaction through intelligent learning material delivery using instructional design modeling. Entropy **23**(6), 668 (2021). https://doi.org/10.3390/e23060668
3. Schwartz, M., Palviainen, Å.: Twenty-first-century preschool bilingual education: facing advantages and challenges in cross-cultural contexts. Int. J. Biling. Educ. Biling. **19**(6), 603–613 (2016). https://doi.org/10.1080/13670050.2016.1184616
4. Papakostas, C., Troussas, C., Krouska, A., Sgouropoulou, C.: User acceptance of augmented reality welding simulator in engineering training. Educ. Inf. Technol. **27**(1), 791–817 (2022)
5. Azuma, R., Baillot, Y., Behringer, R., Feiner, S., Julier, S., MacInTyre, B.: Recent advances in augmented reality. IEEE Comput. Graphics Appl. **21**, 34–47 (2001). https://doi.org/10.1109/38.963459
6. Cheng, K.-H., Tsai, C.-C.: Affordances of augmented reality in science learning: suggestions for future research. J. Sci. Educ. Technol. **22**(4), 449–462 (2013)
7. Akçayır, M., Akçayır, G.: Advantages and challenges associated with augmented reality for education: a systematic review of the literature. Educ. Res. Rev. **20**, 1–11 (2017)
8. Kapetanaki, A., Krouska, A., Troussas, C., Sgouropoulou, C.: Exploiting augmented reality technology in special education: a systematic review. Computers **11**(10), 143 (2022). https://doi.org/10.3390/computers11100143
9. Chen, P., Liu, X., Cheng, W., Huang, R.: A review of using augmented reality in education from 2011 to 2016. In: Popescu, E., et al. (eds.), Innovations in Smart Learning, pp. 13–18. Springer, Singapore (2017)
10. Redondo, B., Cózar-Gutiérrez, R., González-Calero, J.A., Ruiz, R.S.: Integration of augmented reality in the teaching of English as a foreign language in early childhood education. Early Childhood Educ. J. **48**(2), 147–155 (2020)
11. Lee, H.S., Lee, J.W.: Mathematical education game based on augmented reality. In: Pan, Z., Zhang, X., El Rhalibi, A., Woo, W., Li, Y. (Eds.), Technologies for E-Learning and Digital Entertainment, pp. 442–450. Springer, Berlin (2008)
12. Huang, Y., Li, H., Fong, R.: Using augmented reality in early art education: a case study in Hong Kong kindergarten. Early Child Dev. Care **186**(6), 1–24 (2016)
13. Cascales, A., Laguna, I., Pérez-López, D., Perona, P., Contero, M.: An experience on natural sciences augmented reality contents for preschoolers. In: Shumaker, R. (ed.), Virtual, augmented and mixed reality. Systems and applications (VAMR 2013). Lecture notes in computer science, Vol. 8022. Springer, Berlin (2013). https://doi.org/10.1007/978-3-642-39420-1_12
14. Chen, Y., Zhou, D., Wang, Y., Yu, J.: Application of augmented reality for early childhood english teaching. In: 2017 International Symposium on Educational Technology (ISET), pp. 111–115 (2017). https://doi.org/10.1109/iset.2017.34
15. Cheng, K.H., Tsai, C.C.: Children and parents' reading of an augmented reality picture book: analyses of behavioral patterns and cognitive attainment. Computers & Educ. **72**, 302–312 (2014). https://goo.gl/zbVBC1
16. Yilmaz, R.M.: Educational magic toys developed with augmented reality technology for early childhood education. Comput. Hum. Behav. **54**, 240–248 (2016). https://doi.org/10.1016/j.chb.2015.07.040
17. Vosniadou, S.: Children, Schools and Computers. Gutenberg Publications, Athens (2006)
18. Cohen, L., Manion, L., Morrison, K.: Research Methods in Education (6th ed.). Routledge/Taylor & Francis Group, New York, NY, US (2007)
19. Kurt, S.: Kirkpatrick's Model: Four Levels of Learning Evaluation. Educational Technology (2018). 8 Feb 2023. https://educationaltechnology.net/kirkpatricmodel-four-levels-learning-evaluation/

Modeling and Prediction of Meteorological Parameters Using the Arima and LSTM Methods: Sivas Province Case

Aydin Ozan Cetintas(✉) and Halit Apaydin

Department of Agricultural Engineering, Faculty of Agriculture, Ankara University, Ankara 06110, Turkey
ozancetintas@hotmail.com, apaydin@ankara.edu.tr

Abstract. The modeling of meteorological parameters sheds light on the determination of agricultural water needs and dry periods. Predicting precipitation and, therefore, droughts provides various benefits such as increased crop yield and the prevention of harvest losses. The power of prediction plays a crucial role in achieving improved water management, enhanced crop yield, risk mitigation, economic stability, and sustainable agriculture. It is crucial for rural communities that make a living through agricultural production, to benefit from these advantages. From past to present, meteorological parameters have been estimated with many statistical models and machine learning methods. This study aims to determine the prediction success of the statistical method ARIMA and artificial neural network model LSTM by using meteorological data of Zara, Susehri, Ulas, Kangal, Gemerek and Divrigi districts stations of Sivas province, Turkey. The study was conducted using daily data collected over a period of 10 years from all districts. While R^2 ranged between 0.06–0.94 in the ARIMA models, R^2 success was between 0.63–0.96 in the LSTM models. According to the results obtained, LSTM layer with 8-16-32 neurons, epoch value in the range of 100–300, learning rate of $5e^{-4}$ and $146e^{-5}$, between $1e^{-2}$ –$1e^{-6}$ decay, Adam optimization, in which ReLu activation is used in each layer, in the estimation of meteorological parameters in the region. It has been determined that the LSTM method with batch size configuration in the range of 8–256 is the best alternative. Although the ARIMA model is a common model that has been used for many years, it is determined that the LSTM model are superior to the ARIMA model with the diversity and controllability of the variables, and more successful results can be obtained.

Keywords: ARIMA · Deep learning · LSTM · Meteorological parameters · Prediction · Forecasting

1 Introduction

Meteorological factors such as rainfall and temperature impact upon the productivity, quality and irrigation preferences of agricultural crops and production. These meteorological factors can produce various production problems such as drought, drainage problems, nutrient deficiency, ground water and salinity problems. Global warming, a

K. Kabassi et al. (Eds.): NiDS 2023, LNNS 783, pp. 250–261, 2023.
https://doi.org/10.1007/978-3-031-44097-7_27

subject of widespread debate today, causes the emergence of observable extreme dry or extreme wet periods.

Precipitation and the clean water sources that it provides are valuable for determining both the irrigation level and planting/harvesting date in plant cultivation. Successful prediction of precipitation levels allows for a reduction in water waste and irrigation costs, irrigation planning through predicting the dry period, and allowing for high yields by providing the water the plant needs. Taking precautions by knowing in advance the factors that adversely affect crop production and yield, such as instantaneous heavy rainfall or dry periods, which have been increasing in recent years, are factors that protect producers and products.

With the prediction of meteorological parameters in advance, the producer can sufficiently prepare for such problems and even prevent many problems before they occur. To be able to plan the production pattern, to determine the sowing-planting time correctly, to plan irrigation at the right time using the correct and method, to ensure that the producer obtains the product with the targeted quality and yield at the end of the production season are directly related to meteorological factors. For many years, scientists interested in agricultural production and the production sector have been receiving information about meteorological factors through the use of different forecasting methods and planning is based on these data.

With ever-improving computer capacities and software, various methods have been attempted to forecast meteorological parameters. Autoregressive Integrated Moving Average (ARIMA), one of the statistical models that have long been used, is still the traditional method used for forecasting. Artificial neural networks, however, is a method in which the idea and basic structure dates back to earlier years, but with the hardware improvement of computers has become in recent years has become more common. Long-Short Term Memory (LSTM), which is included in machine learning based deep learning, is frequently used in this field due to its multi-layer structure, the ability to deepen learning and the high controllability of variable diversity [1, 2].

The LSTM method is an effective method that has been used for forecasting purposes in various fields, especially in recent years. The ARIMA method has also been used for forecasting purposes for many years and gives successful results. Both methods are frequently used in time series such as meteorological data. Tektas [3] modelled daily temperature, air pressure and wind speed; Bari et al. [4] and Pasaribu et al. [5] modelled precipitation using ARIMA. Akbari Asanjan et al. [6], Duong et al. [7], Kaneko et al. [8] and Kumar et al. [9] modelled precipitation; Salman et al. [10] modelled air pressure, humidity and dew point temperature using the LSTM method. Li et al. [11] compared the results and models by estimating the relative humidity through the ARIMA and LSTM models using meteorological data from the Gansu region of China. They concluded that the ARIMA model is more precise than LSTM model in relative humidity prediction. Poornima and Pushpalatha [12] compared the performance of the LSTM, ARIMA and Holt-Winters methods using rainfall data measured between 1980 and 2014 in Hyderabad, and found the LSTM method to give the best precision. De Saa and Ranathunga [13] compared the performance of the ARIMA and LSTM methods for temperature data measured between 2006 and 2016 in Szeged, Hungary. They concluded that the LSTM performed better than ARIMA method. Sekertekin et al. [14] modelled air temperature, Ouma et al. [15] modelled precipitation and runoff, and Samad et al. [16] modelled precipitation, wind speed, temperature, pressure, humidity and wind direction with LSTM. Dubey et al. [17] modelled the effect of meteorological parameters such as humidity, temperature, cloudiness, wind speed on energy consumption using both LSTM and ARIMA and determined that LSTM was more successful. Wu et al. [18] modelled monthly precipitation in Linjiang, Changchun and Qian Gorlos regions of China using the ARIMA and LSTM methods. In all regions, the LSTM model gave more accurate results.

Obtaining correct estimation results of both methods is crucial in water resources management and agricultural production both in terms of increasing yield and taking early precautions against potential problems that may occur in production and, thereby, eliminating any problems.

Long short-term memory, which is a sub-branch of artificial intelligence methods that have emerged recently, provides accurate results with low error rate on time series, revealing the possibility of accurate predictions of meteorological parameters. In addition, the ARIMA method has also been successfully used in time series forecasting for a long time. For these reasons, this study aims to determine the success of the LSTM and ARIMA methods in the prediction of precipitation, temperature and humidity in the districts of Sivas province.

2 Material and Method

2.1 Material

In Sivas province, which has 17 districts and 1 233 villages, agriculture is the sector with the highest economic importance in the province. 66.5% of the working population is employed in the agricultural sector [19]. Sivas, which has a total of 1 106 085 ha of agricultural land, has 348 845 ha of irrigable land, but 196 573 ha of land is irrigated.

In Sivas, where various crops are grown due to the size of agricultural lands and the presence of agricultural basins with microclimate, triticale, oats, sainfoin, wheat, clover, safflower, sugar beet, potato, barley products are intensively produced. Sivas ranks second in triticale and oat cultivation in Turkey, third in wheat and sainfoin cultivation, sixth in alfalfa cultivation, and seventh in sugar beet and barley cultivation [20].

In this study, rainfall, humidity and temperature data of Zara, Gemerek, Divrigi, Ulas, Susehri and Kangal districts in Sivas province were used (Table 1).

In order to improve the training in LSTM models and to strengthen the input data, precipitation, humidity and temperature data were also included in the input set with the measurement of the day before (t-1). In Matlab, full decomposition of precipitation data was also performed with one-dimensional haar wavelet transform with the wavelet analyzer tool. It is expressed as decompose in the study.

2.2 Method

ARIMA model. One of the best known and most successful linear statistical models for time series forecasting is the ARIMA model. ARIMA is a time series forecasting

Table 1. Geographical information and average climate values of the stations

District	Altitude (m)	Latitude	Longitude	Average Precipitation (mm)	Average Humidity (%)	Average Temperature (°C)
Zara	1338	39.89	37.74	1.14	63.90	9.43
Susehri	1164	40.16	38.07	0.99	59.59	11.18
Ulas	1392	39.44	37.02	1.11	66.53	9.22
Kangal	1521	39.23	37.38	1.01	66.56	7.84
Gemerek	1182	39.18	36.08	0.82	62.83	10.36
Divrigi	1121	39.36	38.11	1.03	53.52	12.48

approach used to predict the future value of a variable by using its past values to find its future value [21, 22].

Time series data can be in two structures, stationary and non-stationary. Stationary time series do not have a time-dependent structure, but non-stationary series have structures also called seasonality. Therefore, the mean and variance of non-stationary data are not in a constant relationship over time.

The ARIMA model is used to express the autocorrelation that determines the relationship between successive values of the data over time and consists of three parts. Stationarity is sought in the ARIMA model consisting of autoregression (AR), difference degree (I), moving average (MA). In time series, the stationarity of the series is checked using the Dickey-Fuller Test. It is a desirable feature that the observation values occurring at equal time intervals consist of a discrete and stationary series. If stationarity is not available, it is aimed to provide stationarity by using the difference degree parameter (I) [23].

ARIMA models are defined as ARIMA (p,d,q).
p: Autoregression degree (Autoregression-AR)
d: Degree of difference (I)
q: Moving Average (Moving Average-MA)

Seasonality or trend formation is often observed especially in natural events and agricultural data. For this reason, they need a model with a degree of difference. The degree of difference usually has a value of '1' and at most '2'.

In non-cyclical models, ARIMA (p,d,q) can be expressed as $d = 0$ and becomes ARMA (p,q). It can also be used as AR (p,d) or MA (d,q) in case $p = 0$ or $q = 0$.

If the data series is non-stationary, it is subjected to the test again by taking the first degree difference. If the series becomes stationary after differencing, $d = 1$ is accepted. If the series is still non-stationary, differencing is applied at second order and continues until the difference becomes stationary. In very complex series, second order differencing may be necessary.

Long-short-term memory (LSTM) model. Artificial neural networks have been applied in various fields such as engineering, hydrology, mathematics, economics, medicine and meteorology for many years. In today's world, where terms such as machine learning and deep learning are becoming more and more important and integrated into life, advanced neural network models are also used in time series forecasting. Artificial neural networks are generally used in simpler conditions with relatively small numbers of data. For situations where the number of data increases and the connection between input and output is difficult to detect, the concept of Deep Learning has emerged. The version of deep learning applications in time series is called Recurrent Neural Network (RNN). The sub-methods of RNN are LSTM, Bidirectional LSTM (BiLSTM), Gated Recurrent Unit (GRU) and Simple Recurrent Network (simpleRNN). LSTM is the most common and well-known model among these methods. Learning the relationship between data, the ability to generalise, the freedom to work with an unlimited amount of variables are the biggest factors in the popularity of these models.

There are 3 gates in an LSTM. The gate that controls the level at which a new value flows in the memory is called the input gate, the gate that calculates the degree of retention of the new value in the memory is called the forget gate, and the gate that calculates the use of the value in memory to calculate the output activation of the block is called the output gate. Despite the artificial neural network structure, which is limited in training and predicting data in time series, the detection of cells that need to be remembered or forgotten in order to overcome this limit is provided through these gates.

An LSTM model can be implemented by using one or more layers in combination. The layer used first should be chosen as the LSTM layer. Subsequent layers act as short-term memories of the information carried by the previous layer. In models where multiple layers can be used, having more layers does not mean that the model will learn better. The use of layers should be at optimum values according to the results given by the model.

In the LSTM method in the study, dense, dropout and batch normalisation layers were tried and used together with LSTM layers.

In order to make the network more stable and regular, batch normalisation layer was used as an intermediate layer to work together with LSTM layers to provide simultaneous learning and to reduce the covariance between hidden layers.

Dense layers were used to provide transition between layers. Different activation functions were tried in LSTM and dense layers, and ReLU was used as the activation function which was found to be successful.

In order to compile the models, the compile method was used to observe the training and performance by taking different parameters. In the study, Adam for optimisation, MSE for loss function and accuracy as a metric were used as model compile parameters.

After compile, the training of the model is done by fit method. In the fit method, the definition of training and test parts, batch size and epoch values are used. The batch size, which allows the training and test data to be trained in groups instead of being handled individually, was tested at values of 16–32-64–128-256 according to the compatibility of the models and the success of the network. In order to calculate how many times the datasets will be trained and the most appropriate weight value, epochs in the range of 10–1000 were tried according to the success of the network.

Different layers and various hyperparameters were tried in the study, and the most successful combinations are given in the LSTM result tables of the relevant district.

When updating the weights in neural networks and LSTM models, a gradient descent algorithm is used to try to minimise the error. Therefore, the derivative of the mean square error is used. Mean Squared Error (MSE), which is a loss function frequently used in regression problems, is used by averaging the squares of the differences between the predicted and actual values. The MSE loss function was used as a loss function in the LSTM model in this study.

In the LSTM method, ReLU (rectified linear unit- RELU) was used as the activation function. ReLU takes values in the range $(0, +\infty)$, negative input gives 0 and positive input gives unchanged output.

Adam optimisation was used for optimisation. Adam optimisation is an optimisation algorithm that provides first order gradient based optimisation and is used to update the weights iteratively. Adam optimisation is simple to implement, requires less memory, and is suitable for non-stationary, noisy and sparse data [24].

In the LSTM model, the training set for each model was divided into 70% training and 30% test data. The alternative parameters in Table 2 were applied to each district and meteorological parameter and the models with the most successful results were used. While the $5e^{-4}$ learning ratio gave the most successful results in precipitation models, the $146e^{-5}$ ratio was found to be successful in humidity models, and $1e^{-2}$, $1e^{-3}$ and $146e^{-5}$ gave the most successful results in temperature data and were used in the study results. In general, it was observed that $5e^{-4}$ and $146e^{-5}$ learning ratios were successful in the models. It was observed and applied that decay rates between $1e^{-2}$ and $1e^{-6}$ were used in the models, low LSTM layers such as 16–32 were sufficient for precipitation and temperature data, but especially humidity models needed higher layers. Intermediate layers were used in all models except for two models and Dense was used as an intermediate layer at different values in some models. Iteration was applied in the range of 50–300 in all models except for two models. Batch normalisation was required in some models, but not in others, and although it did not have a significant effect on the success rate, it prevented the overfit problem in some models. While the batch size was different for each model, it did not follow a clear order and different batch sizes were used in each group. In all models, the output layer was Dense 1, the activation function was ReLU and the loss function was Mean Squared Error.

In ARIMA models, while stationarity was observed in all precipitation and humidity data, non-stationary results were observed in temperature data, and for this reason, models were created by taking "1" difference degree, and models using seasonality were applied. In precipitation data, appropriate model parameters with high success could not be found and models were run according to the most successful results. In temperature and humidity data, models that can give more successful results were used. The results obtained as a result of the study are given in Tables 3, 4, 5, 6, 7 and 8 on the basis of districts.

Table 2. Alternative parameters used in the LSTM model

Parameter	Parameter Range
Learning Rate	$1e^{-1}$, $1e^{-2}$, $1e^{-3}$, $1e^{-4}$, $5e^{-4}$, $1e^{-5}$, $146e^{-5}$, $1e^{-6}$, $1e^{-7}$, $1e^{-8}$, $1e^{-9}$
Decay	$1e^{-1}$, $1e^{-2}$, $1e^{-3}$, $1e^{-4}$, $1e^{-5}$, $1e^{-6}$, $1e^{-7}$, $1e^{-8}$, $1e^{-9}$
Epoch	In the range 10–1000 in multiples of 10
LSTM Layer	In the range of 4–256 as multiples of 2
Activation Function	ReLU, SeLU, LeakyReLU, tanh, sigmoid
BiLSTM Layer	In the range of 4–128 as multiples of 2
Dropout	0.10, 0.20, 0.30, 0.40, 0.50
Dense Layer	4, 8, 16, 32, 64 as intermediate layer, 1 as output layer
Batch Size	In multiples of 2 in the range 4–512
Optimiser	Adam, Adadelta, SGD, RMSprop
Loss Function	MSE, MAE, MAPE, MSLE

Table 3. The best results of LSTM and ARIMA in Zara district

Zara	LSTM Prep.	ARIMA Prep.	LSTM Hum.	ARIMA Hum.	LSTM Temp.	ARIMA Temp.
Training MAE	0.34	-	6.54	-	1.59	-
Test MAE	0.38	1.63	6.56	6.82	1.65	2.18
Training RMSE	0.74	-	8.38	-	2.20	-
Test RMSE	0.89	3.16	8.37	8.90	2.34	2.91
Training R^2	0.97	-	0.68	-	0.94	-
Test R^2	0.96	0.11	0.65	0.60	0.92	0.87

Table 4. The best results of LSTM and ARIMA in Divrigi district

Divrigi	LSTM Prep.	ARIMA Prep.	LSTM Hum.	ARIMA Hum.	LSTM Temp.	ARIMA Temp.
Training MAE	0.44	-	5.55	-	1.43	-
Test MAE	0.40	0.92	6.20	7.05	1.41	1.69

(continued)

Table 4. (*continued*)

Divrigi	LSTM Prep.	ARIMA Prep.	LSTM Hum.	ARIMA Hum.	LSTM Temp.	ARIMA Temp.
Training RMSE	0.66	-	7.49	-	1.94	-
Test RMSE	0.54	1.87	7.96	9.28	1.89	2.22
Training R^2	0.97	-	0.82	-	0.96	-
Test R^2	0.95	0.13	0.82	0.75	0.95	0.94

Table 5. The best results of LSTM and ARIMA in Ulas district

Ulas	LSTM Prep.	ARIMA Prep.	LSTM Hum.	ARIMA Hum.	LSTM Temp.	ARIMA Temp.
Training MAE	0.34	-	6.78	-	1.64	-
Test MAE	0.26	1.02	6.98	7.30	1.69	1.99
Training RMSE	1.05	-	8.75	-	2.25	-
Test RMSE	0.81	2.38	8.89	9.70	2.35	2.66
Training R^2	0.92	-	0.76	-	0.94	-
Test R^2	0.91	0.14	0.70	0.62	0.93	0.90

Table 6. The best results of LSTM and ARIMA in Kangal district

Kangal	LSTM Prep.	ARIMA Prep.	LSTM Hum.	ARIMA Hum.	LSTM Temp.	ARIMA Temp.
Training MAE	0.44	-	4.74	-	1.65	-
Test MAE	0.39	0.84	5.89	6.34	1.70	1.75
Training RMSE	0.70	-	6.22	-	2.34	-
Test RMSE	0.63	2.23	7.61	8.72	2.48	2.54
Training R^2	0.94	-	0.84	-	0.94	-
Test R^2	0.93	0.06	0.80	0.73	0.93	0.92

Table 7. The best results of LSTM and ARIMA in Gemerek district

Gemerek	LSTM Prep.	ARIMA Prep.	LSTM Hum.	ARIMA Hum.	LSTM Temp.	ARIMA Temp.
Training MAE	0.44	-	5.41	-	1.49	-
Test MAE	0.43	0.95	5.60	5.90	1.56	1.61
Training RMSE	0.93	-	6.89	-	2.07	-
Test RMSE	0.87	2.11	7.23	7.90	2.17	2.21
Training R^2	0.88	-	0.80	-	0.95	-
Test R^2	0.86	0.13	0.77	0.72	0.94	0.93

Table 8. The best results of LSTM and ARIMA in Susehri district

Susehri	LSTM Prep.	ARIMA Prep.	LSTM Hum.	ARIMA Hum.	LSTM Temp.	ARIMA Temp.
Training MAE	0.28	-	7.11	-	1.53	-
Test MAE	0.27	0.92	7.49	7.99	1.56	1.64
Training RMSE	0.54	-	9.10	-	2.08	-
Test RMSE	0.50	2.54	9.33	10.39	2.05	2.22
Training R^2	0.97	-	0.65	-	0.94	-
Test R^2	0.96	0.09	0.63	0.53	0.94	0.93

3 Discussion and Conclusion

In this study, general information about the importance of the estimation of meteorological parameters, the important statistical method ARIMA and the deep learning method LSTM models used to estimate these parameters are given. The results obtained as a result of the applications made in the light of this information reveal the prediction success of both methods.

In the ARIMA method, different alternatives other than the ACF and PACF graphs used in the choice of AR and MA parameters were also tried. Dickey-Fuller test was used for stationarity test and stationarity was not detected in temperature data. As a result of the experiments and graphical summaries in ARIMA models, the (p,d,q) values that best represent the data were found and these values were included in the relevant station and parameter in the study. Especially the results found for humidity and temperature parameters in each district revealed that it is still a strong model in time series forecasting. The results of the precipitation parameter did not give successful results.

In the LSTM method, combinations of different layers, optimisation functions, activation functions, learning rates and iteration numbers were tried while searching for the most successful results for precipitation, humidity and temperature parameters. Although most of the experiments yielded successful results, the most successful values and results that were not subjected to overlearning among these combinations were included in the study. While searching for the most successful combination, the hyperparameters given in Table 2 were used and the most successful hyperparameters are given in the results part.

In deep learning methods, it has been observed that learning rates of $1e^{-3}$, $5e^{-4}$ and $146e^{-5}$ converge faster than other learning rates of networks. In this study, these values were also included while searching for the best combination, and it was observed and used in the results section that the best result was achieved with these values in most trials [25].

The results of the applications of both methods in humidity and temperature parameters were satisfactorily successful. However, all LSTM applications were found to be more successful than ARIMA applications. The most successful and close results of the two models in humidity and temperature parameters are 0.95 R^2 for LSTM and 0.94 R^2 for ARIMA in the temperature parameter of Divrigi district. This closeness, which is also seen in the temperature prediction results of Divrigi and Kangal districts, is replaced by larger differences in other districts and parameters. In the MAE and RMSE evaluation metrics, LSTM models were clearly more successful in all results.

In the case of precipitation data, ARIMA models using only precipitation data as input data were outside the acceptable success limits. In the LSTM model, the use of decomposed data together with the data belonging to different parameters enabled the LSTM model to obtain successful results for the precipitation parameter. With this feature, the LSTM model was found superior to the ARIMA model. Table 9 shows the minimum and maximum R^2 result ranges of both methods for all districts.

Table 9. R^2 ranges for the relevant parameters in ARIMA and LSTM models

Method and parameter	R^2 ranges
ARIMA Precipitation	0.06 – 0.14
LSTM Precipitation	0.86 – 0.96
ARIMA Relative humidity	0.53 – 0.75
LSTM Relative humidity	0.63 – 0.82
ARIMA Temperature	0.87 – 0.94
LSTM Temperature	0.92 – 0.95

The most successful results were found in LSTM applications in all evaluation metrics in all parameters of all districts of both models. The LSTM method is successful, consistent and open to improvement. While ARIMA method can only be tested with ARIMA parameters, LSTM method is preferred because it can be used with different models and layers mentioned before, it can get more successful results by using different data together and it is open to improvement. ARIMA seems to be a preferable method in terms of ease of use and accessibility.

In the light of the results obtained from the study, with the successful prediction of precipitation, humidity and temperature parameters by LSTM method, it can be used in terms of irrigation calendar planning, determination of sowing-planting time and arrangement in accordance with the predictions. In addition, the results of the study are expected to have a guiding and precautionary effect on the prediction of seasonal extreme situations and accordingly, to be prepared for agricultural problems such as possible disasters, unfavourable weather events in terms of cultivation and regional and periodic drought. In this respect, it is expected that the outputs of the study will have a positive effect on reducing yield and product losses in agricultural production and increasing agricultural income accordingly.

Acknowledgments. We would like to thank TSMS for meteorological data. The article was developed from the MSc thesis of Aydin Ozan CETINTAS.

References

1. Hochreiter, S., Schmidhuber, J.: Long short-term memory. Neural Comput. **9**(8), 1735–1780 (1997)
2. Gers, F.A., Schmidhuber, J., Cummins, F.: Learning to forget: continual prediction with LSTM. Neural Comput. **12**(10), 2451–2471 (2000)
3. Tektas, M.: Weather forecasting using ANFIS and ARIMA models. Environ. Res. Eng. Manag. **51**(1), 5–10 (2010)
4. Bari, S.H., Rahman, M.T., Hussain, M.M., Ray, S.: Forecasting monthly precipitation in Sylhet city using ARIMA model. Civil and Environ. Res. **7**(1), 69–77 (2015)
5. Pasaribu, Y.P., Fitrianti, H., Suryani, D.R.: Rainfall forecast of merauke using autoregressive integrated moving average model. In: E3S Web of Conferences (73), 12010. EDP Sciences (2018)
6. Akbari Asanjan, A., Yang, T., Hsu, K., Sorooshian, S., Lin, J., Peng, Q.: Short-term precipitation forecast based on the PERSIANN system and LSTM recurrent neural networks. J. Geophysical Res.: Atmospheres **123**(22), 12–543 (2018)
7. Duong, T.A., Bui, M.D., Rutschmann, P.: Long Short Term Memory for Monthly Rainfall Prediction in Camau (2018)
8. Kaneko, R., Nakayoshi, M., Onomura, S.: Rainfall prediction by a recurrent neural network algorithm LSTM learning surface observation data. In: AGU Fall Meeting Abstracts, Vol. 2019, pp. GC43D-1354 (2019)
9. Kumar, A., Islam, T., Sekimoto, Y., Mattmann, C., Wilson, B.: Convcast: an embedded convolutional LSTM based architecture for precipitation nowcasting using satellite data. PLoS ONE **15**(3), e0230114 (2020)
10. Salman, A.G., Heryadi, Y., Abdurahman, E., Suparta, W.: Single layer & multi-layer long short-term memory (LSTM) model with intermediate variables for weather forecasting. Procedia Computer Sci. **135**, 89–98 (2018)
11. Li, Z., Zou, H., Qi, B.: Application of ARIMA and LSTM in relative humidity prediction. In: 2019 IEEE 19th International Conference on Communication Technology (ICCT), pp. 1544–1549. IEEE (2019)
12. Poornima, S., Pushpalatha, M.: Prediction of rainfall using intensified LSTM based recurrent neural network with weighted linear units. Atmosphere **10**(11), 668 (2019)

13. De Saa, E., Ranathunga, L.: Comparison between ARIMA and Deep Learning Models for Temperature Forecasting. arXiv preprint arXiv:2011.04452 (2020)
14. Sekertekin, A., Bilgili, M., Arslan, N., Yildirim, A., Celebi, K., Ozbek, A.: Short-term air temperature prediction by adaptive neuro-fuzzy inference system (ANFIS) and long short-term memory (LSTM) network. Meteorology and Atmospheric Physics, pp. 1–17 (2021)
15. Ouma, Y.O., Cheruyot, R., Wachera, A.N.: Rainfall and runoff time-series trend analysis using LSTM recurrent neural network and wavelet neural network with satellite-based meteorological data: case study of Nzoia hydrologic basin. Complex & Intelligent Systems, pp. 1–24 (2021)
16. Samad, A., Gautam, V., Jain, P., Sarkar, K.: An approach for rainfall prediction using long short term memory neural network. In: 2020 IEEE 5th International Conference on Computing Communication and Automation (ICCCA), pp. 190–195. IEEE (2020)
17. Dubey, A.K., Kumar, A., García-Díaz, V., Sharma, A.K., Kanhaiya, K.: Study and analysis of SARIMA and LSTM in forecasting time series data. Sustainable Energy Technol. Assess. **47**, 101474 (2021)
18. Wu, X., et al.: The development of a hybrid wavelet-ARIMA-LSTM model for precipitation amounts and drought analysis. Atmosphere **12**(1), 74 (2021)
19. Anonymous. Sivas Provincial Directorate of Agriculture and Forestry Annual Report, Sivas (2018)
20. TUIK (2020). https://biruni.tuik.gov.tr/medas/?kn%3D95%26locale%3Dtr Access 20 Dec 2021
21. Box, G.E., Jenkins, G.M., Reinsel, G.C., Ljung, G.M.: Time Series Analysis: Forecasting and Control. John Wiley & Sons (2015)
22. Alabdulrazzaq, H., Alenezi, M.N., Rawajfih, Y., Alghannam, B.A., Al-Hassan, A.A., Al-Anzi, F.S.: On the accuracy of ARIMA based prediction of COVID-19 spread. Results in Physics **27**, 104509 (2021)
23. Dickey, D.A., Fuller, W.A.: Distribution of the estimators for autoregressive time series with a unit root. J. Am. Stat. Assoc. **74**, 427–431 (1979)
24. Kingma, D.P., Ba, J.: Adam: A Method for Stochastic Optimization. ArXiv preprint arXiv:1412.6980 (2014)
25. Jindal, H., Sardana, N., Mehta, R.: Analyzing performance of deep learning techniques for web navigation prediction. Procedia Computer Sci. **167**, 1739–1748 (2020)

A State of Art Review on Testing Open Multi-Agent Systems

Djaber Guassmi, Nour El Houda Dehimi(✉), and Makhlouf Derdour

Computer Science Department, LIAOA Laboratory, University of Oum El Bouaghi, Oum El Bouaghi, Algeria
{djaber.guassmi,dehimi.nourelhouda,derdour.makhlouf}@univ-oeb.dz

Abstract. Open Multi-Agent Systems (MASs) represent a paradigm in which multiple autonomous agents communicate and collaborate to achieve shared objectives in an open and dynamic environment. These systems require specific software engineering capabilities. In the realm of software validation, testing plays a crucial role, but it can be challenging to conceptualize and execute. Designing systems under the Open MAS paradigm poses even greater challenges due to characteristics such as autonomy, reactivity, pro-activity, and social abilities. This paper presents state-of-the-art research on testing Open MAS. The goal is to ultimately provide a new test method that addresses the unique characteristics of Open MAS, along with software and efficient test tools that facilitate the automation of testing activities for these systems.

Keywords: Open multi-Agent Systems · System Level Testing · Quality Assurance

1 Introduction

Multi-Agent Systems (MAS) have spread into more and more crucial application fields, including finance, transportation, utilities, aerospace, and even the military, where the risk of error can result in catastrophic failures. For this reason, there is interest in giving multi-agent system testing more weight.

As these systems grow in complexity, robust testing methodologies become imperative to ensure their reliability, stability, and overall quality. The activity of testing these systems represents an essential task in the process of its quality assurance. However, this activity remains an under researched area despite the rapid evolution of MAS due to the fact that only a few proposals are found in the literature.

System-level testing ensures that all agents in the system operate according to particular specifications and interact correctly. This level of testing has received little interest from researchers compared to the interest in earlier levels of testing [5]; this is related to the testing difficulty of this level. Indeed, at this level of testing, several problems arise. This is particularly visible in the opening of the MAS, which introduces new behaviors in the system [4]. Indeed, it is vital to maintain the test cases generated initially by integrating new test cases to cover the new behaviors. Therefore, our research objectives are

K. Kabassi et al. (Eds.): NiDS 2023, LNNS 783, pp. 262–266, 2023.
https://doi.org/10.1007/978-3-031-44097-7_28

to comprehensively review the state-of-the-art techniques and approaches in multi-agent system testing, highlighting key research contributions and trends.

In the following sections, we will provide an overview of the open MAS, a general explanation of the MAS test's levels, and some previous work on the MAS test.

2 Open Multi-Agent Systems

Both extensibility and scalability define the Open Multi-Agents system. Extensibility is the capacity to flexibly include and/or exclude agents from the system; in this context, we are referring to a totally open MAS. However, openness in terms of evolution allows for the augmentation of existing agents in the system with new services; in this situation, we are referring to an evolutionary MAS. The following are the proprieties of open MAS "Fig. 1" [9]:

- The agent joins or leaves the system regularly and randomly;
- The agent has multiple goals and strategies, which can probably be incompatible;
- The agent is usually self-interested;
- The acquisition of complete information about the agent's existence and resource's availability in the system is always expensive and sometimes unfeasible because the agents do not desire to reveal their complete information;
- The existence of a global scheduler to manage all system agents is as well unfeasible.

3 Levels of Testing Multi-Agent Systems

Unlike the test of other paradigms, which consists of three levels, the MAS test consists of five levels, namely:

1. Unit testing: Tests all units that make up an agent, including blocks of code, implementation of agent units like goals, plans, knowledge base, reasoning engine, rules specification, and ensures that they work as designed;
2. Agent Testing: Consists of testing [8]: (i) the integration of the different units, tested at the unit level, inside an agent and (ii) the ability of the agent to achieve its goals in its environment;
3. Integration or Group Testing: Tests the integration of an agent, tested at the agent level, with the other agents in the development phase; it consists of verifying that: (i) each agent in the company receives the right message from the right agent, provides the correct response and interacts correctly with the environment [8] and (ii) the objective of the society in which the agents interact is fully achieved, in this level of testing, the integration strategy depends on the architecture of the MAS or the dependencies between agents;
4. System-Level Testing involves ensuring that all system agents operate according to particular specifications.At this level, we focus on:
   - Testing of system operation in the environment;
   - Test the expected emergent and macroscopic properties of the system as a whole;

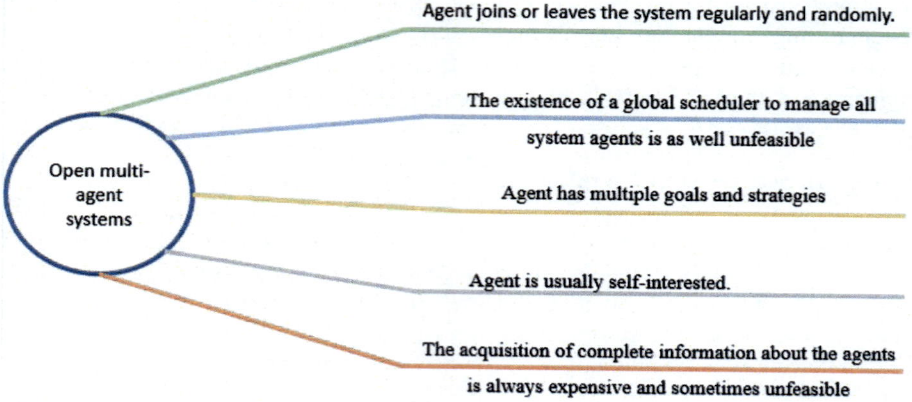

Fig. 1. Properties of open MAS.

- Test the quality properties that the intended system must reach, for instance, adaptation, openness, fault tolerance, and performance;
- At the system test level, it is also possible to verify the main types of agent interactions (cooperation, negotiation, and coordination) or to use other types of tests such as performance tests, conformity tests, and load and/or stress tests.

5. Acceptance Testing: Consists of testing the MAS in the customer's execution environment and verifying that it meets stakeholder goals with the participation of stakeholders.

4 Previous Work on MAS Testing

Only a few approaches have been proposed in the literature dealing with MAS testing in recent years. The following table "Table 1" indicates some Previous work of testing MAS:

The data in the table give an overview of the most recent work in this domain. They are classified according to their relation to the open MAS or the Ordinary MAS, as we notice, most of the testing studies do not target the open systems, and a few focus on open MAS, this gives the motivation to focus on testing this type of systems.

Table 1. Previous work on MAS Testing.

Work/Criteria	Related to ordinary MAS	Related to open MAS
[15] Savarimuthu S., Winikoff M(2013)	yes	no
[17] J. Thaler, P.O. Siebers(2019)	yes	no
[13] N. Nascimento et al.(2020)	yes	no
[14] Nguyen, C.D et al.(2012)	yes	no
[11] Lanus, E et al.(2021)	yes	no
[16] Tang, W., Wang, K (2021)	yes	no
[10] Huang Z et al.(2014)	yes	no
[7] Gonçalves, E.M.N et al.(2022)	yes	no
[3] Dehimi, NEH et al.(2015)	no	yes
[2] Dehimi,NEH, Mokhati,F(2019)	yes	no
[12] Mohajeri Parizi et al.(2021)	yes	no
[1] Dehimi,NEH et al.(2022)	yes	no
[6] Dehimi,NEH et al.(2022)	yes	no

5 Conclusion

Although MAS research is popular, there are few methods available for evaluating these systems in the literature. These studies have made significant contributions to the field by proposing new methods for MAS testing. However, they have not specifically addressed the challenges associated with testing open MAS. Therefore, there is ample scope for further investigation in this area. Promising prospects include the utilization of automated testing methods based on machine learning and artificial intelligence, as well as the establishment of standardized assessment measures for comparative research. Looking ahead, we envision presenting a new test method that effectively addresses the unique characteristics and challenges of testing open MAS in the short to medium term.

References

1. Dehimi, N.E.H., Benkhalef, A.H., Tolba, Z.: A novel mutation analysis-based approach for testing parallel behavioural scenarios in multi-agent systems. Electronics **11**(22), 3642 (2022)
2. Dehimi, N.E.H., Mokhati, F.: A novel test case generation approach based on auml sequence diagram. In: 2019 International Conference on Networking and Advanced Systems (ICNAS), pp. 1–4. IEEE (2019)
3. Dehimi, N.E.H., Mokhati, F., Badri, M.: Testing hmas-based applications: an aspecs-based approach. Eng. Appl. Artif. Intell. **46**, 232–257 (2015)
4. Derdour, M., Alti, A., Gasmi, M., Roose, P.: Security architecture metamodel for model driven security. J. Innov. Digital Ecosyst. **2**(1–2), 55–70 (2015)
5. Derdour, M., Zine, G., Roose, P., Dalmau, M., Alti, A.: Uml-profile for multimedia software architectures. Int. J. Multimedia Intelligence Sec. **1**(3), 209–231 (2010)

6. El Houda, D.N., Soufiene, B., Djaber, G.: Towards a new dynamic model-based testing approach for multi-agent systems. In: 2022 4th International Conference on Pattern Analysis and Intelligent Systems (PAIS), pp. 1–6. IEEE (2022)
7. Gonçalves, E.M.N., Machado, R.A., Rodrigues, B.C., Adamatti, D.: Cpn4m: Testing multi-agent systems under organizational model m oise+ using colored petri nets. Appl. Sci. **12**(12), 5857 (2022)
8. Houhamdi, Z.: Multi-agent system testing: a survey. International Journal of Advanced Computer Science and Applications **2**(6) (2011)
9. Houhamdi, Z., Athamena, B.: Multi-agents collaboration in open system. Int. Arab J. Inf. Technol. **18**(3A), 393–404 (2021)
10. Huang, Z., Alexander, R., Clark, J.: Mutation testing for jason agents. In: Engineering Multi-Agent Systems: Second International Workshop, EMAS 2014, Paris, France, May 5–6, 2014, Revised Selected Papers 2, pp. 309–327. Springer (2014)
11. Lanus, E., et al.: Test and evaluation framework for multi-agent systems of autonomous intelligent agents. In: 2021 16th International Conference of System of Systems Engineering (SoSE), pp. 203–209. IEEE (2021)
12. Mohajeri Parizi, M., Sileno, G., van Engers, T.: Seamless integration and testing for mas engineering. In: International Workshop on Engineering Multi-Agent Systems, pp. 254–272. Springer (2021)
13. Nascimento, N., Alencar, P., Lucena, C., Cowan, D.: A metadata-driven approach for testing self-organizing multiagent systems. IEEE Access **8**, 204256–204267 (2020)
14. Nguyen, C.D., Miles, S., Perini, A., Tonella, P., Harman, M., Luck, M.: Evolutionary testing of autonomous software agents. Auton. Agent. Multi-Agent Syst. **25**, 260–283 (2012)
15. Savarimuthu, S., Winikoff, M.: Mutation operators for the goal agent language. In: Engineering Multi-Agent Systems: First International Workshop, EMAS 2013, St. Paul, MN, USA, May 6–7, 2013, Revised Selected Papers 1, pp. 255–273. Springer (2013)
16. Tang, W., Wang, K.: Automatic generation of test cases of multi-agent systems based on model checking. In: 2021 16th International Conference on Intelligent Systems and Knowledge Engineering (ISKE), pp. 24–29. IEEE (2021)
17. Thaler, J., Siebers, P.O.: Show me your properties: the potential of property-based testing in agent-based simulation. In: Proceedings of the 2019 Summer Simulation Conference, pp. 1–12 (2019)

Internet of Things Based Smart Healthcare System

Sofia Kouah[1,4](✉), Abdelghani Ababsa[2], and Ilham Kitouni[3]

[1] Department of Mathematics and Computer Science, University of Oum El Bouaghi, Oum El Bouaghi, Algeria
sofia.kouah@univ-oeb.dz
[2] Abdelhafid Boussouf University Center, Mila, Algeria
[3] LISIA Laboratory, Abdelhamid Mehri Constantine 2 University, Constantine, Algeria
ilham.kitouni@univ-constantine2.dz
[4] Artificial Intelligence and Autonomous Things Laboratory, University of Oum El Bouaghi, Oum El Bouaghi, Algeria

Abstract. This work deals with healthcare systems where healthcare professionals' and patients' comfort, providing relevant decision in real time and scrupulously respecting barrier gestures in pandemics times are among the important issues. Recently, Internet of Things technology has been rapidly employed by several applications impacting our daily life activities. Special attentions were devoted to healthcare systems, resulting in a sub branch of IoT, named Internet of Medical Things (IoMT). This work intends to propose an IoMT based Smart Healthcare system. The contribution is twofold: Initially, we develop a medical device and a Mobile Application that facilitate diagnosis and reduce proxemics between patient and healthcare professionals. The medical device acquires vital medical information that can be useful for several pathologies' diagnosis. The mobile application offers ability to identify patient, extract his medical e-form and assist diagnosis process. Secondly, we propose a general approach for consultation, that reduces interactions specially in pandemic times and decreases resource and time consummation. The approach is based on Multi-agent System paradigm, Face recognition and IoT technology. The proposal is exemplified by testbed case study on pulmonary embolism diagnosis.

Keywords: E healthcare · Internet of Things · Internet of Medical Things · Multi- Agent Systems · Pulmonary Embolism · Face Recognition

1 Introduction

The Internet of Things (IoT) is a rapidly evolving computing technology that has emerged in the last few decades and has become a widespread research field. It is seen as billions of connected things, which largely hold our daily lives activities and surroundings. The size of global IoT market is expected to increase to 1.6 trillion by 2025 [1]. It can be considered as a system of connected devices, digital and mechanical machines, animals, people or objects. These "Things" are provided with unique identifiers, they can transfer

K. Kabassi et al. (Eds.): NiDS 2023, LNNS 783, pp. 267–281, 2023.
https://doi.org/10.1007/978-3-031-44097-7_29

data over a network without human or computer interaction [2]. The connected things generate an enormous amount of data that should be analyzed, stored, and processed to result in suitable actions in order to improve citizen life quality and reduce negative impacts on the environment. IoT is based on adopting intelligence, which is provided by on-board processing. Due to the versatility of connected things, several IoT real-world applications are being widely used in various fields, such as Industry, Smart Home, Smart Farming, Smart Transportation, Smart City, and many otherers [1, 3–7].

This paper focuses on Internet of Medical Things (IoMT), which is a specialized subset of IoT technologies and frameworks tailored to healthcare and medical domains. The IoMT has numerous applications, including remote patient monitoring and support, tele-consultation, drug management, and medical diagnosis [6, 8–10]. Our research is primarily focused on exploring how IoMT, digitalization, and Artificial Intelligence can contribute to improving healthcare professionals' and patients' well-being, enhancing healthcare management, and facilitating real-time diagnosis and medical data sharing. Previous studies have investigated the potential of advanced technologies in this regard, as evidenced by the various researches carried out in this perspective [11–14]. However, there has been less emphasis on developing a comprehensive consultation management system. Therefore, we aim to propose a general consultation approach that streamlines processes, ensures efficient data sharing, and minimizes healthcare actors' interaction. On the other hand, the COVID-19 pandemic has provoked prevalent loss of human lives across the globe [15]. It has affected more than 213 countries, with a count of 596,313,072 cases and 6,457,440 deaths, as per WHO report on 16 August 2022. The COVID-19 pandemic has ground the world to a standstill and has shed light on the inadequacies of healthcare systems and the weaknesses in epidemic management. The traditional ways of healthcare management have collapsed under these exigent circumstances. In these trying times, there was a dire need for better implementation of available resources and technology to accelerate the management of the pandemic. Among the practical solutions for controlling the spread of the disease and its severity which has been adopted by healthcare establishments was tele-consultation, which is a remote consultation that has reduced contamination and time. Indeed, tele-consultation has shown its limits behind efficiency, especially in urgent situations. Therefore, our study aims to strike a balance between remote and on-site consultation to reduce unnecessary human interactions, avoid proximity, and react as swiftly as possible in emergency situations.

Our proposed consultation approach is based on IoMT and utilizes MAS as an artificial intelligence paradigm that supports intelligence and distribution properties. This study aims to highlight the common technologies and paradigms that can cater to such consultation approaches. The contribution is twofold: Firstly, we provide useful connected medical device and mobile Application. The medical device is designed to collect essential patient information, such as body temperature, blood pressure, oxygen saturation, and heart rate, at the time of their arrival in the physician's office. The collected data is stored in a server for easy access by the physician. The mobile application could be used by the physician to ensure fast and efficient medical consultation, along with the server with minimum physician-patient interaction, particularly in time of pandemics. Secondly, based on this proposal, we provide a general consultation management process. The proposed approach is illustrated through a case study on the diagnosis pulmonary embolism, a critical condition that has become increasingly common during the Covid-19 pandemic. Our proposed solution can be used in conjunction with the connected device and mobile application, or it can be adapted and integrated into other medical systems.

The remainder of the paper is structured as follows: Sect. 2 provides an overview of the key concepts and paradigms employed in this work, including an introduction to IoT, MAS, and relevant medical parameters. In Sect. 3, we review some related works. Section 4 describes our proposed approach, which is structured into two sub-sections. First, we introduce the development of the medical device and mobile application. Second, we present our consultation approach, which is illustrated through a case study focused on the diagnosis of pulmonary embolism. Finally, Sect. 5 provides conclusions and draws prospects for future research.

2 Background

2.1 Internet of Things

IoT is a network of physical objects, called "things", which are embedded with sensors, software, and several technologies in order to connect, collect and exchange data with other devices and systems over Internet. It refers to a state where Things, such as objects, vehicles and clothing, will have more and more information associated with them and may be able to sense, communicate, network and produce new information, becoming an integral part of the Internet [16]. As stated in [17], the main feature of IoT systems is the pervasive presence of a variety of connected things, such as sensors and actuators, which are capable to interact with each other and cooperate with their neighbors to achieve several functionalities. IoT systems have numerous properties, such as distribution, self-configurability, interoperability, heterogeneity, ubiquity and dynamicity, sensing and actuations abilities, communication and embedded intelligence. Regarding IoMT [6], it is an extension of IoT, it is focused on devices and medical technologies, and enables efficient collection of real-time health data and further tailored healthcare delivery, improving patients' life quality.

2.2 Multi Agents System

The MAS paradigm is characterized by agents that are physical or software entities, autonomous, pro-active, reactive, and situated in an environment, interacting with other agents and the environment to achieve their goals [18–20]. Each agent has incomplete information and capabilities to solve problems, there is no global system control; data is decentralized, and computation is asynchronous [20, 21]. Agents exhibit different properties, such as autonomy, reactivity, and proactivity, depending on their functionalities and roles [19]. The interactions among agents are supported by well- structured communication languages, such as KQML and FIPA-ACL, and protocols, such as Contract-Net Protocol, it can take various forms, such as cooperation, collaboration, and negotiation, depending on agents' goals. The applications of MAS are numerous and cover several areas, including healthcare, education, air traffic management, and mobile robotics. MAS modeling benefit from a rich variety of models, platforms, and tools, making their design and implementation more efficient.

2.3 Relevant Medical Parameters

In this subsection, we outline the vital human medical parameters used in this work.

The Body Temperature is the body's heat level which is viewed as an early warning marker of infection and one of the key physiological observations that should be measured and considered whenever a patient is admitted into a healthcare facility. It is also mutual in many diseases particularly for pandemics such as Coronavirus (COVID-19).

- The blood pressure is the force measurement on the arteries walls as the heart pumps blood through our body. Normal pressure is important for the accurate flow of blood from the heart to body's organs and tissues. In fact, high blood pressure is considered as an important world-wide risk factor for cardiovascular disease, leading among others to strokes, ischemic heart disease deaths [22]. High blood pressure has been also identified as a risk factor for coronavirus disease 2019 (COVID-19) and associated adverse outcomes.
- The Respiratory rate is an early, extremely good indicator of physiological conditions, among others [23]: hypoxia, hypercapnia and the metabolic and respiratory acidosis. Respiratory rate changes are usually associated with respiratory conditions, it can also be considered as a good indicator for a patient trying to maintain homeostatic control [23]. Within COVID-19 pandemic, it has been rapidly revealed the importance of respiratory rate because of the severe effects of the virus on the lungs. COVID-19 occurs frequently as a lower-respiratory tract infection, leading to lung tissue inflammation, coughing, and breath shortness.
- The heart rate measures the number of times, per minute, that the heart beats or contracts. It is considered as one of the vital signs for a healthy human body and plays a major role in determining the cardiac function. It is thought as an important physiologic regulator of total cardiac output.
- The oxygen saturation is an important factor that defines blood oxygen content and oxygen delivery. It is vital parameter of human life and particularly crucial for patients with health constraints that can reduces oxygen level in the blood. Such constraints take in among others, asthma, lung cancer, anemia, heart failure, pneumonia and other cardiopulmonary disorders. It has been revealed its importance in monitoring COVID-19 disease evolution and gravity because of the resting SpO2 which has conspicuously found lower in patients with a severe stage of the disease as determined by clinical symptoms and CT scan [24].

3 Literature Review

The use of IoMT has increased significantly in recent years, and its real-life applications have been found to be especially beneficial to healthcare and well-being fields, particularly during the COVID-19 pandemic [25]. The spectrum of applications ranges from simple data collection systems to complex applications that incorporate multiple paradigms and new technologies. As a result, researchers have proposed various systems, tools, and devices, and in this section, we provide a review of some of these solutions.

Authors in [26] have presented a study of numerous digital system projects in the context of healthcare, ranging from Remote Health Monitoring to non-invasive medical care. Examples are given below:

In [27], examples are given for collecting data systems of three basic human vitals parameters, namely body temperature, electrocardiography (ECG), pulse and heart rate, and displaying them on the Thing Speak Server.

In the same context, authors in [28] propose a monitoring system devoted for patients at risk for cardiac arrest. It measures the same parameters by means of wearable sensors. The data is extracted by using a Smartphone and sent to a MySQL server from SQL Lite internal database.

In the context of Smart prediction applications, authors in [29] have developed a predictive system. It consists of two parts. The first one deals with the design of a prototype and the second one deals with classification. The system begins with acquiring four physiological parameters: heart rate, body temperature, ECG and blood oxygen. Then, the data are transmitted to a gateway node. Concerning the prediction, it operates Support Vector Machine (SVM) machine learning algorithm on the received data in order to predict patients' healthy state.

Another interesting IoMT application relates to automatic Insulin Injection [30], where authors provide a system that collects four body vitals and visualize them on web page along with online streaming by means of webcam to ensure continuous monitoring. Two kinds of account are distinguished patient account and doctor account. Whenever instability is detected, a notification will be sent to the doctor. This later analyzes insulin injection necessity and reacts remotely. Thus, the required insulin dose will be released in the patient by means of solenoid valve involved once pressing button. In another study, an alarm system integrated with a pulse oximeter and WLAN router, was developed that alarms the patients on detecting a drop in oxygen saturation below the critical level [31].

Although presented work address various health contexts and utilize numerous technologies, paradigms and materials, they do not deal with consultation approach to pursue that re duces human interaction between healthcare professionals and patients and therefore preventing contamination during pandemics, facilitating and accelerating consultation process. The purpose of this paper is to cope with such drawbacks.

4 Proposed Approach

The proposed approach is illustrated in Fig. 1. It aims at providing a successful solution for consultation by using smart devices and advanced information and communication technologies. It is focused on IoT, MAS and Face recognition paradigms. Thus, it involves two main stages (See Fig. 1): Pre-Processing and Consultation. Pre-Processing stage, assisted by a professional healthcare, consists of acquiring personal and medical information by using a connected medical device. Consultation stage, supported by a doctor, intends to provide patient diagnosis or therapeutic advice through mobile application means. MAS is implied in both stages; it represents the main intelligent core of the system.

The system behaves as follows: Upon start-up, the patient comes before a healthcare professional (emergency physician or nurse), the latter checks his existence in the healthcare system database. If the patient does not exist, a medical e-form will be created and sent to the data server located in MAS. This sheet includes his photo and personal information as well as medical information provided by the patient and acquired from

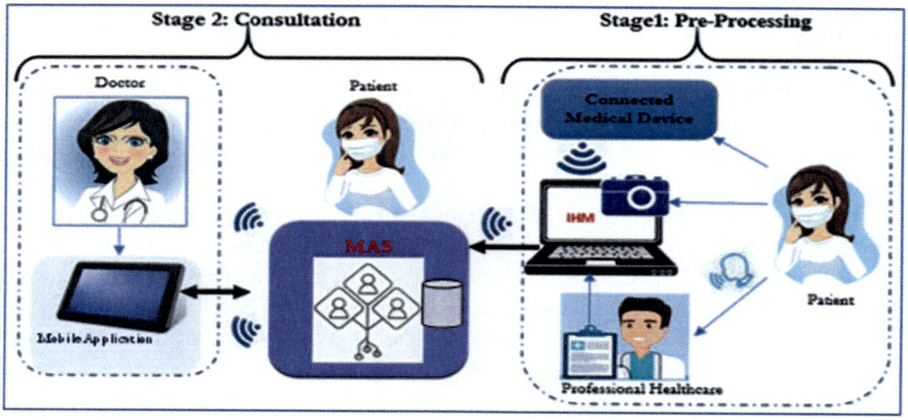

Fig. 1. IoT based Healthcare Architecture.

the connected medical device. The later can measure some physical values related to patient medical constants, such as: body temperature, blood pressure, heart rate and oxygen saturation. If the patient already exists, the healthcare professional only collects medical data and sends the e-form to the database server. In this phase, patient's capture is gathered by means of a connected camera either to the medical device or to the desktop application. This pre-processing step is followed by the consultation carried out by a specialist physician who uses mobile application. During the consultation, the physician performs a facial recognition of the patient. To that end, he can use the camera fitted into mobile device or any connected camera. Once identified, patient's medical e-form is extracted automatically from the database and displayed on the mobile screen. Thus, he can make the appropriate decision and prognostics by invoking the diagnosis process. It should be noted that this process is assisted by a decision support system (i.e., MAS application). In addition, the physician can question the patient, ask for a medical check-up or perform clinical examinations.

4.1 Connected Medical Device

The development of the medical devices is presented by giving the required components, as well as software parts, which ensure data exchange and communication between system's components. This device aims at measuring the following medical parameters: body temperature, blood pressure, breathing rate, heart rate and oxygen saturation. As stated above, these parameters could be useful for many diseases diagnosis and should respect typical values. Thus, the medical device requires the following hardware components (Fig. 2): Body Temperature, Blood Pressure, Breathing Rate, Pulse and Oxygen Saturation Sensors, a Microcontroller, Connectivity Module such as Bluetooth or WIFI and wires.

The use cases diagram (Fig. 3) specifies the functional requirements of the medical device. Four kinds of actors are distinguished:

- Administrator (Primary actor): controls and manages the medical device.

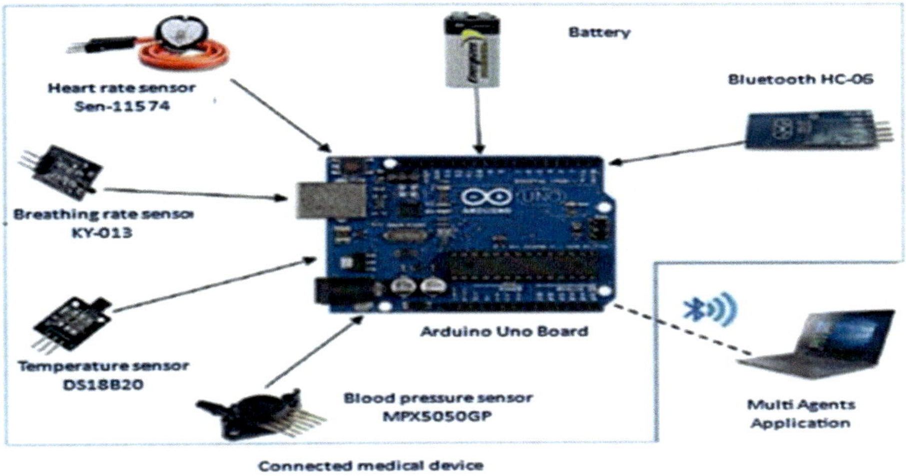

Fig. 2. Medical Device Hardware Architecture.

- Professional healthcare (secondary actor): takes cares of the patient and places the medical device on the patient.
- Patient: person of which we need to collect his medical parameters.
- Multiagent application: to stock and treat data in order to assist decision making about patient's diagnosis.

Three main use cases are distinguished in this diagram: Data collection which consist of collecting the medical parameters from the sensors, Control and management of the medical device and Data Transmission. By following sensors' connection schemas, we obtain the physical part of the medical device, illustrated in Fig. 4.

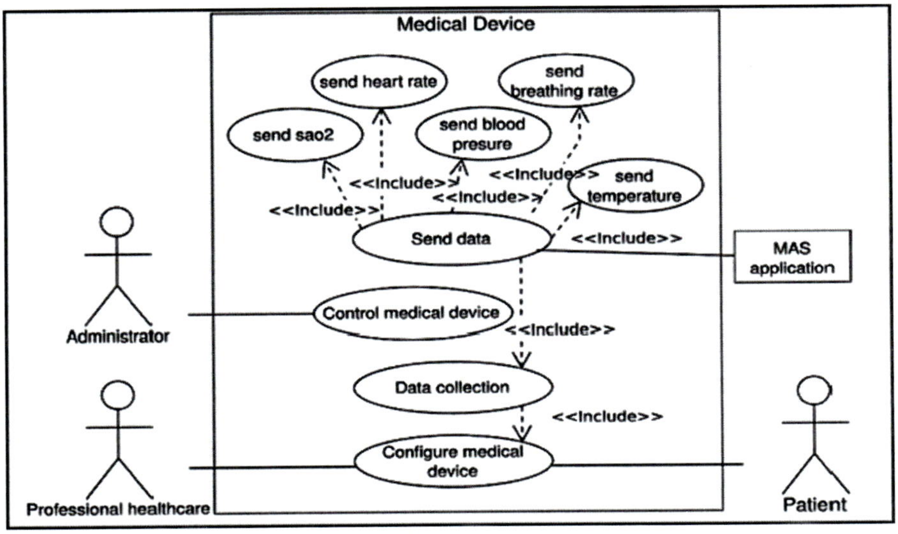

Fig. 3. Medical Device Use Cases Diagram.

Fig. 4. Medical device physical part.

4.2 Mobile Application

The mobile application uses camera in order to enables patients' recognition and identification. It contributes to the following facilities: the extraction of patient medical files from database server, the exchange of useful information between healthcare professionals in order to take relevant decisions at real time, such as responding, in a timely manner, to an emergency event. It helps to ensure better coordination between staffs, optimize resources, save time, avoid unnecessary displacing and make consultation easier. Its functional requirements are shown by the use cases diagram depicted in Fig. 5, where two kinds of actors are distinguished:

- Physician (Doctor): person who uses the mobile application to ensure consultation.
- Multiagent application: the application which interact with both mobile application and the device to assist healthcare professionals in their decisions. It should be noted that there is no direct interaction between mobile application and the connected medical device.

Three main use cases are distinguished in this diagram: Facial recognition which includes facial capture, medical form display and diagnosis result display. The mobile application main screen looks like depicted in Fig. 6. Regarding the face recognition functionality, it is organized based on two complementary phases: face detection and face recognition, and ensured by MAS application whenever requested as well as diagnosis.

4.3 Agent-based System for Pulmonary Embolism Diagnosis

This section is devoted to the development of diagnosis IoT system that integrates both mobile application and the medical device. It is based on Multi agent paradigm which is suitable for modeling IoT systems properties, especially interactions, communications and intelligence features which are key-stones for both fields. As medical pathology,

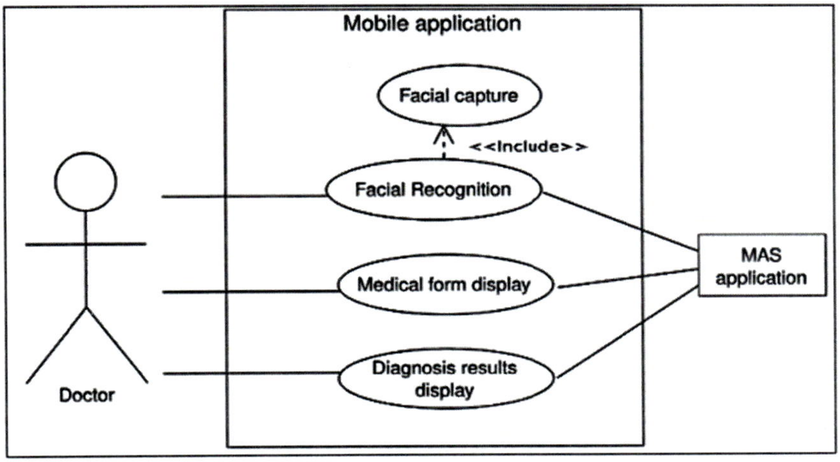

Fig. 5. Mobile application use cases.

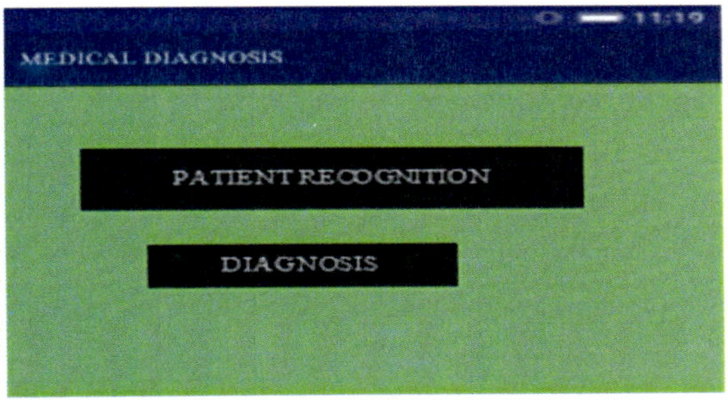

Fig. 6. The main of mobile application interface.

Pulmonary Embolism (PE) have chosen. Therefore, the multi-agent application is composed of several kinds of agents whose overall aim is to analyze medical parameters and specify potentially PE critical level. The medical device is used for collecting the necessary parameters to PE diagnosis; however, the mobile application ensures consultation. It should be noted that the MAS application can be adapted to several diagnosis pathologies.

Pulmonary Embolism Description. The PE is defined as a sudden obliteration, which can be total or partial, of the trunk or a branch of the pulmonary artery by an embolus, most often cruoric [32]. It remains a major challenge in medicine. Within COVID-19 pandemic, PE is considered among the essential concerns of professional healthcare, which should be predicted and prevented. It has been reported that a high prevalence of PE in patients with COVID-19 at time of hospital admission has been noticed. As well, it has been found 31% incidence of thrombotic complications in Intensive Care Unit (ICU) Hospitalization in patients with COVID-19 infections that is remarkably high. Several materials are practicable by healthcare professionals in order to assist PE diagnosis, including clinical scoring systems, laboratory data, and imaging studies. If healthcare professionals suspect that patient have a pulmonary embolism, they should look-up for the most appropriate test for PE diagnosis. These tests could cover Blood tests, Electrocardiogram Test, Imaging Test, etc. Since the clinical polymorphism, PE is often unrecognized. Likewise, because of its presentation variability, the diagnosis depends on the PE like-hood and patient stability. Scoring systems could be used to assist delineating PE like-hood and the underlying thromboembolic events. Wells and Geneva scores are examples of the mostly used Scoring systems, which will be adopted int this research. If healthcare professional suspects that patient has PE, he deals with medial algorithms, recommended by the European Society of Cardiology (ESC). We distinguish two different clinical situations: High risk and non-high risk. These two situations do not have the same prevalence. Nevertheless, they require different response.

Description of MAS Architecture. Recall that the global functioning of PE diagnosis multi-agent system is specified as follows: First, information is collected from connected medical device and further data are introduced by the healthcare professional. These Data are then stored by MAS in order to be processed whenever a patient consult a physician. The later uses mobile application, to ensure patient consultation. The PE diagnosis decision making is achieved by the MAS and returned back to the doctor for validation and display. Thus, MAS ensures the following main functionalities: Facial recognition, PE Diagnosis, PE criticality level computation, calculate PE Probability, Data management and Storage as well as gathering Data, which are dispatched between MAS' agents. Now, we describe MAS by means of Voyles approach components [33], given by: agents involved in MAS, environment, organization and interactions between agents.

Agents Identification. The MAS is composed of:

- Agent of Things (several instances, 1 instance per medical object) is an intelligent agent which supervises the connected medical device and ensures the following tasks: gather medical constants from the medical device, validate medical constants (legitimate values), create new medical form, recognize patient and gather patient information.
- Local Coordinating Agent (several instances, 1 instance per a set of mobile applications) is an intelligent agent which manages mobile applications related to several doctors and ensures the following tasks: recognize patient, ensure PE diagnosis, inform PE diagnosis results and prognostic. The PE diagnosis is based on probability assessments upon Wells and PESI scores within clinical decision rules. Wells score is a clinical prediction score based on simple, noninvasive clinical parameters. It is used conjointly with d-dimer to avoid unnecessary tests whenever clinical suspicion is not high. On the other hand, the pulmonary embolism severity index (PESI) is a score that has been validated most extensively to date, and has been in use for the risk assessment of pulmonary embolism (PE), as well as to determine the severity of disease.
- Global Coordinating Agent (unique instance in a given hospital), its role is summarized as follows: manage and store patient information, deliver the medical report to patients and assist them, it can also communicate with the outside world.

Environment Identification. The physical environment supports agents' actions and have many resources; it consists mainly of all physical materials (connected medical device and mobile application) and surrounding agents.

Organization Identification. Agents are organized according to the hierarchical organization structure depicted in Fig. 7. They interact together according to the existing relationship.

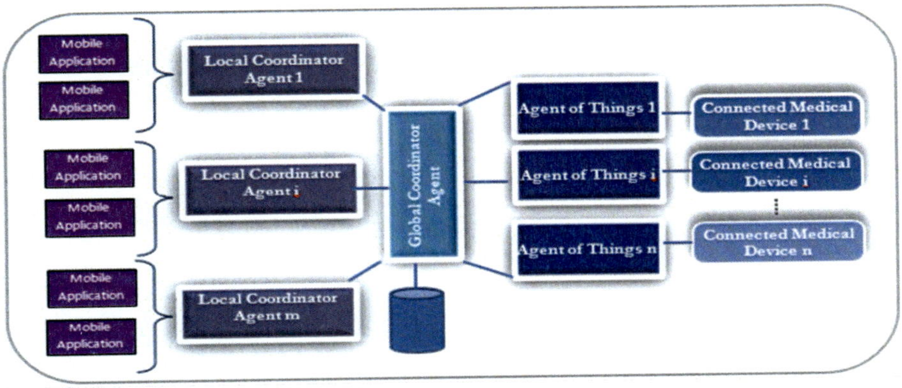

Fig. 7. PE Diagnosis MAS.

Agents Communication. It is described by means of sequences diagrams given in the Fig. 8.

Figure 9 and Fig. 10 show some interfaces of PE Diagnosis MAS.

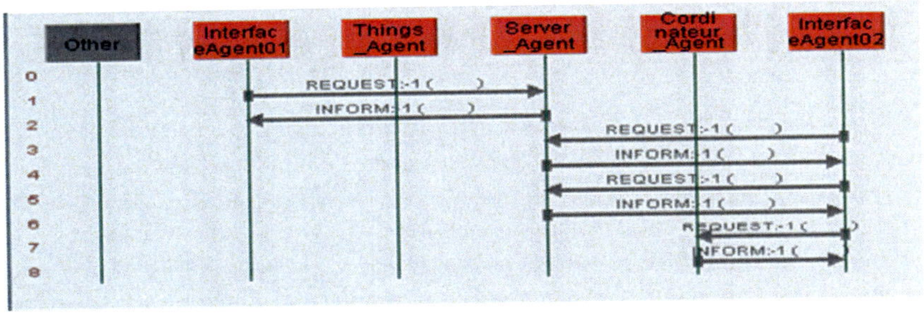

Fig. 8. Sniffer (interactions between agents).

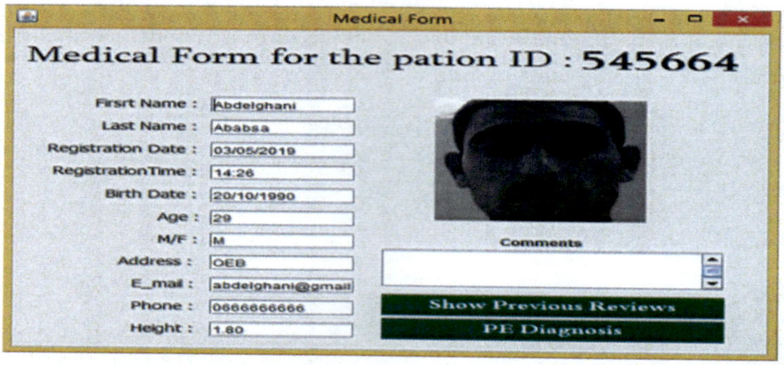

Fig. 9. Interface of Medical form for recognized patient.

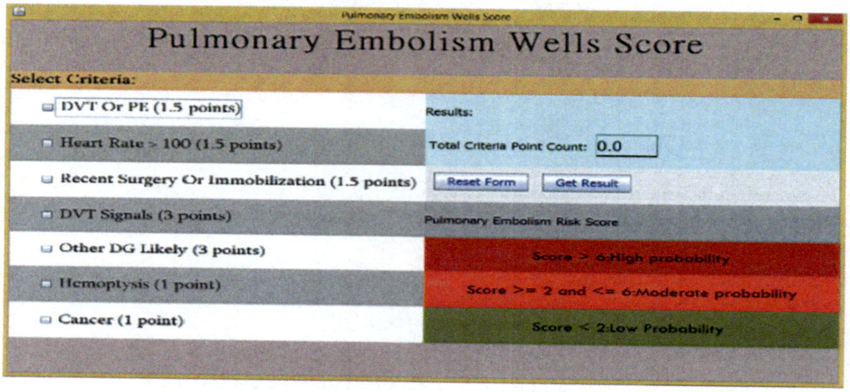

Fig. 10. Pulmonary Embolism Interface.

5 Conclusion

In this work, we proposed an innovative approach for healthcare system management during consultation. The use of IoMT and intelligent systems has marked a significant change in the development of healthcare systems by offering cost-effective, noninvasive, and swift facilities for patients. Our main goal was to provide a tradeoff between remote and on-site consultation. The proposed approach has great potential to ensure consultation and system management that reduces proxemics and unnecessary interactions, and reacts in real-time in emergencies. The approach offers a wide range of benefits in terms of healthcare professional assistance, keeping track of patient medical data, and personal information related to their healthy state. Data are exchanged naturally between healthcare services. Our future progress work consists of adopting machine and deep learning for agents' reasoning to enhance the intelligence feature of diagnosis results based on the patient's data. In the future, we plan to improve our approach by incorporating advanced technologies such as Cloud Computing, augmented reality, Big

Data, and Fog/Edge Computing, as well as wearable devices with increased miniaturization, power, and cost-effectiveness. Additionally, we plan to apply this approach to other pathologies to further benefit patients.

References

1. Abdulrazak, B., Codjo, J.A., Paul, S.: Self-healing approach for IoT architecture: AMI platform. Participative Urban Health and Healthy Aging in the Age of AI: 19th International Conference, ICOST 2022, Paris, France, June 27–30, 2022, Proceedings (2022)
2. Laghari, A.A., Wu, K., Laghari, R.A., et al.: A review and state of Art of internet of things (IoT). Arch Computat Methods Eng. **29**, 1395–1413 (2022). https://doi.org/10.1007/s11831-021-09622-6
3. Atitallah, S.B., Driss, M., Boulila, W., Ghézala, H.B.: Leveraging deep learning and IoT big data analytics to support the smart cities development: review and future directions. Computer Science Rev. **38**, 100303 (2020). ISSN 1574-0137. https://doi.org/10.1016/j.cosrev.2020.100303
4. Kiran, S., Neelakandan, S., Reddy, A.P., Goyal, S., Maram, B., Rao, V.C.S.: Chapter 11 - Internet of things and Wearables-Enabled Alzheimer Detection and Classification Model Using Stacked Sparse Autoencoder, Editor(s): Hemanth
5. Jude, D., Gupta, D., Khanna, A., Khamparia, A.: Wearable Telemedicine Technology for the Healthcare Industry, Academic Press, pp. 153–168 (2022), ISBN 9780323858540. https://doi.org/10.1016/B978-0-323-85854-0.00012-5
6. Wagan, S.A., Koo, J., Siddiqui, I.F., Attique, M., Shin, D.R., Qureshi, N.M.F.: Internet of medical things and trending converged technologies: a comprehensive review on real-time applications. J. King Saud University - Computer and Inf. Sci. ISSN 1319–1578 (2022). https://doi.org/10.1016/j.jksuci.2022.09.005
7. Hemmati, A., Rahmani, A.M.: The internet of autonomous things applications: a taxonomy, technologies, and future directions. Internet of Things, **20**, 100635 (2022). ISSN 2542-6605, https://doi.org/10.1016/j.iot.2022.100635.
8. Wang, I., Mu, Z.: Risk monitoring model of intelligent agriculture Internet of Things based on big data. Sustainable Energy Technologies and Assessments, **53**(Part C), 102654 (2022). ISSN 2213-1388, https://doi.org/10.1016/j.seta.2022.102654.
9. Shreya, S., Chatterjee, K., Singh, A.: A smart secure healthcare monitoring system with Internet of Medical Things. Computers and Electrical Engineering, **101**, 107969 (2022). ISSN 0045-7906, https://doi.org/10.1016/j.compeleceng.2022.107969.
10. Kakhi, K., Alizadehsani, R., Kabir, H.M.D., Khosravi, A., Nahavandi, S., Acharya, U.R.: The internet of medical things and artificial intelligence: trends, challenges, and opportunities, Biocybernetics and Biomedical Eng. **42**(3), 749-771 (2022). ISSN 0208-5216, https://doi.org/10.1016/j.bbe.2022.05.008.
11. Khan, I.A., et al.: XSRU-IoMT: explainable simple recurrent units for threat detection in Internet of Medical Things networks. Future Generation Computer Syst. **127**, 181–193 (2022). ISSN 0167–739X, https://doi.org/10.1016/j.future.2021.09.010
12. Su, X., An, L., Cheng, Z., Weng, Y.: Cloud–edge collaboration-based bi-level optimal scheduling for intelligent healthcare systems. Future Generation Computer Syst. **141**, 28–39 (2023). ISSN 0167–739X, https://doi.org/10.1016/j.future.2022.11.005
13. Dao, N.-N.: Internet of wearable things: advancements and benefits from 6G technologies. Future Generation Computer Syst. **138**, 172–184 (2023). ISSN 0167–739X, https://doi.org/10.1016/j.future.2022.07.006

14. Kumar, P., Kumar, R., Gupta, G.P., Tripathi, R., Jolfaei, A., Islam, A.K.M.N.: A blockchain-orchestrated deep learning approach for secure data transmission in IoT-enabled healthcare system. J. Parallel and Distributed Comput. **172**, 69–83 (2023). ISSN 0743–7315, https://doi.org/10.1016/j.jpdc.2022.10.002
15. Hu, H., Xu, J., Liu, M., Lim, M.K.: Vaccine supply chain management: an intelligent system utilizing blockchain, IoT and machine learning. Journal of Business Res. **156**, 113480 (2023). ISSN 0148-2963. https://doi.org/10.1016/j.jbusres.2022.113480
16. Kaur, S., Hasija, Y.: Role of computational intelligence against COVID-19. In: Raza, K. (eds) Computational Intelligence Methods in COVID-19: Surveillance, Prevention, Prediction and Diagnosis. Studies in Computational Intelligence, **923**. Springer, Singapore (2021). https://doi.org/10.1007/978-981-15-8534-0_2
17. Sofia, K., Ilham, K. : Multi-layer agent-based architecture for internet of things systems. J. Information Technol. Research (JITR) **11**(4), 32–52 (2018)
18. Kouah, S., Saïdouni, D.E., Ilié, J.M.: Synchronized petri net: a formal specification model for multi agent systems. JSW **8**(3), 587–602 (2013). https://doi.org/10.4304/jsw.8.3.587-602
19. Kouah, S., Saidouni, D.E.: Fuzzy labeled transition refinement tree: application to stepwise designing multi agent systems. In: Fuzzy Systems: Concepts, Methodologies, Tools, and Applications, pp. 873–905 ((2017)). IGI Global, Hershey, PA
20. Kouah, S., Kitouni, I., Saidouni, D.E.: Towards agent oriented methodology: a fuzzy based formal refinement approach for multi-agent. In: Proceedings of the International Conference on Engineering & MIS 2015, p. 50 (2015). ACM. https://doi.org/10.1145/2832987.2833060
21. Kouah, S., Saïdouni, D.E., Kitouni, I. : Open fuzzy synchronized petri net: Formal specification model for multi-agent systems. International J. Intelligent Information Technologies (IJIIT) **12**(1), 63–94 (2016)
22. Sheppard, J.P., et al.: Association between blood pressure control and coronavirus disease 2019 outcomes in 45 418 symptomatic patients with hypertension: an observational cohort study. International journal of hyertension. https://www.ahajournals.org/doi/suppl/https://doi.org/10.1161/HYPERTENSIONAHA.120.16472
23. Rolfe, S.: The importance of respiratory rate monitoring. British Journal of Nursing **28**(8). https://doi.org/10.12968/bjon.2019.28.8.504
24. Sheppard, J.P., et al.: Association between blood pressure control and coronavirus disease 2019 outcomes in 45 418 symptomatic patients with hypertension: an observational cohort study. Hypertension **77**(3), 846–855 (2021)
25. Seshadri, D.R., et al.: Wearable sensors for COVID-19: a call to action to harness our digital infrastructure for remote patient monitoring and virtual assessments. Journal Frontiers in Digital Health (2020). https://doi.org/10.3389/fdgth.2020.00008
26. Ashfaq, A., et al.: A review of enabling technologies for Internet of Medical Things (IoMT) Ecosystem. Ain Shams Engineering J. **13**(4), 101660 (2022). ISSN 2090–4479
27. Aloulou, H., Abdulrazak, B., De Marassé-Enouf, A., et al. : Participative urban health and healthy aging in the age of AI: 19th International Conference, ICOST 2022, Paris, France, June 27–30, 2022, Proceedings (2022)
28. Ruman, M.R., Barua, A., Rahman, W., Jahan, K.R., Jamil Roni, M., Rahman, M.F.: IoT based emergency health monitoring system. In: 2020 International Conference on Industry 4.0 Technology (I4Tech), pp. 159–162 (2020)
29. Kakria, P., Tripathi, N., Kitipawong, P.: A real-time health monitoring system for remote cardiac patients using smartphone and wearable sensors. Int. J. Telemed Appl. (2015), pp. 1-11 (2015)
30. Athira, A., Devika, T.D., Varsha, K.R., Bose, S.S.: Design and development of IOT based multi-parameter patient monitoring system. In: 2020 6th International Conference on Advanced Computing and Communication Systems (ICACCS), pp. 862–866 (2020)

31. Tripathi, V., Shakeel, F.: Monitoring health care system using internet of things - an immaculate pairing. In: 2017 International Conference on Next Generation Computing and Information Systems (ICNGCIS), pp. 153–158 (2017)
32. Pessinaba, S., et al.:L'embolie pulmonaire au centre hospitalier universitaire Campus de Lomé (Togo): étude rétrospective à propos de 51 cas. Pan Afr Med J. **27**, 129 (2017). Published online 2017 juin 18. French. https://doi.org/10.11604/pamj.2017.27.129.6855
33. Ricordel, P.M., Demazeau, Y.: Volcano, a vowels-oriented multi-agent platform. In: From Theory to Practice in Multi-Agent Systems: Second International Workshop of Central and Eastern Europe on Multi-Agent Systems, CEEMAS 2001 Cracow, Poland, September 26–29, 2001 Revised Papers 2, pp. 253-262. Springer Berlin Heidelberg (2002)

Integrating CAD to University's Social Enterprise to Promote Local Weavers' Livelihood and Traditional Craft Preservation

Jorelyn P. Concepcion[1](✉), Aurelio P. Vilbar[2], Lynnette Matea S. Camello[1], and Charmaine Lee M. Cabrera[1]

[1] Cebu Technological University- Argao, Cebu, Philippines
jorelyn.concepcion@ctu.edu.ph
[2] University of the Philippines Cebu, Cebu, Philippines

Abstract. This study addresses the need to revive the traditional handloom weaving industry in Argao, Cebu, Philippines. It examines the impact of the Cebu Technological University-Argao Campus' social enterprise that uses Computer-Aided Design (CAD) software in the handloom weaving design, on the local weavers' livelihood and in the preservation of the traditional handloom weaving industry in Argao. Using the exploratory research design, the research utilized various data-gathering procedures, such as the analysis of the weaver's pay slips to determine the increase in income and semi-structured interviews to assess the impact of the project on the weavers and designers. The findings reveal that the use of CAD technology has had a positive impact on reducing production time and costs, on archiving designs, increasing market demand for value-added products, and, consequently, increasing income for the weavers. The project also improved the income of weavers, upscaled their technical skills, and positively impacted the community's socio-economic status. The study recommends the continued use of CAD technology to ensure efficiency, consistency, and accuracy in the production of handwoven products, which is essential for the preservation of the traditional craft.

Keywords: handloom weaving · CAD technology · university as a social entrepreneur · craft preservation

1 Introduction

Loom weaving is the craft of interlacing two sets of yarns, threads, or fibers called the warp and the weft to create a fabric, which is a century-old industry in the Philippines [1]. In Argao, Cebu, Philippines, handloom weaving, called 'hablon,' was a major source of income for the town during the Spanish colonial period. The local weavers' designs were supplied to the elite market and outside the municipality.

However, as a traditional craft, handloom weaving is considered an almost dying industry [2] because of the popularity and availability of cheaper machine-made textiles

K. Kabassi et al. (Eds.): NiDS 2023, LNNS 783, pp. 282–291, 2023.
https://doi.org/10.1007/978-3-031-44097-7_30

[3]. In addition, the specialized and time-consuming handloom weaving techniques cannot meet the demands of commercial production [4]. Industrialization, mass production, and digitization threatened the retooling and upscaling of the local weavers' skills [1]. What was once a personal activity, handloom weaving, has evolved into a source of income for women and, over time, a cottage industry. The impact of industrialization, mass production, globalization, and digitization put pressure on this industry, which brought about changes in production technology and market access but also left many behind, including the weavers [1].

Cognizant of the local weavers' needs, the Cebu Technological University-Argao Campus (CTU) established the Hinablon sa Cebu (Loom weaving) Project to revive the handloom weaving industry by training the local weavers on different weaving patterns, color harmony, and fabric designs. It further used the Computer-Aided Design (CAD) software Hatch Embroidery Digitizer, Adobe Illustrator, and Thunder Laser, which was the technology that allowed the designers to create designs for handwoven textile manufacturing and increase production.

In collaboration with the local government unit, and public and private organizations, the project became the university's extension flagship to fulfill its function as a social entrepreneur whose aim is to strengthen the hablon weaving industry while supporting the local weavers. This paper seeks to determine the impact of the university's social entrepreneurship project on the local weavers' income and to what extent the use of CAD increased the efficiency of handloom weaving production.

1.1 Review of Related Literature

The University as a Social Entrepreneur

This study is anchored on the Framework for Effective Institutional Arrangements for Livelihood and Income-Generation Interventions, which states that social enterprises are businesses set up mainly for the benefit of marginalized low-income groups. The framework highlighted that a social entrepreneur might provide production centers for community beneficiaries, capability-building activities for the workers, and resources to run the business [5].

CTU is a higher education institution (HEI) whose vision-mission is to provide extension programs addressing the community's needs [6]. It continuously works toward sustainable economic, social, and environmental development through relevant community engagements and industrial partnerships toward functional, productive, resilient, and entrepreneurial communities [7].

This strategic objective is achieved through the Hinablon sa Cebu project of CTU-Argao Campus, which provides livelihood through technology adoption, production, and marketing of research-based product innovation. In this research, livelihood is a means that enable people to earn a living [8]. Eventually, some of these livelihood activities are developed into a social enterprise.

This project was conceived to address the decline of handloom weaving in Argao. To achieve this goal, the university engaged donors, the Department of Trade and Industry, who granted a Shared Service Facility (SSF) for Handloom Weaving. The SSF provided twenty-five handloom machines and four CAD equipment: RICOMA FHT-1204,

IKONIX FHT 1204, RICOMA RCM 1201TC-7S, and HTJ1390 CNC Laser Cutting Machine. On the other hand, the university provided the capital expenditures for the construction of the building and weaving center, the working capital, and capacity-building of the local weavers.

Using CAD in Preserving Traditional Craft

CAD technology has become an indispensable tool in product design and engineering. This technology employs specialized software and hardware that allows professional designers, architects, and engineers to create and modify designs on a computer. CAD technology has been widely adopted in various industries and fields, including product design, such as in the development of handwoven products [9].

Hinablon sa Cebu utilized CAD technologies to develop value-added products. Value-adding is the process where product salability is enhanced by adding some incentives [10]. In the Indian handloom sector, value addition and embellishment of products using embroidery and applique work are being done to increase marketability [11]. Among these are digital embroidery machines RICOMA FHT-1204 and the IKONIX FHT-1204, both with four heads, 12 needles, a maximum speed of 1000 stitches per minute (spm), and an embroidery area of X400 mm Y680mm. The RICOMA RCM-1201TC-7S was also used, which has one head, 12 needles, a max speed of 1200 spm, and an embroidery area of 560 mm × 360 mm. Additionally, the study employed the HTJ-1390 100w CO2 CNC laser cutting machine for wood, acrylic, and plastic with an engraving area of 1300 × 900 mm and an engraving speed of 8000 mm/min.

The single-head embroidery machine is utilized for prototyping new designs because it works with small textile quantities. Once the design is approved, mass production is carried out using the 4-head digital embroidery machines. In addition, the laser cutter machine is employed to print company logos on MDF boards attached to handwoven leis.

In Argao, weavers traditionally use pedal-frame looms made of wooden and bamboo parts, which have a lifespan of about three years. These looms typically support a 24-inch to 36-inch-wide metal reed [12]. However, to accommodate the innovative 60-inch metal reed of the SSF for Handloom Weaving and ensure durability and stability, most wooden parts were replaced with metal using CAD technology. This innovation has not only allowed for the production of the standard 60-inch textile commonly found in the market today but has also opened up opportunities to create many other handwoven products. The use of CAD technology has allowed weavers to explore new designs and techniques that were not possible with traditional looms.

In Argao, traditional handwoven textile designs were created using pen and paper. However, these weaving patterns are now rare and can only be seen in personal towels, blankets, and other handwoven products that were popular during the time when weaving was still thriving in the municipality [12]. As a result, there is a risk of losing these traditional textile and weaving designs. To preserve these weaving designs, CAD is used as a tool for a renovation that can sustain the age-old traditional designs, thus, preserving the cultural heritage [13].

Designing and prototyping textiles can be a tedious and time-consuming process, as evidenced by the experiences of designers and weavers in Hinablon sa Cebu. To address this issue, 3D CAD software is utilized to draw cross-section diagrams of the ends and picks and generate perspective projections of the weave from these cross-sections [14]. Their study proved that CAD software could generate the perspective projection of a given weave structure directly by inputting the point paper design of the weave. This can significantly reduce prototyping time and cost. Similarly, the implementation of technology in India's rapidly growing traditional carpet-weaving industry has yielded positive outcomes, including increased revenue and enhanced competitiveness, resulting from the adoption of technology [15].

The Department of Textiles and Apparel Designing in India conducted a study highlighting the significant role of CAD in facilitating the representation of designs. By digitizing designs, designers can smoothly transition from initial concepts to photorealistic fabric representations. This innovative approach reduces the complexity involved and saves time in the fabric drafting process, ultimately streamlining the overall design process. The study underscores the transformative impact of CAD technology in the realm of textile design, enabling more efficient and visually accurate representations of fabric designs [16].

CAD technology has found another use in the handloom weaving industry, specifically in product value-adding. Adding features to the basic product can increase its value and competitiveness in the market. They argue that various surface enrichment techniques, such as fabric and screen painting, hand embroidery, and applique work, can enhance handwoven products' value. These surface enrichment techniques fulfill the modern-day consumer's demand for made-ups with new and unique designs [17].

In this project, value addition to handwoven products is accomplished through various methods such as hand painting, digital embroidery, and digital fabrication. These techniques enhance the exclusivity and market value of handwoven products like fashion apparel, accessories, and home accents. In addition to these methods, laser printing on medium density fiber (MDF) boards attached to handwoven leis are also utilized to promote company brands and increase branding presence.

2 Methodology

This impact assessment research used a descriptive design in which the researchers used various data gathering procedures to answer the research questions. It employed data analysis of the weavers' pay slips to determine their increase of income. It also used semi-structured interviews to determine the impact of the project on the weavers and the designers. This research utilized a focus group discussion to triangulate the data. The independent FGD facilitator is not affiliated with CTU-Argao.

Ms. Debbie Palao is the managing director of the Holistic Coalition of the Willing, Inc. (HoliCOW), a group of designers, manufacturers, and crafters who are dedicated to sustainability. She has been a designer since 1990 and has provided mentorship to Hinablon sa Cebu in various areas, such as quality assurance, product development, and marketing strategies. Ms. Kaye Batiquin, the current manager of HoliCOW, Inc., is currently a core member of the Creative Cebu Council UNESCO. She has won for HoliCOW the prestigious Good Design Award from the Systems of Design Centre Philippines. Her mentorship in the Hinablon sa Cebu project is in visual merchandising, design digitization, and archiving.

3 Findings

3.1 CAD in Designing Traditional Craft

The three CAD software, coupled with marketing and positive customer satisfaction, increased the gross production sales of the Hinablon sa Cebu Project from Php 5,396,421.40 in 2021 to Php 11,182,086.03 in 2022, according to sales record and the project's financial statements.

Figure 1 shows the CAD software Hatch Embroidery Digitizer creating the design while Fig. 2 shows the actual output.

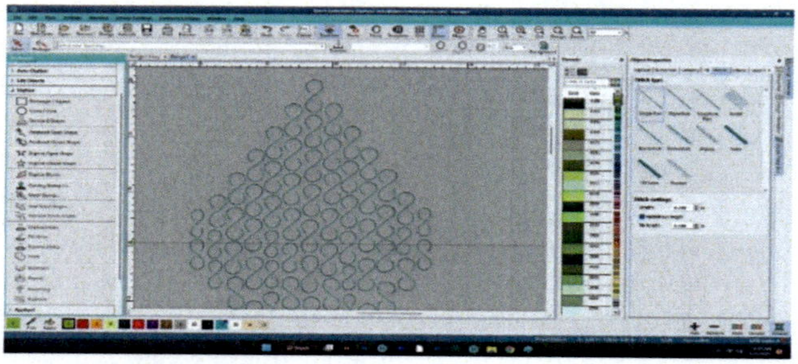

Fig. 1. Rendered design using CAD software Hatch Embroidery Digitizer

Fig. 2. Actual embroidered bed runner

Bed runners are integral to bedroom accouterment in the hospitality industry because they provide a pop of personality and aesthetics to the otherwise all-white, uniform standard of hotel amenities but also protect the end of the bed from luggage dust or shoe dirt for guests. This table runner, designed by Ms. Palao, employs a "combi" (combination thread) weave in green hues embroidered with an orange stylized 'S' monogram to create an example of the dynamic variety one can employ for a custom order.

In a semi-structured interview with Ms. Palao, she claimed that the use of CAD technology had a positive impact on reducing production time and costs in designing and prototyping. She explained that the technology assisted the designer in conceptualizing color schemes and graphics more effectively to better serve buyers requiring specific corporate colors or marketing directions. This also provides the designer with the ability to present the design to the buyer before it is executed, thereby facilitating the decision-making process. Without CAD technology, this would be challenging.

Ms. Batiquin emphasized that CAD technology can be used in archiving designs. This is crucial for preserving traditional crafts like handloom weaving. With CAD, weaving patterns can be archived, and new designs can be created based on archived ones. This helps in preserving the Cebuano weaving tradition, which is currently endangered. According to her, archiving not only helped in preserving the tradition but also enabled analyzing the production process to identify areas for improvement such as cost and time reduction. CAD technology can also help in saving resources.

The use of CAD technology directly impacts the income of weavers by increasing market demand for value-added products. When market demand for a product increases, production time for weaving the product also increases, resulting in increased income for the weavers. Therefore, the integration of CAD technology can potentially contribute to the economic empowerment of weavers and their communities.

Table 1 shows that the weaver's average gross annual income significantly increased by 89.92% from 2019–2022. The weavers' average annual income in 2019 is Php 71,257.62 2019 and Php 105,100.49 in 2022. However, the significant decrease in 2020 was due to the COVID-19 pandemic, which halted the orders and production of hablon products.

The local weavers' significant income increase proved the community's economic advantages when Hinablon sa Cebu Project adopted the social enterprise model [5]. The growth was attributed to the purchase of Hatch Embroidery Digitizer CAD software that promoted mass production of embroidered graduation sash, "sablay". From the 2021 gross sales of 3.88 M, and 2022 sales of 7.38 M, it is projected to increase to 8.80 M in 2023.The software has enabled the mass production of embroidered graduation sashes, known as "sablay," with a daily output of 64 pieces (from a 4-head embroidery machine) compared to the manual production rate of approximately 2 pieces per day. In the interview, the weavers said, "We earned a lot from the greater demand for the production of the sablay."

From the interview results with the weavers as shown in Table 2, the project empowered the weavers to send their children to school, purchase basic needs, start their financial savings, and upscale their weaving technical skills. As expressed by Weaver A, "My life is good because I was able to send my children to school. She is about to finish her college degree." Weaver B added, "Working here is a big help for my family, especially now that I don't have a husband anymore. My children are in grade 5 and grade 2. My income here is what I use to provide for them."

The weavers' statement provides evidence of the significant economic impact that Hinablon sa Cebu has had on its beneficiaries. This is supported by the framework developed by Dacanay [18] for understanding how social enterprises can serve the poor and address poverty. This study sheds light on the effects that social enterprises have on poverty in developing countries, from the perspective of the poor as stakeholders.

Table 1. Weaver's Income and Percentage Increase from 2019–2022

	2019	2020	% decrease	2021	% increase	2022	% increase
Weaver A	59,299.90	17,679.70	−70.19%	40,759.10	130.54%	67,320.66	65.17%
Weaver B	181,910.95	58,195.50	−68.01%	143,154.80	145.99%	168,644.55	17.81%
Weaver C	70,687.10	22,196.20	−68.60%	74,737.10	236.71%	162,245.78	117.09%
Weaver D	59,847.90	24,414.60	−59.21%	53,740.20	120.12%	70,270.81	30.76%
Weaver E	49,660.00	2,700.00	−94.56%	–		175,435.10	
Weaver F	82,266.10	34,088.50	−58.56%	79,765.20	133.99%	116,399.52	45.93%
Weaver G	50,378.90	20,891.30	−58.53%	42,766.20	104.71%	106,341.76	148.66%
Weaver H	51,836.30	17,578.20	−66.09%	35,113.40	99.76%	67,069.84	91.01%
Weaver I	71,183.10	14,315.00	−79.89%	66,489.00	364.47%	106,676.56	60.44%
Weaver J	41,375.90	28,013.60	−32.29%	47,935.90	71.12%	103,234.74	115.36%
Weaver K	65,387.70	23,136.00	−64.62%	41,982.80	81.46%	60,407.60	43.89%
Weaver L	–	–		29,761.40		85,920.28	188.70%
Weaver M	–	–		32,026.10		81,539.14	154.60%

According to this, social enterprises that empower the poor as transformational partners tend to have a significant qualitative impact on poverty reduction [18].

Another positive impact of the project is the improvement of the living conditions of the weavers. Weaver C shared, "I have bought many things with my salary. I was able to install a faucet and buy foam. Now, I will renovate my kitchen which was damaged by the last typhoon." Agreeing to Weaver C, Weaver D said, "My weaving skills here have also helped me with my struggles and our daily needs. My child is a Person with Disability. My income is used to buy food and other necessities."

The weavers' testimonials confirmed the findings [18] on social enterprise in the Philippines, which revealed that among the poor, women are one of the significant groups served. SEPPS, which stands for social enterprise with the poor as primary stakeholders, aims to reduce poverty or improve the quality of life of specific segments of the poor.

Table 2. Impact on the community.

Themes	Example	Sample Statements
Improved income	Sent children to school	"My weaving skills have greatly improved here. I was able to send my child to school here, and he even went to college. I also have a senior high school student who will start studying soon."
	Could afford to purchase basic needs	"My life here is very good because my earnings are big. It also helped with the allowance of my students before. They were able to finish their studies here. I can also buy things for the house that can be used for a long time, like tables and chairs."
	Learned the value of savings	"My weaving skills have been a big help for me. Before, when I was still newly married, I did not have work. When there is a need like when my child got sick and we didn't have money in our pocket. Since I started weaving, I now have some savings. We can even afford to buy a little something extra."
Upscaled in technical skills	Learned new weaving techniques	"In our job, I learned different designs. Before, it was just plain, made of sinamay and polyhemp. But now, there are already many designs."
	Improved productivity	My skill in weaving has improved because I weave anything that needs to be woven. I don't pick and choose. I also weave fast now. Recently, I even tried weaving with a new kind of thread

Dacanay's study showed that a large percentage of the poor targeted by SEPPS are marginalized or socially and economically challenged women. The study reported that SEPPS had positive impacts on the poor, such as increased, diversified, and sustainable sources of income, improved access to social and community services, and movement out of poverty.

The project also upscaled the technical skills of the weavers. According to Weaver E, her skill improved because she weaved different materials and designs. She added that her speed in weaving also improved. Weaver F shared, "I improved a lot here. It's different from before when I used to weave alone at home. In here, I can interact with fellow weavers and I can see the CAD software and the process of designing."

Weaver F's statement supports the importance of integrating CAD technology into traditional craft design to increase productivity. Similar to the findings of Gulati, et al., this study proved that CAD can create new products that promote the convergence of traditional craft and technology with efficiency [19].

4 Conclusion

The Hinablon sa Cebu project has had a three-pronged positive impact on the local community, improving the livelihood of weavers, preserving a traditional craft, and contributing to economic development. Through this initiative, the university has acted as a livelihood intervenor and a social entrepreneur, empowering the community and achieving its mission of economic development. The project has revived a dying industry, sustaining a living cultural tradition and benefiting both weavers and the local community. To preserve and enhance the traditional craft of handloom weaving, future initiatives include CAD documentation of traditional designs. This ensures their preservation and future manipulation, maintaining efficiency, consistency, and accuracy in handwoven product designs.

The implementation of CAD technology to enhance the value of handwoven fabric presents a challenge in terms of the technical expertise required for its utilization. In order to promote sustainability, the university trained both faculty members and enterprise personnel in the digitization of designs and operation of CAD machines to ensure that the knowledge and expertise will be available to future users.

CAD technology plays a crucial role in supporting product development by adding value to handwoven products and streamlining production. Integration of intricate designs and innovative patterns enhances the visual appeal and market value of handwoven products, leading to increased income for the weavers. CAD technology empowers designers and weavers to efficiently incorporate these value-adding elements, elevate the quality of their handwoven creations, benefiting their economic prospects.

Acknowledgment. The authors extend their heartfelt appreciation to the Hinablon sa Cebu communities, advocates, DTI-VII, and CTU Administration whose unwavering commitment and collective efforts have been instrumental in establishing and sustaining a sustainable social enterprise for handloom weaving in Argao, Cebu, Philippines. Their shared vision and selfless dedication have been invaluable in advancing a cause greater than ourselves.

References

1. Leh-Arcena, L.L., Castro, R., Misa, A., Abola, K., Subido, D.: Sustaining handloom weaving in the Philippines, p. 67 (2020). https://www.britishcouncil.ph/sites/default/files/british_council_-_crafting_futures_-_full_report.pdf
2. Sales, T.L.: Argao 400 Years in Legend and History (2008)
3. Respicio, N.A., Yoshikawa-Zialcita, G.E.: Filipino Styles and Techniques: Weaving Ways (2020)

4. Philippine Textile Research Institute, TELA NATION, ONE ASEAN: PTRI 2016 STAKEHOLDERS CONFERENCE (2016). https://ptri.dost.gov.ph/technical-services/86-new-articles/169-tela-nation-one-asean-2
5. Morato, E.A.: Programs, A Guidebook for Designers adn Developers of Livelihood (2016)
6. University, C.T.: CTU: Vision, Mission and Goals. https://www.ctu.edu.ph/vmgo/
7. C. T. University, EXTENSION COMMUNITY Manual of Operations and Policies (2023)
8. van Woller, H.G.G., et al.: Livelihood & Food Security Conceptual Framework Lift – Lift project. Usaid, p. 36 (2011). https://www.marketlinks.org/sites/default/files/resource/files/LIFTESFramework_8172011.pdf
9. Bimber, O., Raskar, R.: Spatial Augmented Reality: Merging Real and Virtual Worlds, 1st edn. A K Peters/CRC Press (2005)
10. Malik, T., Parmar, S.: Value Addition To Textiles (2007). https://www.fibre2fashion.com/industry-article/1830/value-addition-to-textiles#:~:text=VALUEADDITION is the process, i.e. something new %26 unique
11. Dahiya, R., Yadav, S.: Impact of training on value addition in handloom weaving. Int. J. Educ. Manag. 342–345 (2020)
12. Olaguir, L.: Personal Interview (2023)
13. Hani, U., Das, A.K.: Digital technology as a tool for the democratization of powerful mathematical ideas. Rev. Colomb. Educ. **65**(74), 109–132 (2018)
14. Panneerselvam, R.G., Balamurugan, D., Prakash, C., Raja, D.: A study on generating the perspective projection of different weave structures using 3D computer-aided designing tools. J. Textile Inst. 1–13 (2022). https://doi.org/10.1080/00405000.2022.2145449
15. Bowonder, B., Swamy, J., Mastakar, N.: Regaining competitiveness using an ICT platform in a traditional industry: adoption of computer aided design for carpet weaving. Int. J. Serv. Technol. Manag. **6**(3–5), 402–415 (2005). https://doi.org/10.1504/IJSTM.2005.007417
16. Yasmin, S.S., Boruah, R.R., Gogoi, P., Tamuli, P., Mandal, A., Baruah, S.: Significance of CAD software in the development of traditional handloom motifs, vol. 12, no. 6, pp. 1618–1621 (2023)
17. Yadav, S., Verma, M., Rose, N.M.: Consumer Preferences for Development of Textile Products for Entrepreneurial Activity, vol. 11, no. 2, pp. 289–296 (2021)
18. Dacanay, M.L.: Social Enterprises and the Poor Enhancing Social Entrepreneurship and Stakeholder Theory (2012)
19. Gulati, V., Singh, K., Katyal, P.: A CAD paradigm for generating woodworking motifs. Int. J. Comput. Appl. **47**(5), 38–40 (2012). https://doi.org/10.5120/7188-9932

Intelligent Air Quality Monitoring System: A Comprehensive Review

Halima Mahideb[1,2](✉), Sofia Kouah[3], and Ahmed Ahmim[1]

[1] Department of Mathematics and Computer Science, Faculty of Science and Technology, Mohamed-Cherif Messaadia University - Souk Ahras, Souk Ahras, Algeria
mahideb_h@outlook.fr, ahmed.ahmim@lrs-annaba.net

[2] University of Constantine 3, Salah Boubnider, Constantine, Algeria

[3] Department of Mathematics and Computer Sciences, University of Oum El Bouaghi, Oum El Bouaghi, Algeria
sofia.kouah@univ-oeb.dz

Abstract. The Smart City utilizes new information and communication technologies to provide, implement, and promote development strategies that address urban concerns and establish a coordinated, technologically enabled, and sustainable infrastructures. Environmental sustainability is a key indicator that impacts Smart cities advancements. Air quality monitoring and regulation are among the important environmental challenges that contemporary societies face. The World Health Organization stated in a March 2022 publication that air pollution causes the premature deaths of 13 people every minute. Thus, an air quality monitoring system is necessary for smart cities to assist the government in choosing the best course of actions to reduce pollution, as well as to make it easier for residents to avoid exposure to pollutants. Our PhD research aims to provide an intelligent system that deals with air quality monitoring in Smart City based on the Internet of Things [14], optimization algorithms, and deep learning algorithms. This paper reviews some relevant aspects and works related to the air quality field.

Keywords: Smart City · Internet of things · Air Quality · Learning Algorithms · Optimization Algorithms

1 Introduction

A Smart City is considered as an urban setting where technology enables efficient links between data and its application, resulting in a responsive, adaptable, and healthy environment [1]. This integration increases the Smart City's responsiveness, predictability, adaptability, and capacity for learning. We can infer from the foregoing that the concept of the smart city is to develop infrastructure within the city that make use of new technologies in order to enhance the quality, effectiveness, and interactions of urban services while ensuring cost effectiveness. On the other hand, the internet of things (IoT) [15] is obviously connected to several Smart City domains like environment [2]. Thus, developing intelligent frameworks is essential for promoting a smarter environment [3].

K. Kabassi et al. (Eds.): NiDS 2023, LNNS 783, pp. 292–298, 2023.
https://doi.org/10.1007/978-3-031-44097-7_31

Air pollution refers to the modification of the chemical makeup of the atmosphere caused by hazardous gases, mainly anthropogenic in nature and generated by human endeavors in current times [4]. As a result, the air pollution can harm the environment or materials. Gases, particles, and biological molecules are only a few examples of the various forms of air pollution [3]. The World Health Organization has reported that Air pollution causes the premature deaths of 13 people every minute, according to a March 2022 World Health Organization statistics[1]. The presented arguments show the air quality importance and makes it among the primary challenges in Smart City. These statistics highlights the significant environmental health risk presented by air pollution to public health worldwide. Therefore, it is crucial for smart cities to adopt air quality monitoring systems in order to provide citizens with a means to avoid exposure to pollutants and to convince government to make informed decisions to minimize their impact [5]. This paper focuses on the field of air quality monitoring and aims to provide a comprehensive review for the development of an intelligent air quality monitoring system. The remainder of the paper is structured as follows: Sect. 2 outlines the objectives. Section 3 presents related works. Section 4 explains the methodology and expected outcomes. Finally, Sect. 5 concludes the paper and provides an overview of ongoing works.

2 Outline of Objectives

The aim of our PhD research work is to provide contributions related to the intelligent control and monitoring of air quality within the framework of the Smart City. To this end, we aim to propose innovative solutions, including models, methods and strategies, by taking into account the following aspects:

- The various goals of monitoring sites, which have an impact on the responsibilities and the functionalities required as well as the nature of decision-making.
- Decisions made at various levels of the monitoring system, which are typically classified as multilevel decisions since they are dependent on one another.
- Data analysis is an important key for the continuous improvement of the decision-making process as well as for studying and understanding of the phenomena and factors contributing to air quality.
- Considering the organizational structure and distributed nature of the control sites and their governance as well as the inherent characteristics of those systems.

To achieve this, our plan involves utilizing optimization algorithms and Artificial Intelligence techniques including machine learning and deep learning. This choice is motivated by the following requirements: gathering and analyzing control data, facilitating intelligent control decisions, predicting air quality, improving performance across different levels, and taking into account the aforementioned factors.

Furthermore, the performance of the control system can be improved by integrating new technologies among others Internet of Things and Cloud Computing, as well as services managing large volumes of data.

[1] https://www.who.int/multi-media/details/air-pollution-kills-13-people-every-minute.

3 Related Work

As mentioned above, one of the key stones of a smart city management involves providing of air quality control. Thus, many studies focus on new technologies-based air quality system. Herein, we give a summary of some relevant works:

In [6], the authors proposed a domesticated IoT based framework for monitoring air pollution. Two sensors were installed in the University of Rwanda, namely CO2 and PM2.5 sensors. The gathered data is delivered to the cloud through LoRaWAN Protocol-based gateway which ensure interaction between sensors and the cloud part of the network. The results revealed significant amounts of contaminants in the environment. However, additional automated solutions to improve the proposed system is needed, due to the important presence of contaminants in the environment. As a result, the authors' upcoming work will focus on creating and deploying such systems an automated IoT-based cooling systems that will be triggered when pollutant levels exceed recommended limits.

In [3], for monitoring the indoor environment in homes where people have respiratory conditions people, the authors presented an IoT-based monitoring system. This system monitors carbon dioxide (CO2) levels, Particulate Matter (PM2.5 and PM10), as well as daily household activities. The proposed work aims to give feedback of the link between asthma and indoor air pollution. The study will also concentrate on human activities and data-driven behavioral changes to enhance indoor air quality and, consequently, the asthma condition utilizing IoT-based LCS indoor air quality monitoring system.

In [7], the authors developed an IoT powered sensor monitoring system, designed for citizens to have affordable access to indoor air quality data, as well as to educate them on methods to improve air quality in their personal dwellings. As an integral component of two European Union-funded initiatives, the endeavor affords researchers the prospect of amassing quality data for investigating the correlation between the quality of atmospheric air and everyday pursuits in the region of Bradford, situated in the United Kingdom. As perspective, the authors intend to utilize data-based analytics in various ways, such as visualization, interviews, and workshops, to track how citizen behavior changes in response to poor air quality.

In [12], an outdoor air quality and noise surveillance system based on the IoT is proposed. The system collects various data from disparate sources, including crowdsourcing, citizen surveys, social media analytics, mobile and static wireless sensor networks, and to continuously combine objective estimations with subjective evaluations of noise and air quality. Through the application of cutting-edge AI and deep learning methods. The suggested method is able to estimate air pollutant concentration and noise levels at small scales and with sufficient accuracy in big metropolitan areas because to the variety of data sources. The authors also looked at how air quality and noise affected users' everyday routines and how they interacted.

In [8], based on the IoT, a framework for monitoring air pollution in smart cities is proposed. To measure humidity, carbon emissions, temperature, smoke, sound, and other hazardous particles in the atmosphere, the proposed system gathers data from the city's central office, processes it, and sends it forth. A data bank is created using the gathered data so that it can be used later and possibly shared with other academic institutions and environmental organizations. The proposed framework should be evaluated using simulations and deploying multiple sensor nodes throughout the city in order to gather real-time data.

In [9] Researchers created an IoT smart device using affordable sensors to track the air quality in an indoor setting called (IoT-based AQM). They increased the measurement precision of inexpensive sensors and calculated the air quality index using a type 1 fuzzy inference system. The authors intend to employ this method in the future to estimate the air quality model using machine learning techniques and data collected over a 12-month period by an IoT-based AQM device. Furthermore, they plan to apply type-2 fuzzy sets to improve the BI-CO solution.

In [2] the researchers proposed a system for monitoring air pollution in smart cities, based on the IoT and cloud computing. They have designed and implemented a smart mobile application that monitors the air pollution index and through the data obtained provides real-time recommendations to citizens. Despite the fact that the application provides a significant novelty in the form of real-time recommendations based on the parameters measuring air pollution, it requires expansion in the areas of intelligent environmental monitoring. Moreover, the authors also aim to enhance the application by integrating sensors and cameras into the city's already-existing street poles that are powered by those circuits. Sensors should be small, adaptable, inexpensive, able to be calibrated, and Internet access enabled. The authors' future directions focus on security concerns by integrating cloud computing services, and also on enhancing the application through weather and pollution forecasting, in addition to providing recommendations.

The authors in [10] described a design for a platform to monitor air quality and analyze data called AirQ platform. The authors are required to demonstrate this platform. The AirQ platform, as mentioned by the authors, is an affordable alternative for real-time air quality monitoring at specified locations. The AirQ can deliver air quality data that is strongly correlated with the data from the NEA's air quality station at a minuscule fraction of its cost by means of sensor calibration technique. As a result, the AirQ platform can be used to create extensive and dense networks for monitoring air quality in smart cities. Various locations such as schools, daycare facilities, hospitals, community centers, bus stops, parks, etc. can benefit from the implementation of AirQ devices in the future.

In [11], the authors suggested a system for tracking vehicles that produce pollution above a set limit, monitoring ambient air quality on highways, and controlling traffic signals and deflection intelligently to cut down on emissions of pollutants and save time and money. This system based on the IoT and big data technologies. They used a visual real time Air Quality Meters (AQM), Electrochemical Toxic Gas Sensors and AQI to monitor pollution levels. Due to the high energy consumption and the associated operating and maintenance costs of the proposed solution, the authors intend to run out of solar energy sources in the future to alleviate these challenges.

In [13] the authors proposed an innovative model for smart cities based on Social IoT. The system monitors traffic congestion and air quality in the city by tracking all modes of transportation as well as pedestrians and measuring air quality using both mobile and fixed sensors. For the purpose of advising citizens on traffic and improving it, real-time data is gathered and analyzed. Various challenges are encountered at each level of the proposed method. Further efforts are needed to improve the sensing nodes' accuracy, reduce their energy requirements, simplify deployment procedure, and minimize their environmental impact.

Table 1 recapitulates the presented works and highlights their feature keys.

Table 1. Comparison of air quality monitoring systems in smart city

Reference	Considered Parameters	Environment	Data access	Attention
[6]	CO_2, And PM2.5	Indoor	Cloud server	Necessitate incorporating remedies rather than just monitoring
[3]	CO_2, and PM	Indoor	machine learning models	Give the relationship between asthma and indoor air pollution
[7]	BME680 SGP30, PM CO, NO_2 and NH_3	Indoor	Cloud server	Explore the relationship between air quality and daily activities
[9]	CO and PM	Indoor	Cloud server	Estimate the proposed model using machine learning and data gathered
[8]	CO_2 and NO_2	Outdoor	Web and mobile application	the suggested system did not simulated
[2]	PM,CO,O_3, NO_2 and SO_2	Outdoor	Cloud server	Extension in the aspect of smart environmental monitoring and user assistance
[10]	PM,CO_2, CO,NO_2 and SO_2	Indoor	Cloud server	Real-time air quality data and alerts are provided by a smartphone app
[11]	CO,NO and SO_2	Outdoor	Web server	Employ solar energy, which would be less expensive to use and maintain

(continued)

Table 1. (*continued*)

Reference	Considered Parameters	Environment	Data access	Attention
[12]	PM, Uv And IR	Outdoor	Cloud server	Implemented in a specific area of Cagliari
[13]	SO2, CO, CO2, NOx and PM	Outdoor	Web server	Obstacles to overcome: energy requirements, lessening the impact of the devices on the environment, and enhancing sensor accuracy

4 Methodology and Expected Outcomes

In our PhD research, we aim to develop an advanced intelligent air quality monitoring system by employing several technologies such as Artificial intelligence, particularly machine learning algorithms, and Deep Learnings algorithms as well as optimization techniques, Cloud Computing and Internet of Things. The review of existing literature has identified key requirements that should serve as the basis of the system to be proposed. These requirements include: gathering and analyzing air quality data, decision support for intelligent control, air quality forecasting, and performance optimization across multiple levels while considering the aforementioned considerations.

5 Ongoing Work and Future of the Research

A comprehensive review of relevant literature has revealed key requirements that serve as the central elements of our envisioned system. It aimed at pinpointing existing gaps in knowledge within our research area. An ongoing work focus on the study of additional pertinent parameters and features, that meaningfully impact system efficiency, to be gathered and analyzed rigorously. As perspective, after recognizing such parameters, we intend to employ machine learning, Deep Learning ang optimization algorithms to existing air quality datasets, as well as proposing an intelligent air quality monitoring system for smart city leveraging the IoT.

References

1. Carlos Augusto, J. (ed.): Handbook of Smart Cities
2. Jovanovska, E.M.: No pollution Smart City Sightseeing Based on WSN Monitoring System; No pollution Smart City Sightseeing Based on WSN Monitoring System (2020)

3. John, R., Kureshi, R.R., Thakker, D., Mishra, B.K.: Internet of Things (IoT) and Indoor Air Quality (IAQ) monitoring in the health domain. In: ACM International Conference Proceeding Series, Association for Computing Machinery, pp. 215–218 (2021). https://doi.org/10.1145/3494322.3494704
4. Tayeb, M.S., Benallal, M.A., Benabadji, M.S., Houari, A.: IoT monitoring system for air quality assessment and collecting data. Indonesian J. Electric. Eng. Comput. Sci. **28**(3), 1592–1600 (2022). https://doi.org/10.11591/ijeecs.v28.i3.pp1592-1600
5. Rhaiem, M., Chaabane, M., Bouassida Rodriguez, I.: Air Quality System in Smart City: Systematic Literature Review (2022). http://ceur-ws.org
6. Twahirwa, E., Mtonga, K., Ngabo, D., Kumaran, S.: A LoRa enabled IoT-based Air Quality Monitoring System for Smart City. In: 2021 IEEE World AI IoT Congress, AIIoT 2021, pp. 379–385. Institute of Electrical and Electronics Engineers Inc. (2021). https://doi.org/10.1109/AIIoT52608.2021.9454232
7. Kureshi, R.R., Thakker, D., Mishra, B.K., John, R.: AQ-SCIENCE: air quality - smart cities with iot-enabled citizen engagement approach. In: ACM International Conference Proceeding Series, Association for Computing Machinery, pp. 177–180 (2021). https://doi.org/10.1145/3494322.3494354
8. Ali, A.: A framework for air pollution monitoring in smart cities by using IoT and smart sensors. Informatica (Slovenia) **46**(5), 129–138 (2022). https://doi.org/10.31449/inf.v46i5.4003
9. Zareb, M., Bakhti, B., Bouzid, Y., Batista, C.E., Ternifi, I., Abdenour, M.: An intelligent IoT fuzzy based approach for automated indoor air quality monitoring. In: 2021 29th Mediterranean Conference on Control and Automation, MED 2021, pp. 770–775. Institute of Electrical and Electronics Engineers Inc. (2021). https://doi.org/10.1109/MED51440.2021.9480313
10. Lim, H.B., Iqbal, M., Yao, Y., Wang, W.: A Smart e-Science Cyberinfrastructure for Cross-Disciplinary Scientific Collaborations, pp. 67–97 (2010). https://doi.org/10.1007/978-1-4419-5908-9_3
11. Padmanabha Reddy, Y.C.A., Parameswaran, T., Sathiyaraj, R.: A smart environment monitoring framework using big data and IoT. In: 2021 IEEE Mysore Sub Section International Conference, MysuruCon 2021, pp. 399–404. Institute of Electrical and Electronics Engineers Inc. (2021) https://doi.org/10.1109/MysuruCon52639.2021.9641609
12. Fadda, M., Anedda, M., Girau, R., Pau, G., Giusto, D.D.: A social Internet of Things smart city solution for traffic and pollution monitoring in cagliari. IEEE Internet Things J **10**(3), 2373–2390 (2023). https://doi.org/10.1109/JIOT.2022.3211093
13. Bardoutsos, Filios, G., Katsidimas, I., Krousarlis, T., Nikoletseas, S., Tzamalis, P.: A multidimensional human-centric framework for environmental intelligence: air pollution and noise in smart cities. In: Proceedings - 16th Annual International Conference on Distributed Computing in Sensor Systems, DCOSS 2020, pp. 155–164. Institute of Electrical and Electronics Engineers Inc. (2020). https://doi.org/10.1109/DCOSS49796.2020.00036
14. Sofia, K., Ilham, K.: Multi-layer agent based architecture for Internet of Things Systems. J. Inform. Technol. Res. **11**(4), 32–52 (2018)
15. Sofia, K., Ilham, K.: Internet of Things agents diagnosis architecture: application to healthcare IoT system. In: 3rd International Conference on Advanced Aspects of Software Engineering (ICAASE'18). Constantine 2 University, on December 01–02, 2018

One Dimensional Fingerprinting as an Alternative to the Free Space Path Loss Equation for Indoor Positioning

Dimosthenis Margaritis(✉), Helen C. Leligou, and Dimitrios G. Kogias

University of West Attica, Egaleo, Greece
dimomarg@gmail.com

Abstract. This paper proposes an alternative approach to indoor navigation using One Dimensional Fingerprinting (1DFP) as an alternative to the Free Space Path Loss (FSPL) equation. The method involves producing a distance/RSSI table for each beacon and using linear interpolation to estimate distances for unknown RSSI values. Trilateration is then used to convert these distance estimates into a position in space. The proposed method was tested in a rectangular room using three beacons and compared to expected values. Results showed significantly better accuracy than comparable studies using FSPL, with mean errors in distance estimation of 170, 145 and 146 cm. The study highlights the unsuitability of FSPL for indoor environments and the potential for further improvements in accuracy.

Keyword: Bluetooth Low-Energy · Indoor Positioning · Trilateration

1 Introduction

The rapid growth of the Internet of Things (IoT) has brought about a need for indoor navigation systems that can accurately locate and track devices and users within indoor environments. While satellite navigation systems are effective outdoors, they are largely unusable indoors due to signal attenuation and obstruction by walls and other structures. Therefore, the conventional approach to indoor positioning is to use Bluetooth Low Energy (BLE) beacons[1]. This is most often implemented by using the Free Space Path Loss (FSPL) formula to convert Received Signal Strength Indices (RSSI) to beacon distance and trilateration to convert distances from known beacons to a set of coordinates. This, however, yields highly inaccurate results, as the FSPL formula has been shown to be inadequate at modelling signal propagation in closed spaces [1].

An alternative approach to using FSPL and trilateration is fingerprinting [5]. Fingerprinting involves collecting a large dataset of signal strength values at different locations within the indoor environment and then producing a location estimate from a set of RSSIs using the k Nearest Neighbors algorithm. While fingerprinting can be significantly more

Supported by the University of West Attica.

[1] The techniques described below can also be implemented using Wi-Fi. [6] Such implementations yield higher accuracy at the cost of exponentially worse battery life.

K. Kabassi et al. (Eds.): NiDS 2023, LNNS 783, pp. 299–303, 2023.
https://doi.org/10.1007/978-3-031-44097-7_32

accurate, it requires the creation of a large dataset, a very labor-intensive process which is often prohibitive for practical applications.

Attempting to strike a balance between accuracy and feasibility, we propose using One Dimensional Fingerprinting (1DFP) as an alternative to the FSPL equation. 1DFP is implemented by producing a distance/RSSI table for each beacon and using linear interpolation to produce a distance estimate for unknown RSSI values.

2 Related Studies

Expected results in setups similar to our own in terms of room geometry and beacon count can be found in [4] and [5], with the former reporting an average error value of 896 cm and 797 cm at 5 and 10 measuerments using the conventional FSPL-trilateration method. Assuming the error value follows simple exponential decay, sampling a beacon producing 10 advertisements per second for 1 min. would yield 320 cm of mean error. The latter study was done in the same arrangement using Fingerprinting, resulting in significantly lower error values (233 cm and 194 cm across the width of the room and 345 and 259 cm across the length of the room for one and five measuremenets, respectively).

A theoretical maximum value for accuracy was set in [1]. In that study, an error value of 3–7% was achieved with sampling durations exceeding 1 h and using Taylor series to model the distance/RSSI relation, however, the study does not account for the possibility of that being non injective. Methods for sampling and accuracy enhancement in envioronments with more than three beacons were tested in [2]. Lastly, also notable is [3], being a study of the effects of increased beacon count on accuracy.

3 Proposed Method

Our method for estimating a user's position given a set of beacons of known coordinates is as follows.

- Measure RSSI as a function of distance for each beacon.
- As the produced tables are intended for use in linear interpolation, the possibility of non injective RSSI/Distance correlations is a problem. A simple method of ameliorating this is eliminating the least amount of measurements needed to produce an injective function.
- Given a set of unknown RSSI values, use linear interpolation on the corresponding beacons to produce a set of distance estimations.
- Use trilateration to convert these values to a position in space. Trilateration is implemented as in Pakanon et al., 2020 [4].

$$A_1 = 2 * (x_{b2} - x_{b1})$$
$$A_2 = 2 * (x_{b3} - x_{b1})$$
$$A_3 = 2 * (x_{b3} - x_{b2})$$
$$B_1 = 2 * (y_{b2} - x_{b1})$$
$$B_2 = 2 * (y_{b3} - x_{b1})$$
$$B_3 = 2 * (y_{b3} - x_{b2})$$
$$C_1 = d_{b1}^2 - d_{b2}^2 + x_{b2}^2 - x_{b1}^2 + y_{b2}^2 - y_{b1}^2$$
$$C_2 = d_{b1}^2 - d_{b3}^2 + x_{b3}^2 - x_{b1}^2 + y_{b3}^2 - y_{b1}^2$$
$$C_2 = d_{b2}^2 - d_{b3}^2 + x_{b3}^2 - x_{b2}^2 + y_{b3}^2 - y_{b2}^2$$
$$\begin{bmatrix} x \\ y \end{bmatrix} = \begin{bmatrix} A_1^2 + A_2^2 + A_3^2 & A_1B_1 + A_2B_2 + A_3B_3 \\ A_1B_1 + A_2B_2 + A_3B_3 & B_1^2 + B_2^2 + B_3^2 \end{bmatrix}^{-1} * \begin{bmatrix} A_1C_1 + A_2C_2 + A_3C_3 \\ B_1C_1 + B_2C_2 + B_3C_3 \end{bmatrix}$$

Where x_{bn} and y_{bn} represent the x and y coordinates of beacon n and d_{bn} the distance to it.

Linear interpolation is implemented thus:

$$d = d_{\downarrow} + (s - s_{\downarrow}) \frac{d_{\uparrow} - d_{\downarrow}}{s_{\uparrow} - s_{\downarrow}}$$

With $s_{\downarrow}$ and $s_{\uparrow}$ representing the largest smaller and smallest larger known RSS indices and $d_{\downarrow}$ and $d_{\uparrow}$ their respective associated distances. d is the yielded distance and s is the input signal. The result is clamped between the largest and smallest values in the dataset.

4 Measurement Method

All measurements were conducted at the University of West Attica's Ancient Olive Grove Campus, Hall ZA216, a rectangular room measuring 8.1 over 8.02 m. Three beacons were arranged around the room. Their advertising frequency and transmission power were set to maximum (10 Hz and 4d BmW, respectively). Subsequently, the coordinates of the beacons were determined. Then, for each beacon, its Received Signal Strength Index was measured as a function of distance along a line starting from the beacon and crossing the center point of the room, yielding the results shown in Fig. 1.

The measurements were taken using a smartphone which was held by the operator during the entire process. Each data point had a sampling time of one minute, during which the operator moved their arm in a circular motion to average out any potential resonance spots that could have impacted the readings. This methodology was chosen to ensure the accuracy and consistency of the measurements, as any fluctuations or inconsistencies in the received signal strength index could have affected the overall results.

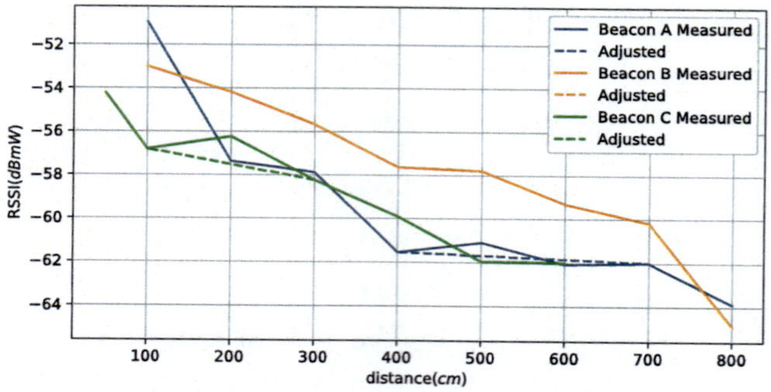

Fig. 1. RSSI as a function of distance

5 Accuracy Study

Twelve measurement points were designated, as shown in Fig. 1. Following the measurement method detailed in the previous section, each beacon was sampled at each point for one minute. Distance estimates to each beacon were produced at each point, resulting in mean error values of 170 cm, 145 cm and 146 cm. The sets of estimated distances at each point were then used for trilateration, which resulted in 166 cm of mean error. Results and comparisons to expected values are presented in Fig. 2.

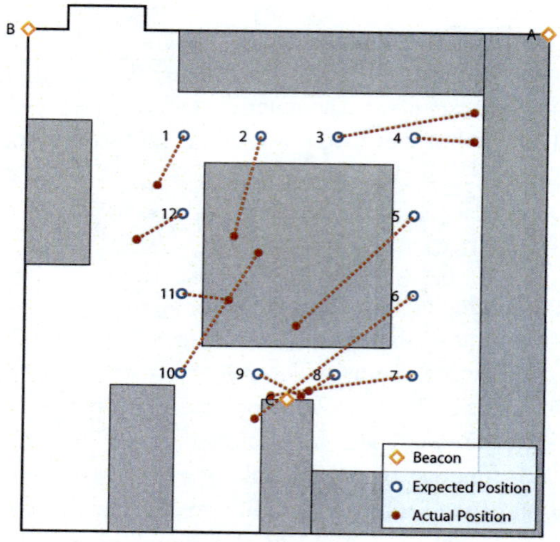

Fig. 2. Actual positions vs expected positions

6 Conclusions

Our method yielded significantly better results than comparable studies utilizing BLE beacons and the FSPL formula 4, 3]. Our results are indicative of the FSPL formula being obsolete for distance estimation in indoor environments and set a new baseline for accuracy in trilateration. Further improvements in accuracy appear possible, however a study on the reliability of commercial BLE beacons in terms of providing consistent transmission strength is warranted, as well as an accuracy study on the BLE transceivers of modern smartphones.

Acknowledgements. This study was conducted at the University of West Attica's Laboratory of Computational Intelligence and Intelligent Systems and partially financed by University of West Attica and by the European Regional Development Fund (ERDF) within the Regional Operational Program "Attica" under the call "Research and Innovation Synergies in the Region of Attica" (Project name: OTPORER-PPE, Project code: ATTP4-0353772).

References

1. Astafiev, A., Zhiznyakov, A., Privezentsev, D.: Development of indoor positioning algorithm based on bluetooth low energy beacons for building RTLS-systems, pp. 1–5 (2019). https://doi.org/10.1109/RUSAUTOCON.2019.8867751
2. Bembenik, R., Falcman, K.: Ble indoor positioning system using rssi-based trilateration. J. Wirel. Mob. Networks Ubiquitous Comput. Dependable Appl. **11**(3), 50–69 (2020)
3. Ji, M., Kim, J., Jeon, J., Cho, Y.: Analysis of positioning accuracy corresponding to the number of ble beacons in indoor positioning system. In: 2015 17th International Conference on Advanced Communication Technology (ICACT), pp. 92–95. IEEE (2015)
4. Pakanon, N., Chamchoy, M., Supanakoon, P.: Study on accuracy of trilateration method for indoor positioning with BLE beacons, pp. 1–4 (07 2020). https://doi.org/10.1109/ICEAST50382.2020.9165464
5. Phutcharoen, K., Chamchoy, M., Supanakoon, P.: Accuracy study of indoor positioning with bluetooth low energy beacons, pp. 24–27 (2020). https://doi.org/10.1109/ECTIDAMTNCON48261.2020.9090691
6. Raza, A., Lolic, L., Akhter, S., Liut, M.: Comparing and evaluating indoor positioning techniques, pp. 1–8 (2021). https://doi.org/10.1109/IPIN51156.2021.9662632

Check for updates

The Interaction of Disabled People and New Forms of Packaging: A Holistic Approach

Maria Poli[1](✉), Konstantinos Malagas[2], and Spyridon Nomikos[3]

[1] Department of Interior Architecture, University of West Attica, Aigaleo, Greece
mpoli@uniwa.gr

[2] Laboratory of Informatics and New Technologies in Shipping, Transport, and Insular Development (LINTSTID), Department of Shipping Trade and Transport, University of the Aegean, Chios, Greece
Kmalagas@aegean.gr

[3] Department of Graphic Design and Visual Communication, University of West Attica, Aigaleo, Greece
nomic@uniwa.gr

Abstract. Disabled people require particular services and a friendly environment to achieve accessibility, autonomy, and inclusion in every aspect of their daily lives. All related stakeholders should contribute to these targets. The shopping experience and the new forms of packaging that incorporate new interactive technologies are highly important for disabled people. The study aims to emerge a conceptual framework that includes beyond disabled people, the issues of packaging, purchase, and product management, and how the various stakeholders may impact those. The study sample included people with mobility, vision, and hearing problems and their caregivers (n = 12). The Soft Systems Methodology (SSM) was applied to collect the required information. Initially, a detailed discussion in a focus group took place about those people's problems when interacting with packaging and shopping; furthermore, based on the relevant literature, all agreed and designed a rich picture and the study's outcome. In addition, interviews with three (3) persons led to a related stakeholder analysis and then answered some general specific questions (PQR). The close cooperation of people with disabilities and private and state-owned organisations based on the real needs of these people is strongly recommended. This cooperation leads to providing higher quality services to disabled persons, and specific actions should be taken to facilitate their shopping experience, mainly through advanced forms of packaging. The study offers valuable insights to the disabled people community, practitioners, policy-makers and academicians.

Keywords: Persons with disability (PwD) · new technologies · intelligent packaging

1 Introduction

People with disabilities (PwD) consist of a significant part of the population. On a global basis, more than 1.3 billion people live with some form and degree of disability, representing at least 17% of the world's population and making up the world's largest minority

K. Kabassi et al. (Eds.): NiDS 2023, LNNS 783, pp. 304–315, 2023.
https://doi.org/10.1007/978-3-031-44097-7_33

group. They posit more than $13 trillion in annual disposable income [1]. In particular, 360 million people worldwide have moderate to profound hearing loss, 285 million people are visually impaired (39 million of whom are blind), and 75 million people require a wheelchair to move (of which only 5–15% have access to one) [2]. Shopping is vital for PwD, requiring a facilitative environment to manage this. Thus, disabled people need additional assistance in shopping, particularly accessible information, as this enforces their independence, and technology offers valuable services towards this direction. Thus, incorporating technology (such as RFID (Radio Frequency Identification)) in packaging can further improve the provided services. Intelligent packaging monitors the condition of packaged foods by providing information regarding the quality of the packaged food during transport and storage [3]. In addition, this form of packaging, through the use of time-temperature indicators (T.T.I.s) or food freshness indicators, reacts immediately to significantly changing conditions that affect the state of packaged food [4, 5]. All consumers may benefit considerably from these functions, including PwD, where providing the required information encourages their inclusion in the economy and society [6].

However, PwD live in a broader environment and require advanced services in every aspect of their daily lives. All related stakeholders and public and private organisations should closely cooperate in providing a facilitative environment that satisfies the evolving needs of PwD.

Soft System Methodology (SSM) was applied to collect the required information to respond to the study's objectives. Participants with problems in mobility, vision, and hearing problems and their caregivers ($n = 12$) participated in a focus group discussion about the issues and the difficulties they faced in their shopping experience. They designed a rich picture and the outlook of the current study. In addition, interviews with 3 participants ($n = 3$ - people with mobility problems and a caregiver) were done. Those led to related stakeholder analysis and answered general specific questions (PQR). Consequently, the study suggested adopting a more holistic approach where all the associated stakeholders should closely cooperate with the PwD community (PwD, their families, caregivers and PwD associations) and monitor these people's real needs. Based on these, public and private organisations must offer PwD the appropriate services and introduce specific actions to facilitate their shopping experience, mainly through advanced forms of packaging. The study provides valuable insights to the PwD community, practitioners, policy-makers, and academicians.

2 Literature Review

In this part of the study, the packaging and disability issues are briefly presented.

2.1 Packaging

Packaging is vital for marketing and consumers, preserving food, extending its shelf-life, and leading to the highest quality and safety along the production chain and during storage [7]. Thus, companies based on attractive and helpful packaging can increase sales, providing the services consumers require [8]. Incorporating advanced technologies in shopping and packaging is associated with the smart economy and the increased use of e-commerce, leading to higher productivity, advanced services, and sales of products [9].

New advanced technologies based on the developments in information and communication technologies (ICT) offer new intelligent services to consumers, in particular, in shopping and packaging (where packaging consists of a "smart object" communicating with other "smart objects" and the surrounding environments) [10]; in addition, new materials creating packages that can carry readable data and information [11]. Thus, e-commerce is gaining strength by using RFID tags against counterfeit products, which also positively contributes to their distribution [12]. Furthermore, the technology offers innovative solutions to facilitate the transition from conventional packaging to intelligent packaging. Innovative packaging through indicators and sensors ensures the safety of the products, but also through the provided information enabling the user to manage their stocks and waste [13]. Innovative packaging has communication capabilities, uses information systems, and facilitates a new communication behaviour through traceability in the supply chain [14]. However, intelligent packaging extends to smart packaging. It includes information about the product condition, date, place of production, storage or cooking method, and origin, and all these may be able to transmit to consumers [15]. Thus, intelligent packaging aims to continuously monitor food quality during the whole chain and access and reduce the insurgence of food-borne diseases and food waste [7]. Both forms of packaging positively contribute to product recycling, improve waste management and protect the environment, and assist consumers in using more accessible products and services, particularly PwD, who need these services.

2.2 Disability

PwD often need more in their environmental living options. According to the State of the World's Cities 2010/2011 [16], PwD, along with older people and slum dwellers, are the three main groups of people who appear to be the most systematically excluded in various cities in Asia, Africa, Latin America, and the Caribbean.

Disability is defined as "any condition of the body or mind (impairment) that makes it more difficult for the person with the condition to do certain activities (activity limitation) and interact with the world around them (participation restrictions)" (CDC, nd). According to World Health Organization [17] figures, about 1.3 billion people experience significant disability (16% of the world population or 1 in 6 of us). These figures will rise, and it is estimated that over 20% of the global population will be over 65 by 2050, and most of them will have health problems. PwD usually experience more limitations in their daily life than others. Thus, those people find difficulties using mainly public services; in particular, they find 15 times more inaccessible and unaffordable transportation than those without problems. The main types of disabilities impact a person's: vision, movement, thinking, remembering, learning, communication, hearing, mental health, and social relationships [18].

Recently, a shift has been observed from the 'medical model' to the 'social model' where PwD are viewed as being disabled by society rather than their bodies [19]. However, these models are presented as dichotomous [20]; a more balanced approach should be adopted, giving the appropriate emphasis to the different aspects of disability [21, 22].

In addition, a facilitative regulatory framework is required to promote PwD rights [23]. At the global stage, the United Nations Convention on the Rights of Persons with Disabilities (UNCRPD) [24] is the general standard to which regions and countries adjust their conditions. Currently, areas and countries implement regulations that protect the rights of inclusion and accessibility of PwD in social activities. Still, the main problem is the level of implementation of these regulations, as there are cases where they still need to be fully applied.

Greater access of PwD to the level of social protection can improve their capabilities but also potentially contribute to the reduction of poverty for the entire population [25]. Therefore, the target should be the establishment of an environment where products and services are more accessible, and this allows more inclusion and participation of citizens in society. High accessibility to websites, products and services is critical, and the role of technology in these targets is indisputable. Smart and intelligent packaging facilitates the shopping experience of PwD. The easy interaction of packaged foods with PwD provides significant benefits to all stakeholders, such as those people, the rest of the consumers, the governments, the economy, and the society in total, and this should be encouraged.

Packaging and disability issues are examined in the specific study, using the appropriate methodology to respond to the study objectives.

3 Methodology

The Soft Systems Methodology (SSM) was applied in the current study. SSM was initially presented by Checkland (1981) [32] to examine complicated issues that arise in everyday life. The stakeholder theory focuses on the interconnected relationships that should be developed between a company and its customers, suppliers, employees, investors, communities, and others with interests in an organisation [26]. Initially, this theory has applied in Project Management. The key stakeholders mainly impact a project's outcome and are further classified as internal and external [27]. In particular, this theory is applied in a research or project to investigate:

- the beneficiaries involved,
- the stakeholders who will be affected either positively or negatively, as well as.
- the stakeholders who have the potential to positively or negatively influence the outcome and results of a project system [28–30].

The particular theory focuses on the prosperity of stakeholders [31] and the increased role of cooperation and shared values in improving performance and proposes that a firm should create value for all stakeholders, not just shareholders [26].

The detailed analysis of the stakeholders is essential, and the creation of a relevant map assists in quickly identifying and managing project stakeholders who can come from every level of an organisation and outside of this; those stakeholders that have a high and low impact and those that have a high or low interest in the specific work [27].

Finally, the detailed stakeholder analysis leads to the following benefits: i) gain more support and resources through the identification of the real stakeholders, ii) increase project visibility, iii) prevent costly obstacles that can arise later on in the project life,

iv) encourage the communication through the proper channels at the right time, and v) share the right level of information with stakeholders [27].

The starting point of the SSM examination is a so-called problematic situation in which one of the involved persons (probably a researcher) feels that there is room for improvement. Still, they are determining what measures should be taken to achieve this. Thus, based on the worldviews (the opinions of various involved persons for a specific subject), a detailed insight into how people involved in the situation should act and for what purpose is achieved. These worldviews can be further discussed to find the best solution for change that is arguably desirable and culturally feasible [33]. Applying SSM, many novel methods, tools, and techniques have been developed, such as rich pictures, the PQR formula, CATWOE (Customer, Actor, Transformation, Worldview, Owner and Environment), and purposefully activity model. The PQR formula is helpful, focusing on the functions a system should perform and remaining at least functioning on a basic level, answering the following questions: what should be done? (P), how it should be done? (Q), and why it should be done? (R) [34].

SSM was selected as the researchers found it important to examine the problems that PwD face in their shopping experience (the problematic situation). In particular, when PwD are using packaged foods and products, and to identify the related stakeholders and how these may contribute to solving the problematic issues. A focus group detailed discussion about the problematic situation, the study's outlook including a rich picture, some interviews about the matters examined, and a PQR formula were utilised to collect the required information.

Focus group is a non-standard technique of information gathering, based on apparently informal discussion and interaction among a group of people, where the roles of moderator and observer are essential to lead the discussion according to 'the cognitive purposes outlined', observe non-verbal behaviours and collect non-verbal information [35]. Thus, both the moderator (the first author of the current paper) and the observer (an assistant to the moderator) had a pivotal role in collecting all the necessary information in the focus group discussion, with the participation of twelve (12) PwD (n = 12). In particular, three (3) participants and one (1) caregiver/escort from the groups of people with mobility, vision, and hearing problems participated in the focus group discussion about the problematic situation. After the detailed discussion, all the participants agreed and designed a rich picture and the study's framework with the moderator's assistance. Then, two (2) PwD with mobility problems and one (1) caregiver, through unstructured interviews, identified the relevant main stakeholders, their role, and their impact (stakeholder analysis). Finally, the PQR formula was applied, and the participants (n = 3) answered three specific questions (PQR). The above sources collected the required information to respond to the study's objectives successfully.

4 The Study's Findings

In this part of the paper, the description of the main problems that PwD face when visiting shopping areas and using packaged products, followed by the design of a rich picture and the outlook of the current study, then the stakeholder analysis, and finally, the answers to the specific general questions (PQR) are presented.

4.1 The Description of the Problematic Situation

The focus group participants pointed out the following issues:

- All the participants agreed that they have significant problems with accessibility and autonomy in using packaging and, more broadly, in purchasing products.
- Private businesses should invest in offering services that satisfy the PwD needs; their employees should assist these people and focus on the relevant staff training.
- Packaging designers should use the appropriate technology to facilitate the shopping experience of PwD.
- Public Administration should give importance to the PwD-offered services, creating the public infrastructures that facilitate these people's accessibility and autonomy and should create the corresponding protective regulatory framework.

4.2 The Rich Picture

With the participants' contribution (n = 12), the focus group moderator and her assistant designed the following figure (Fig. 1) that contains the study's framework. Thus, based on the detailed discussion in the focus group and the relevant literature, the problematic situation was identified that PwD meet when visiting shopping areas and using the new packaging forms. In addition, this figure includes a rich picture that shows the complex shopping environment for PwD, and the involved people, organizations and PwD associations can positively contribute to this problem.

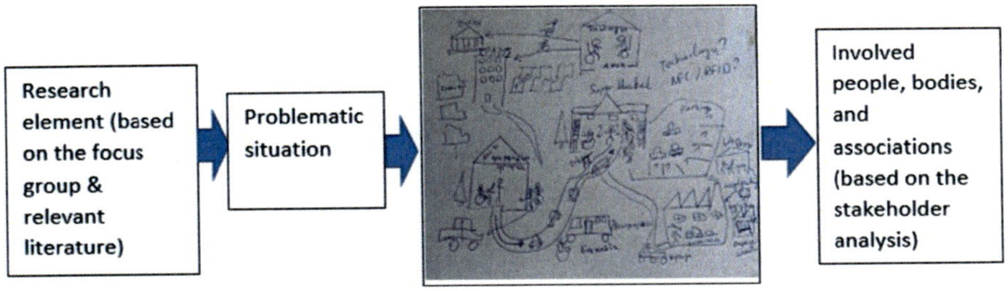

Fig. 1. The outlook of the study includes a rich picture

4.3 Stakeholder Analysis

In addition, the knowledge and experience of some of the participants in the focus group were used to analyse the involved stakeholders. In particular, two (2) PwD with mobility problems and one caregiver from this group were interviewed and asked to examine the related stakeholders. Thus, in the following Table 1, the interviewees identified the main stakeholders, their primary tasks, the level of importance of their operation and intervention, and their expectations from them.

Table 1. Stakeholder Analysis

Involved	Main tasks	Importance (high, medium, low)	Expectations from them
Government	Develop the 'appropriate legislation', issue regulations-circulars, monitor international developments, and deal with the whole life of the PwD	High	Improve the standard of living of the PwD, and solve their problems (financial, social, etc.)
Parliament	Approve and control legislation	High	Improve the standard of living of PwD and oversee the government's work
Authorities/Public Administration	Execution of the relevant legislation	High	Provide high-quality services to PwD
Commercial enterprises	Providing friendly PwD services and attracting them as consumers	Moderate	Improve PwD quality of life by providing satisfactory services, and through this, can achieve economic benefits to all (economy, enterprises, societies)
Employees of commercial enterprises	Providing friendly PwD services and attracting them as consumers	High	Provide satisfactory services to PwD as consumers (improving their satisfaction)
Manufacturing companies	Providing friendly PwD services and attracting them as buyers of products	Moderate	Attract PwD as their product buyers, and increase their customer base and revenues
Packaging production companies	Attracting the PwD as consumers (use of appropriate packaging)	Moderate	Attract PwD as their product buyers, and increase their customer base and revenues

(*continued*)

Table 1. (*continued*)

Involved	Main tasks	Importance (high, medium, low)	Expectations from them
Packaging designers	Provide friendly to PwD package design and attract them as consumers	Moderate	Attract PwD as their product buyers, and increase their customer base and revenues
Associations/Organizations/Non-Governmental Organizations (NGOs)	Support PwD needs	High	Focus on the improvement of the accessibility – autonomy, quality of the PwD life, and protect their rights
Scientific community	Support & research PwD needs and suggest solutions	High	Identify new ways to improve the accessibility – autonomy, and quality of the PwD life
Caregivers/Family environment	Support PwD's daily living	High	Assist PwD in their daily activities

4.4 The PQR Formula

Table 2 below presents the participants' answers (n = 3) to the specific questions (PQR).

Table 2. Answers to the general questions

Question	Answer
What should be done? (P)	The top priorities are accessibility and autonomy improvement in the product market for PwD
How it should be done? (Q)	Improving product management through packaging (on a practical and environmental level) can be achieved through the close cooperation of PwD, the industry and public authorities. In addition, the research community can contribute to this
Why it should be done? (R)	These people will be satisfied by providing those services that PwD need; the customer base will increase with significant benefits for all (PwD, the economy and society)

5 Discussion and Conclusions

Disabled persons, their families, and caregivers consist a significant proportion of the population, and public and private enterprises should not ignore them. This group requires a facilitative environment to satisfy its primary needs and achieve a satisfactory standard of life. The shopping experience is significant for PwD, and the incorporation of advanced technologies in the packaging (smart packaging, intelligent packaging), makes their shopping experience easier and friendlier.

In the broader environment where PwD live, there are barriers, such as attitudinal, architectural, ecological, interactive, communication, and intrinsic, which depend on each person's physical, psychological, or/and cognitive functions [36]. The entire community is responsible for minimising these barriers and implementing the appropriate strategies and policies [37]. Many stakeholders are involved in providing high-quality services to PwD to enjoy their shopping using advanced forms of packaging (smart packaging, intelligent packaging). Private and state-owned enterprises are engaged in the provision of high-quality services to PwD. The first ones should find out the real needs of PwD, train their staff to provide good services to these people, and use the appropriate technology, particularly in packaging, to facilitate the shopping experience of PwD. The state-owned organisations should create the proper infrastructure and introduce a facilitative regulatory framework that improves accessibility and autonomy, including using technology in packaging. Moreover, the research community can assist in this direction by making relevant studies that lead to finding the services in shopping and packaging that ideally fit the real needs of PwD. PwD families' and caregivers' involvement is also essential, as both know these people's needs and behaviours well. The above issues were proved in the current study, and the following Fig. 2 conclusively presents specific measures that offer solutions to the problematic situation.

Conclusively, identifying the real needs of PwD and the close cooperation between the PwD community, the industry, and the state is highly recommended. Based on these two issues, improvements in shopping areas providing higher accessibility through constructing the appropriate infrastructure, training employees, and introducing intelligent packaging and a facilitative regulatory framework and processes are essential for PwD. The stakeholder analysis shows that disabled people seem to trust the government, the public authorities, the employees of private enterprises, the research community, and their families and caregivers more to improve their shopping experience. In addition, innovations in packaging friendly to PwD needs should be encouraged. Therefore, a more holistic approach should be adopted, wherein the center is the real needs of PwD and the required services in shopping and packaging. Around these, all the related stakeholders (private and public) that their actions may impact the PwD shopping experience must be located.

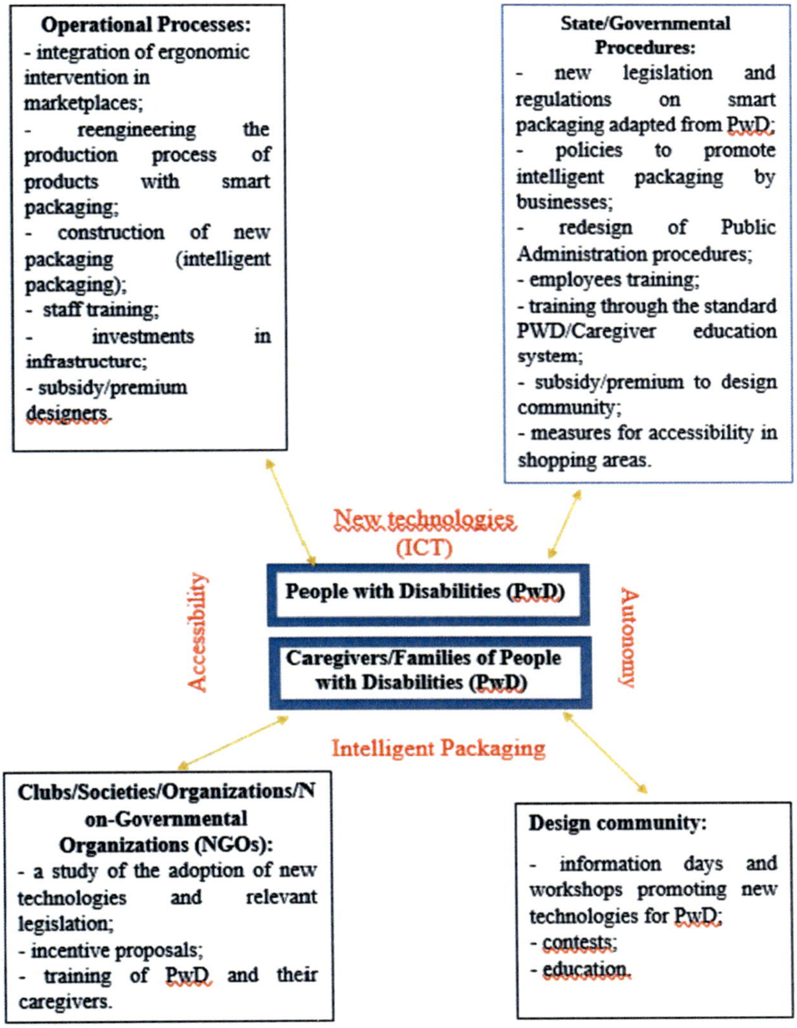

Fig. 2. A proposed a framework to handle the problematic situation better

6 Limitations of the Study and Suggestions for Future Studies

The particular study was applied in Athens, Greece; thus, studies in other places may give different results. In addition, the participation of more PwD with other disabilities (intellectual, aged people, etc.) in focus groups and interviews is essential in future studies. Finally, the implementation of quantitative research with the participation of a large number of participants representing all disabilities is also recommended.

References

1. World Economy News - Forum. 3 ways companies can support disability inclusion and boost innovation (2021). https://www.hellenicshippingnews.com/3-ways-companies-can-support-disability-inclusion-and-boost-innovation/. Accessed 18 April 2023
2. Assistive technology, Debating Europe. How will new technology improve accessibility for people with disabilities? (2016). https://www.debatingeurope.eu/2016/02/18/will-new-technology-improve-accessibility-people-disabilities/#.YktsgCi_xPZ. Accessed 29 March 2023
3. Nicoletti, M., Serrone, P.D.: Intelligent and Smart Packaging, Open Access IntechOpen London (2017). https://doi.org/10.5772/intechopen.68773. Accessed 21 Nov 2022
4. Mariusz, T.: The potential of intelligent packaging in the reduction of food waste. In: Sieć Badawcza Łukasiewicz. S.B. (eds.) Commodity Science and Research – Management and Quality Science in the Face of Sustainable Development Challenges, pp. 121–130. Instytut Technologii Eksploatacji, Radom, Poland (2019)
5. Schaefer, D., Cheung, W.M.: Smart packaging: opportunities and challenges. In 51st CIRP, Conference on Manufacturing Systems. Procedia of CIRP, vol. 72, pp. 1022–1027 (2018)
6. Howells, R, Chemers M.M.: Midget Cities: Utopia, Utopianism and the Vorschein of the 'Freak' Show. Disabil. Stud. Quar. **25**(3) unpaged (2005)
7. Dobero, A., Escher, A., Bertucci, S., Castellano, M., Lova, P.: Intelligent packaging for real-time monitoring of food-quality: current and future developments. Appl. Sci. **11**, 3532 (2021)
8. Mazhar, M., Daud, S., Bhutto, S., Mubeen, M.: Impact of product packaging on consumers buying behavior: evidence from Karachi. J. Market. Consum. Res. **16**, 35–42 (2015)
9. Connolly, D., Lund, H., Mathiesen, B.V.: Smart Energy Europe: the technical and economic impact of one potential 100% renewable energy scenario for the European Union. Renew. Sustain. Energy Rev. **60**(C), 1634–1653 (2016)
10. Nomikos, S., Politis, A., Darzentas, J., Spyrou, T., Darzentas, J.: Exploring cross-media concepts for future packaging – challenges for the printing industry. In: 32nd IARIGAI International Conference 2005, Proceedings, pp. 317–319. 4–7 September 2005, Porvoo, Finland
11. Siegemund, F., Krauer, T.: Integrating handhelds into environments of cooperating smart everyday objects. In: Markopoulos, P., Eggen, B., Aarts, E., Crowley, J.L. (eds.) EUSAI 2004. LNCS, vol. 3295, pp. 160–171. Springer, Heidelberg (2004). https://doi.org/10.1007/978-3-540-30473-9_16
12. Das, R.: RFID Explained (2004). http://www.cnepc.net/downloading%5CRFID_WhitePaper.pdf. Accessed 24 Dec 2022
13. Verghese, K., Lewis, H., Lockrey, S.: Packaging's role in minimizing food loss and waste across the supply chain. Packag. Technol. Sci. **28**, 603–620 (2015)
14. Butler, P., Harrop, P.: Highlights of IDTechEx/ PIRA Conference Intelligent & Smart Packaging USA, ID TechEX. Smart Pack. J. 1–13 (2005)
15. Lee, S.Y., Lee, S.J., Choi, D.S., Hur, S.J.: Current topics in active and intelligent food packaging for the preservation of fresh foods. J. Sci. Food Agric. **95**(14), 2799–2810 (2015)
16. UN. United States Nations Human Settlements Program UN-HABITAT (2011). Global Report on Human Settlements 2011: Cities and Climate Change (2011). https://unhabitat.org/global-report-on-human-settlements-2011-cities-and-climate-change. Accessed 18 March 2023
17. World Health Organization (WHO) (2023). Disability – Key Facts. https://www.who.int/en/news-room/fact-sheets/detail/disability-and-health. Accessed 11 April 2023
18. CDC. Impairments, Activity Limitations, and Participation Restrictions, Centre for Disease Control and Prevention (2020). https://www.cdc.gov/ncbddd/disabilityandhealth/disability.html. Accessed 11 Dec 2022
19. Oliver, M.: The Politics of Disablement. Macmillan and St Martin's Press, Basingstoke (1990)

20. Thomas, C.: Female Forms: Experiencing and Understanding Disability. Open University Press, Buckingham (1999)
21. Shakespeare, T.: Disability Rights and Wrongs. Routledge London (2006)
22. Forsyth, R., Colver, A., Alvanides, S., Wooley, M., Lower, M.: Participation of young severely disabled children are influenced by their intrinsic impairments and environment. Dev. Med. Child Neurol. **49**(5), 345–349 (2007)
23. Bekiaris, E., et al.: Research for TRAN Committee – Transport and tourism for persons with disabilities and persons with reduced mobility. European Parliament, Policy Department for Structural and Cohesion Policies, Brussels (2018)
24. UN (ND) Convention On The Rights Of Persons With Disabilities (CRPD). https://social.desa.un.org/issues/disability/crpd/convention-on-the-rights-of-persons-with-disabilities-crpd. Accessed 23 Feb 2023
25. European Community. The 2010 European Report on Development: Social Protection for Inclusive Development - A New Perspective in EU Cooperation with Africa, European University Institute, San Domenico di Fiesole, p. 1 (2010)
26. Stakeholdertheory.org, About the Stakeholder Theory. http://stakeholdertheory.org/about/. Accessed 19 April 2023
27. Martins, J.: What is stakeholder analysis and why is it important? (2023) https://asana.com/resources/project-stakeholder. Accessed 11 April 2023
28. Grimble, R., Wellard, K.: Stakeholder methodologies in natural resource management. A review of principles, contexts, experiences and opportunities. Agricul. Syst. J. **55**(2), 173–193 (1997)
29. Grimble, R.: Stakeholder methodologies in natural resource management. Socio-economic Methodologies. Best Practice Guidelines. Chatham, UK: Natural Resources Institute (1998). https://core.ac.uk/download/pdf/42389820.pdf. Accessed 30 April 2023
30. Στρατηγέα, Α.: Θεωρία και Μέθοδοι Συμμετοχικού Σχεδιασμού. Αθήνα: Σύνδεσμος Ελληνικών Ακαδημαϊκών Βιβλιοθηκών (2015). https://repository.kallipos.gr/bitstream/11419/5428/1/00_master_document_FINAL_21_3_2016-KOY.pdf. Accessed 3 May 2023
31. Freeman, R.E., Dmytriyev, S.D., Phillips, R.A.: Stakeholder theory and the resource-based view of the firm. J. Manag. **47**(7), 1757–1770 (2021)
32. Checkland, P.: Systems Thinking, Systems Practice. Wiley, New York (1981)
33. Checkland, P.: Soft systems methodology: a thirty-year retrospective. Syst. Res. Behav. Sci. **17**, 11–58 (2000)
34. Svensson, O., Zafar, A.: Synergizing Systems – Using SSM in prototypical system development, Thesis-Information Systems, Masters, Linnaeus University, Sweden (2020)
35. Acocella, I.: The focus groups in social research: advantages and disadvantages. Qual. Quant. **46**, 1125–1136 (2012)
36. McKercher, B., Packer, T., Yau, M.K., Lam, P.: Travel agents as facilitators or inhibitors of travel: perceptions of people with disabilities. Tour. Manage. **24**(4), 465–474 (2003)
37. Preston, J., Raje, F.: Accessibility, mobility and transport-related social exclusion. J. Transp. Geogr. **15**(3), 151–160 (2007)

Emotion Recognition Through Accelerometer and Gyroscope Sensors: A Pilot Study

Michael Dela Fuente[1,2](✉), Carlo Inovero[1,2], and Larry Vea[2]

[1] Polytechnic University of the Philippines, Manila, Philippines
mbdelafuente@pup.edu.ph
[2] Technological Institute of the Philippines, Manila, Philippines

Abstract. Human social interaction and behavior influence the ability to perceive emotions from people. Capturing these emotions through non-intrusive means will allow data collection to be more natural and free flowing. Recent trends used mobile phone sensors to collect data through non-intrusive means. This study explored using accelerometer and gyroscope sensors to capture human emotion during gameplay. We tested our dataset with J48, Random Forrest, Naïve Bayes, Neural Network, and Support Vector Machine (SVM) using 10-fold cross-validation. We have identified that Random Forest proved to be a better model for identifying emotions using the sensors, even with a not so good-quality data.

Keywords: Emotion Recognition · Mobilephone Sensors · Accelerometer · Gyroscope

1 Introduction

The advancement in computing made it possible for people to use smartphones beyond the communication functions for which they were initially designed. People utilize their devices for various tasks not limited to ordering food, shopping, or entertainment. Mobile phones progressively become more complex over time to suit increasing demands and needs. With mobile phones having a variety of sensors, the utilization of these integrated sensors enables the deployment of new applications that previously were impractical.

Human social interaction and behavior are influenced by the ability to perceive emotions from people [8]. Different approaches were introduced to recognize emotion and behavior using technology. Often one issue is that emotion captured is unnatural, especially if the subject is aware of the activity. Capturing these emotions through non-intrusive means will allow data collection to be more natural and free flowing, removing uneasiness from participants.

Recent trends used mobile phone sensors to collect data through non-intrusive means [2]. Mainly, data from accelerometers and gyroscopes are being studied to recognize human emotions [7, 9]. This study explored using accelerometer and gyroscope sensors to capture human emotion during gameplay.

K. Kabassi et al. (Eds.): NiDS 2023, LNNS 783, pp. 316–326, 2023.
https://doi.org/10.1007/978-3-031-44097-7_34

2 Emotion Recognition

Emotion recognition is simply a process of identifying human emotion. In psychology, emotion recognition is an attribution to observable emotional states, whether visual or auditory nonverbal cues [1]. Visually, Ekman introduced six basic emotions of fear, anger, joy, sadness, disgust, and surprise [4].

With the advancement of technology, several techniques have been developed to recognize emotions [2]. Most familiar to these is the use of the camera to identify visual cues, mainly in the face. But a lot of studies currently look beyond the face and recognize emotion through other physiological signals.

In [3], depth map video sequences recognized emotional states from body motion patterns. Another made predictions about emotions while a user is walking with a mobile phone in the pocket [7]. Piskioulis, Tzafilkou, and Economides investigated the recognition of enjoyment and frustration while using a prototype gaming app [9]. Rumiantcev, on the other hand, investigated the use of data from wearable device sensors to recognize emotion and human activity [10]. Sensor data, while a user is typing on a mobile device's keyboard, was used to infer the user's emotional state [11]. Lastly, [6] utilizes sensors in smartphones to estimate affect based on the link between affect and movement.

With these recent studies, interest in multimodal and non-intrusive approaches to recognizing emotion is gaining popularity. Notably, using mobile phones and other wearable devices with sensors has become of interest to researchers.

3 Methodology

This pilot study utilizes a quasi-experimental design. In this experiment, participants were asked to play a mobile game, and during gameplay, the actions and movements were recorded using a third-party application. The facial expressions are recorded using another device, preferably a computer with a camera. The entire duration of the activity is about 10–15 min.

The data source for this experiment will come from the gameplay recordings of participants from the SHS/College. The following materials/resources were used in the conduct of the investigation:

- Android phone – that is capable of recognizing movements through accelerometer and gyroscope sensors
- Mobile game – in this case, Mobile Legend, a popular MMORPG mobile game played by almost all age groups and very familiar with the target participants.
- Sensor Record – a free Android app designed by Martin Golpashin to easily track sensor data of smartphones in high frequency with the capability to store information in external storage as a.csv file.
- Computer with a camera – this will be used to capture the participant's facial expressions while in gameplay.

This work is guided by [5], which presents the general emotion recognition process (Fig. 1).

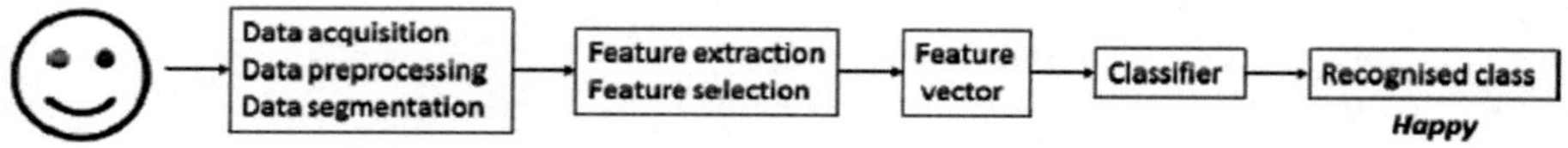

Fig. 1. General emotion recognition process

3.1 Sensors

The identified sensors for this research are the accelerometer and gyroscope. The accelerometer measures the force of acceleration caused by movement, vibration, or gravity, while the gyroscope measures the orientation or angular velocity. For this experiment, two types of accelerometer information were recorded, one considering gravity with motion and position detection and the other without gravity with motion and position detection (acceleration linear). Thus, three pieces of information from sensor data were recorded, accelerometer (with gravity), accelerometer linear (without gravity), and gyroscope.

3.2 Ethical Considerations

Participants were oriented about the study, the information required, and the expected output of the research. A consent form was given since the experiment must record the participant using a camera. Participation in the study is voluntary, and only those who provided their consent were requested to perform the task. In total, eight participants voluntarily joined the activity.

3.3 Procedures

In the conduct of the activity, some of the participants were observed in the presence of the researchers, while others were asked to submit the recordings online via Google Drive following a list of steps presented below:

Pre-experiment Stage

1. Install Sensors Toolbox, Sensor Record, and Mobile Legends
2. Verify if your phone can do the experiment by running Sensors Toolbox. Check if the device has the following sensors: accelerometer and gyroscope.
3. You can also verify it through Sensor Record. There should be Accelerometer, Accelerometer Linear, and Gyroscope enabled.
4. Set up your computer for video recording. Make sure that your face is visible.

Experiment Stage

1. Start the video recording. Your face should be visible during the recording, so do not block your look with the mobile phone while playing.

2. Open Sensor Record and choose Accelerometer, Accelerometer Linear, and Gyroscope with 100 ms intervals, then press recording (red button at the bottom of the screen).
3. Open your Mobile Legend and begin playing. A 10–15-min gameplay is enough for this experiment.

Post-experiment Stage

1. After gameplay, stop the recording for video and sensor recording.
2. Go to bit.ly/ER-Game and create a folder with your full name. Upload the video recording and the.csv files created in Sensor Record (there should be three.csv files).

3.4 Data Acquisition

The primary data considered for this study are the generated data from mobile phones' accelerometers and gyroscope sensors. For this purpose, a third-party application, Sensor Record, was used. The participants were asked to play a game while the app was recording their actions and bodily movement. At the same time, facial expressions are recorded using the computer and camera.

At the end of this activity, three.csv files and a video recording serve as data for the analysis. The.csv file contains the time stamp, the time in milliseconds with a 100 ms interval, and the values for the X, Y, and Z coordinates.

The video recordings for each participant are viewed and analyzed, particularly the facial expressions made during the gameplay. In the analysis, we identify the time when a participant completed the emotion and the length of time it lasted. We then compared this from the time in the.csv and added an annotation indicating the specific emotion we observed. For those portions where no emotion was observed, we annotate them as "neutral."

Only the data coming from six participants were processed. The data coming from the other two participants are with faults. Emotion or any change in facial expression was not observed in one, while the other had missing recorded data as provided by the app.

3.5 Data Pre-Processing

Once the raw.csv data from the sensors were obtained, the data must be adjusted based on different interval sizes of i. This adjustment in data requires taking every i line of raw data and simplifying them into single lines corresponding to their means and standard deviations.

Before this, we remove the first and last 10 s (10000 ms) in the.csv file to accommodate the actions of the participants before and after gameplay (e.g., switching windows from Sensor Record to the mobile game).

Concerning the emotion after simplification, we determine the majority emotion for every i line, which will serve as the dominant emotion. In this study, we have considered three emotions based on what was observed, namely happy, frustrated, and neutral.

3.6 Feature Engineering

In this study, we have collected data from three sensors and computed the variance and distance for every fifty rows. We prepared a separate file for each sensor data combining the information collected from the six participants that will contain the emotion, the variance for each coordinate, and the total distance. This will serve as the dataset for the model and testing.

We eliminated instances from the dataset with 0 distance to account for errors in the recordings where the third-party app stopped recording the actions of the participant. From these, we established the following number of instances per sensor: accelerometer (134), accelerometer linear (102), and gyroscope (102). Table 1 presents a summary of the number of instances per emotion for each sensor.

Using Weka 3.8.6, we balanced the instances in each sensor data so that each class has the same total weight using the CLASSBALANCER filter with ten discretization intervals.

Table 1. Number of Instances per Emotion for Each Sensor

Emotion	Accelerometer	Accelerometer Linear	Gyroscope
Happy	45	31	31
Frustration	39	36	36
Neutral	50	35	35
Total	**134**	**102**	**102**

3.7 Model Development

In this experiment, we are interested in knowing the accuracy of the classification as well as the Precision, Recall, F-Measure, and Kappa values.

We use Weka 3.8.6 to create the models for this experiment. In this study, we tested our dataset with J48, Random Forrest, Naïve Bayes, Neural Network, and Support Vector Machine (SVM). For our test option, we used cross-validation with ten folds. We applied the models to the three datasets and investigated which provided more substantial data.

4 Results and Discussion

This section presents the results of our model development and experimentation. We divided the presentation into sections to represent the results of the experiment per sensor data. In the end, we compare the performance of the models selected for this study.

4.1 Accelerometer

The computed values for accuracy, precision, recall, f-measure, and kappa statistics for the accelerometer data are presented in Table 2.

Here we can see that for accelerometer data, Random Forest and J48 provide higher accuracy ratings as compared with the other models used. Unfortunately, the values computed for precision, recall, f-measure, and kappa statistics are low indicative that there is a high incidence of misclassified instances in the dataset.

Table 2. Results of Model Testing for Accelerometer Data

Model	Accuracy	Precision	Recall	F-Measure	Kappa
J48	60.9573%	0.589	0.610	0.595	0.4144
Naïve Bayes	35.5271%	0.367	0.355	0.303	0.0329
Neural Network	44.4217%	0.467	0.444	0.407	0.1663
Random Forest	64.8148%	0.630	0.648	0.637	0.4722
Support Vector	33.2821%	0.374	0.333	0.270	−0.0008

4.2 Accelerometer Linear

In the accelerometer linear data, we can establish that Random Forest and J48, like in accelerometer data, gave the highest accuracy ratings as compared with other models tested.

Table 3. Results of Model Testing for Accelerometer Linear Data

Model	Accuracy	Precision	Recall	F-Measure	Kappa
J48	49.2260%	0.466	0.492	0.460	0.2384
Naïve Bayes	40.6503%	0.460	0.407	0.355	0.1098
Neural Network	40.9754%	0.422	0.410	0.379	0.1146
Random Forest	59.1091%	0.555	0.591	0.556	0.3866
Support Vector	34.3088%	0.366	0.343	0.285	0.0146

Also, like the results of the accelerometer data, a high percentage of misclassification of instances occurred, resulting in low values for precision, recall, f-measure, and kappa statistics. Table 3 summarizes the data collected from the experiment using accelerometer linear sensor data.

4.3 Gyroscope

Regarding the result generated from the data from the gyroscope sensor, we have identified Random Forest and J48 as the two models that provided a high accuracy rating with 67.6805% and 42.0601%, respectively.

Table 4. Results of Model Testing for Gyroscope Data

Model	Accuracy	Precision	Recall	F-Measure	Kappa
J48	42.0601%	0.407	0.421	0.407	0.1309
Naïve Bayes	39.9505%	0.467	0.400	0.354	0.0993
Neural Network	39.6604%	0.463	0.397	0.401	0.0949
Random Forest	67.6805%	0.667	0.677	0.659	0.5152
Support Vector	32.9638%	0.402	0.330	0.324	−0.0055

Values generated for precision, recall, f-measure, and kappa statistics are again low, indicative of high misclassification of instances. Table 4 summarizes the data generated from the experiment using gyroscope sensor data.

4.4 Model Comparison

In Table 5 we present a comparison of the models across the different sensors utilizing the accuracy data derived from the experimentations. Here, it is evident that Random Forrest and J48 performed well across the various sensors. Both models are based on Decision Trees.

Table 5. Model Comparison Across Sensors

Model	Accelerometer	Accelerometer Linear	Gyroscope
J48	60.9573%	49.2260%	42.0601%
Naïve Bayes	35.5271%	40.6503%	39.9505%
Neural Network	44.4217%	40.9754%	39.6604%
Random Forest	64.8148%	59.1091%	67.6805%
Support Vector	33.2821%	34.3088%	32.9638%

The performance of the other three models, namely Naïve Bayes, Neural Network, and Support Vector Machine, can be attributed to the quality and type of data used, as all three best perform with high dimensional data. In this case, the dataset used for the study contains very limited dimensionality, and thus, we conclude they have significantly affected the models' performance. It is said that neural networks are prone to overfitting and may not fully utilize their representation learning capabilities. Naïve Bayes, though it works well with low dimensional data, may have trouble capturing complicated feature relationships. Similarly, SVMs can still function well in low-dimensional data but may be less critical in finding optimal decision boundaries.

4.5 Emotion Recognition

After the model development and testing, it was observed that among the emotions under study, frustration is the most recognized by the models using different sensor data. This is evident for accelerometer and accelerometer linear data, as presented in Fig. 2.

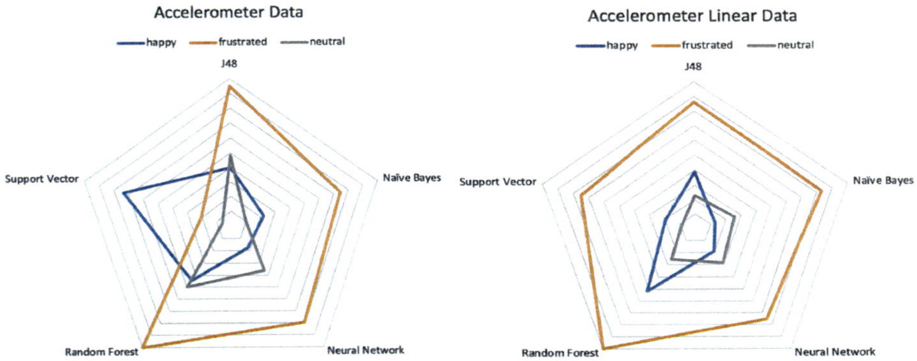

Fig. 2. Dominant Emotion using Accelerometer and Accelerometer Linear Data

Results using Gyroscope data gave varying classifications (Fig. 3). For Naïve Bayes and Support Vector, the dominant emotion is happiness. J48 and Random Forest identified frustration, while Neural Network identified neutral emotion using gyroscope data.

4.6 Feature Selection

In the feature selection, we use the two best model based on the results of the experiment, which is Random Forest and J48. To identify the features that contribute to the performance, we performed attribute evaluation using Learner Based Feature Selection, particularly the use of the WrapperSubsetEval technique with BestFirst as the search method. The evaluation mode was done using 10-fold cross-validation.

The result of the feature selection has identified, for the accelerometer data, the variance of the Z axis for J48 and the variance of the Y axis for Random Forest to have contributed to the performance of emotion recognition.

For the accelerometer linear data, the variance of the Z axis is the best feature for J48, which appeared in 9 evaluations of the 10-fold cross-validation. Likewise, Random Forest has identified the variance of the Z axis as the best attribute.

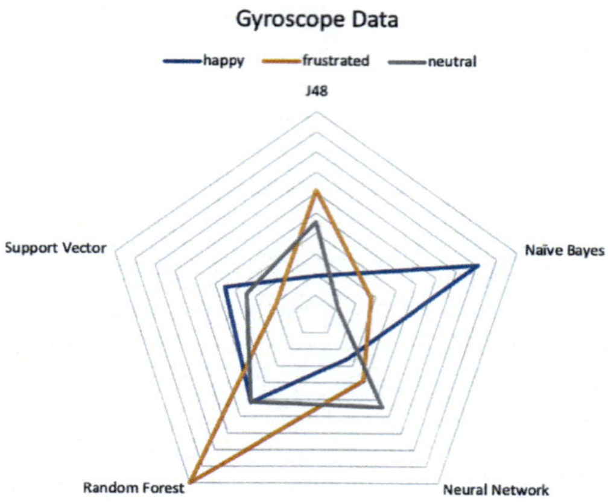

Fig. 3. Dominant Emotion using Gyroscope Data

Table 6. Features Contributing to Emotion Recognition

	J48	Random Forest
Accelerometer Data	VarZ (10 folds)	VarY (10 folds)
Accelerometer Linear Data	VarZ (9 folds)	VarZ (10 folds)
Gyroscope Data	VarZ(10 folds)	VarX (8 folds)

Finally, the gyroscope data identified the variance of the Z axis as a contributing attribute in recognition of emotion under J48, while the variance of the X axis is the contributing attribute for Random Forest, which appeared in 8 folds of the cross-validation. Table 6 summarizes the features contributing to emotion recognition.

4.7 Overall

We compare the accuracy of each sensor with the five models (see Fig. 4). Across all sensors, Random Forest provides a better rating in terms of correctly classifying the instances, whereas Support Vector provides the least feasible rating.

Data from the accelerometer (both with and without gravity) provides a useful basis for recognizing emotion. Although, the rating is not too far away as compared to the gyroscope.

The results of the experiments are less promising. This can be attributed to the following:

1) The amount of data used. We were only able to gather very little data for this experiment. This can also be seen in the video recordings, where identified emotions were very few.
2) The quality of data used. We noticed that the use of the third-party application is not advisable as the recordings stop at times. This is evident from the data collected.

Also, the posture of the player while playing the game affected the movements (e.g., playing the game with arms resting on a table).

3) The type of game played. We believe that the type of game used for the experiment contributed to the quality of data collected. The sensors used for this study are very much suited to games that will require a lot of movement.

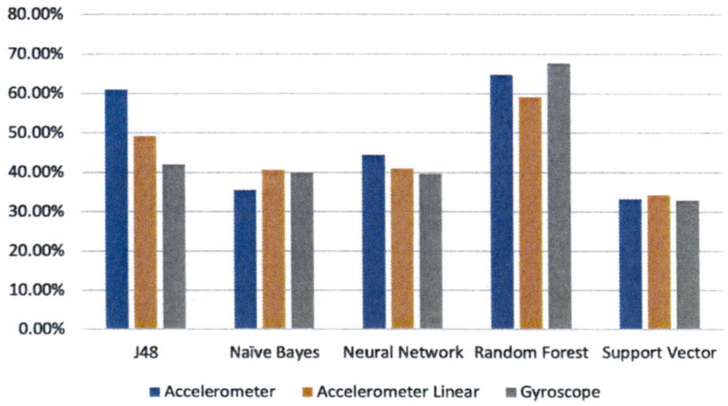

Fig. 4. Comparison of Accuracy for Each Sensor Using Different Models

Although the results are less promising, we have identified that Random Forest proved to be a better model for identifying emotions using the sensors, even with a not so good-quality data. The second best is identified to be J48.

5 Conclusion

In this study, we investigated using accelerometers and gyroscope sensors to recognize emotions, particularly happiness, and frustration. We recorded the data from the sensors using a third-party application while participants played a mobile game. The facial expressions of each participant were also recorded while in gameplay. Emotions were identified from the video recordings, and the sensor data were annotated with the identified emotions. The collected data from the participants are then pre-processed in preparation for model development. We used Weka 3.8.6 to create the models for this study and tested our dataset with J48, Random Forrest, Naïve Bayes, Neural Network, and Support Vector Machine (SVM). We have identified that Random Forest proved to be a better model for identifying emotions using the sensors, even with a not so good-quality data.

From the three datasets coming from the accelerometer, accelerometer linear, and gyroscope, J48 has identified the variance of the Z axis as a contributing attribute in recognition of emotion. Random Forest, on the other hand, has the variance of the Y axis for accelerometer data, the variance of the Z axis for the accelerometer linear data, and the variance of the X axis for the gyroscope data.

The result of the experiments is less promising due to several factors. Thus, we recommend further improving the study by increasing the dataset and identifying more suitable materials for the experiments.

References

1. Banziger, T.: Measuring emotion recognition ability. In: Encyclopedia of Quality of Life and Well-Being Research, pp. 3934–3941. Springer, Dordrecht (2014). https://doi.org/10.1007/978-94-007-0753-5_4188
2. Dzedzickis, A., Kaklauskas, A., Bucinskas, V.: Human emotion recognition: review of sensors and methods. Sensors (Switz.) **20**, 592 (2020). https://doi.org/10.3390/s20030592
3. Elfaramawy, N., Barros, P., Parisi, G.I., Wermter, S.: Emotion recognition from body expressions with a neural network architecture. In: HAI 2017 – Proceedings of the 5th International Conference on Human Agent Interaction, pp. 143–149. Association for Computing Machinery, Inc. (2017). https://doi.org/10.1145/3125739.3125772
4. Simeng, G., Wang, F., Patel, N.P., Bourgeois, J.A., Huang, J.H.: A model for basic emotions using observations of behavior in Drosophila. Front. Psychol. **10**, 781 (2019). https://doi.org/10.3389/fpsyg.2019.00781
5. Kołakowska, A., Szwoch, W., Szwoch, M.: A review of emotion recognition methods based on data acquired via smartphone sensors. Sensors (Switz.) **20**(21), 1–43 (2020). https://doi.org/10.3390/s20216367
6. Mottelson, A., Hornbæk, K.: An affect detection technique using mobile commodity sensors in the wild. In: UbiComp 2016 – Proceedings of the 2016 ACM International Joint Conference on Pervasive and Ubiquitous Computing, pp. 781–792. Association for Computing Machinery, Inc. (2016). https://doi.org/10.1145/2971648.2971654
7. Olsen, A.F., Torresen, J.: Smartphone accelerometer data used for detecting human emotions. In: Proceedings of the 2016 3rd International Conference on Systems and Informatics, Shanghai Jingyue International Aviation Hotel, Shanghai, China, 19–21 November 2016. ICSAI 2016, pp. 410–415 (2016)
8. Picard, R.W.: Affective Computing (1995). Retrieved from https://affect.media.mit.edu/pdfs/95.picard.pdf
9. Piskioulis, O., Tzafilkou, K., Economides, A.A.: Emotion detection through smartphone's accelerometer and gyroscope sensors. In: UMAP 2021 – Proceedings of the 29th ACM Conference on User Modeling, Adaptation and Personalization, pp. 130–137. Association for Computing Machinery, Inc (2021). https://doi.org/10.1145/3450613.3456822
10. Rumiantcev, M.: Emotions and activity recognition system using wearable device sensors. In: Conference of Open Innovation Association, FRUCT, IEEE Computer Society (2021). https://doi.org/10.23919/FRUCT50888.2021.9347652
11. Zualkernan, I., Aloul, F., Shapsough, S., Hesham, A., El-Khorzaty, Y.: Emotion recognition using mobile phones. Comput. Electr. Eng. **60**, 1–13 (2017). https://doi.org/10.1016/j.compeleceng.2017.05.004

Internet of Everything Based Intelligent System for Sleep Recording and Analysis

Chaima Hannachi[1](✉), Sofia Kouah[1], and Meryem Ammi[2]

[1] Department of Mathematics and Computer Sciences, University of Oum El Bouaghi, Route de Constantine BP 321, 04000 Oum El-Bouaghi, DZ, Algeria
{chaima.hannaachi,sofia.kouah}@univ-oeb.dz

[2] Forensic Science Department, Naïf Arab University for Security Sciences, Riyadh, Saudi Arabia
mammi@nauss.edu.sa

Abstract. Human health is highly affected by sleep, as it plays a crucial role in different physiological processes. Healthcare studies, such as polygraphy, are commonly used to assess and monitor sleep quality. Recent technological advancements have made it more effective and less-cost to use smart devices for that purpose, including the integration of people and processes, referred to as Internet of Everything (IoE). Systems IoE-based can be less obtrusive and more cost-effective. To obtain insightful evaluations of various sleep parameters such as: sleep duration, stage and quality on a large scale, intelligent methods to analyze data are necessary. This paper presents a comprehensive overview of sleep monitoring systems that use smart techniques such as IoE, highlighting prior research on IoE-based systems, identifying critical components, characteristics, and challenges.

Keywords: Sleep Study and Analysis · Artificial Intelligence · Cloud Computing · Big Data · Internet of Everything

1 Introduction

Sleep is a crucial biological process that significantly impacts our physical and mental health. During this biological activity, our body undergoes a series of vital physiological changes that are necessary for repairing and rejuvenating our body and brain. Thus, lack of sleep or poor-quality sleep can lead to a variety of health problems, including decreased cognitive function, mood disturbances, and overall reduced well-being [1].

Sleep studies are tests used to diagnose and identify sleep disorders and quality. They are conducted to monitor various physiological parameters such as brain waves, eye movements, heart rate, and breathing to better understand sleep patterns and identify abnormalities [2–4]. By analyzing the data collected, personalized treatment plans for sleep disorders such as sleep apnea, insomnia, and restless leg syndrome can be developed [5].

Recently, the integration of the Internet of Everything (IoE) with Artificial Intelligence (AI) including among others machine learning and deep learning techniques has

K. Kabassi et al. (Eds.): NiDS 2023, LNNS 783, pp. 327–337, 2023.
https://doi.org/10.1007/978-3-031-44097-7_35

shown great potential to improve the diagnosis and treatment of sleep disorders [6–9]. IoE can provide a more comprehensive understanding of sleep patterns by collecting and analyzing data from multiple sources, such as wearable devices, sensors, and electronic health records. By integrating these data sources, IoE can offer a general view of the patient's health, which is essential for identifying sleep disorders. This can help healthcare professionals to make more accurate diagnoses and develop customized therapy plans.

Moreover, the use of intelligent techniques leveraged from artificial intelligence such as machine learning is inevitable to deal effectively with the large amounts of data collected during sleep studies. Deep learning is among machine learning techniques that uses neural networks and has proven promising results in detecting and classifying sleep disorders. With its ability to analyze large amounts of data from multiple sources and recognize complex patterns in the data, deep learning can revolutionize the field of sleep study. The use of IoE and deep learning techniques can lead to more accurate diagnoses, better treatment options, and improved overall health outcomes for individuals with sleep disorders.

The aim of this paper is to focus on exploring the sleep recording and analysis research field. The paper is structured as follows: Section 2 provides some basic concepts, Section 3 reviews related works, Section 4 outlines the problem and objective of our research, Section 5 presents the methodology and current progress, and Section 6 concludes the paper.

2 Background

This section introduces a background on sleep studies, machine learning techniques and IoE.

2.1 Sleep Study

Sleep Definition: According to Walker [1], sleep is a crucial factor of both physical and mental health, marked by reduced consciousness and muscle activity. It comprises five separate stages, with the first stage being a transitional phase between wakefulness and sleep, characterized by reduced muscle tone and slow eye movements. Stage 2 is a deeper phase, where muscle tone and breathing slow down even further, and the heart rate decreases. The deepest stage of sleep, stage 3, is characterized by slow and synchronized brain waves and minimal muscle activity, also known as slow-wave sleep (SWS). During Rapid Eye Movement (REM) sleep, rapid eye movements occur, and the body becomes essentially paralyzed, with no muscle activity except for breathing and eye movements. This stage is associated with the consolidation of memories, as dreams take place. Lastly, the final stage is awakening, where one transition from sleep back to a state of consciousness (Fig. 1).

Impact of Sleep: Sleep has numerous impacts on both individuals and society as a whole. For individuals, poor sleep can lead to physical health problems such as obesity, diabetes, and cardiovascular disease [10, 11]. It can also increase the risk of mental health issues such as depression and anxiety, impair cognitive function and reduce the overall

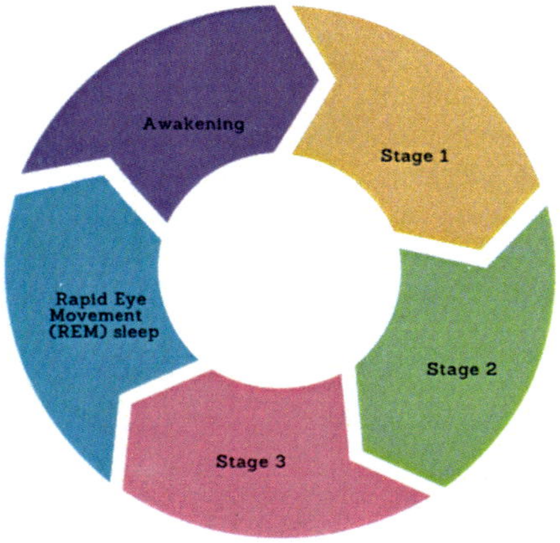

Fig. 1. Sleep stages

quality of life [12]. Sleep deprivation can also increase the risk of accidents, injuries, and other safety hazards., for society, sleep-related issues can lead to decreased work productivity, increased healthcare costs, and reduced economic growth [13]. Chronic sleep deprivation can impact public safety by increasing the risk of accidents, errors, and other safety hazards in critical professions.

Sleep Pathologies: Sleep Pathology refers to specific abnormalities or disturbances that may occur during sleep and can be detected by various sleep monitoring techniques. Insomnia is a common sleep disorder where individuals experience persistent difficulty initiating or maintaining sleep, leading to daytime impairment [14]. Sleep apnea is characterized by brief interruptions in breathing during sleep, causing excessive daytime sleepiness and loud snoring [15]. Restless legs syndrome is a neurological disorder resulting in uncomfortable sensations in the legs and an urge to move them [16]. Parasomnias refer to a group of sleep disorders that involve abnormal behaviors, movements, or experiences during sleep, including sleepwalking, night terrors, and sleep-related eating disorder [17].

Sleep Study: A sleep study is a medical test used to diagnose and determine the severity of sleep disorders. During this test, the patient is connected to multiple sensors and electrodes that measure brain waves, eye movements, heart rate, breathing patterns, and muscle activity, the data collected during the examination is evaluated by a sleep specialist to develop an individualized treatment plan.

2.2 Internet of Everything

The Internet of Everything (IoE) is a network that connects everything, including devices, people, and data, enabling seamless communication and collaboration across different domains. IoE expands on the Internet of Things (IoT) [18] concept by adding people and their interactions with technology. IoE has four key elements: networks, people, data, and processes [19] which work together to create an IoE architecture. The architecture comprises data acquisition, transport, processing, and analysis layers, allowing for the seamless flow of information and intelligent decision-making. IoE has various application domains, including smart homes, smart cities, industrial automation, and healthcare [20]. In healthcare, IoE can integrate medical devices, patient monitoring systems, electronic health records, and other data sources to enhance patient outcomes, reduce costs, and enable remote monitoring and telemedicine [21] (Fig. 2).

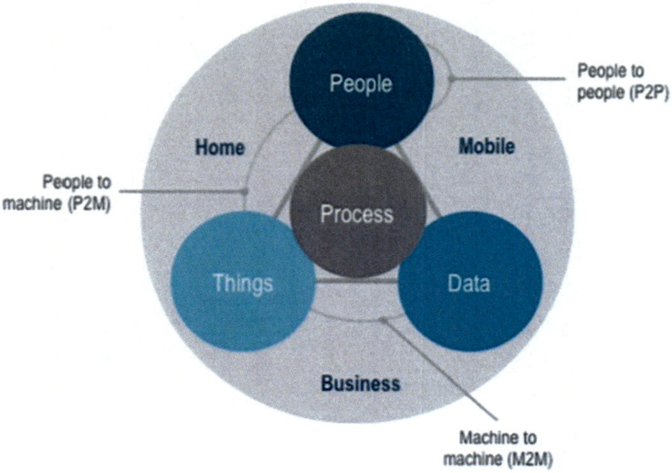

Fig. 2. Internet of Everythings [19]

2.3 Artificial Intelligence/Machine Learning

Numerous definitions of artificial intelligence have emerged since John McCarthy first introduced the term in 1956 [22]. It is the science and engineering of making intelligent machines, especially intelligent computer programs [23]. Artificial intelligence is a broad field that comprises Natural Language Procession (NPL), Computer Vision, Robotic, Expert Systems, Cognitive Computing, and Machine learning. Machine learning is a field of study that gives computers the ability to learn without being explicitly programmed [24] and is often categorized into three main classes: supervised learning, unsupervised learning, and reinforcement learning. These classes of ml allow for the development of models that can be used to analyze data, make predictions, and decision-making processes. AI and ML have diverse applications across different industries such as healthcare, finance, manufacturing, transportation, marketing, and more. In healthcare, for example, machine learning algorithms assist with medical diagnosis, patient data analysis, identifying disease risks, and developing personalized treatment plans in healthcare [25]. In finance, AI and ML are used for fraud detection [26], predicting market trends, and optimizing investment portfolios. Additionally, AI is used in manufacturing for quality control and production optimization, and in transportation, machine learning is used for traffic prediction, route optimization, and autonomous vehicles. The potential for innovation and growth in AI and machine learning is unlimited, and their applications continue to evolve.

3 Related Work

This section provides a review of scientific literature related to sleep recording and analysis systems. We have categorized these systems based on the technology employed, as outlined below:

- **Systems Using Deep Learning**: Zhang et al. [27] proposed a sleep monitoring model based on a single-channel electrocardiogram (ECG) using a CNN-LSTM network, The input data is processed through convolutional neural Network (CNN) layers to extract relevant features, which are then combined with Long-Short-Term-Memory (LSTMs) layers to support sequence prediction. The proposed model attained an accuracy of 96.1%. Chang et al. [26] introduced a one-dimensional (1D) CNN model that uses a single-lead 1D ECG to identify sleep apnea. The system was trained and validated using ECG recordings from the MIT PhysioNet Apnea-ECG Dataset. The proposed methodology demonstrated an accuracy of 87.9%.
- **Systems Using Machine Learning:** Yücelbaş et al. [6] presented an automatic sleep staging system using a morphological method based on ECG signal and MITBPD dataset. The classifier model utilized 15 features extracted from HRV (Heart Rate Variability) to identify 3 sleep stage classes: Wake, NREM and REM. The system achieved an accuracy of 77.02% for healthy individuals and 76.79% for the patient. Memar and Faradji [7] proposed a system that utilizes EEG signals to classify sleep and wake stages. The system decomposes the EEG sample into eight sub-bands with varying frequency contents and utilizes a random forest classifier to classify the extracted features. The overall accuracy rates obtained are 95.31% and 86.64% for nested 5-fold and subject cross-validation, respectively.
- **Systems Using the Internet of Things:** Saleem et al. [8] proposed a sleep monitoring system that can be used for patients at home or in hospitals. The system utilizes three sensors: an accelerometer, a microphone, and a pulse oximeter, which are connected to a microcontroller. The microcontroller communicates with a computer system, where the data is stored and used for training and testing an intelligent random forest method to predict the person's sleeping quality. The proposed system achieved an accuracy of 95,67%. Mendonça et al. [9] proposed a non-invasive home monitoring device for the diagnosis of obstructive sleep apnea (OSA). The device uses a pulse oximeter to measure blood oxygen saturation. An automatic scoring algorithm was developed to assess OSA minute by minute using statistical and frequency-based features that are fed to a classifier The proposed system achieved an accuracy of 88%. Lin et al. [28] proposed a Polysomnography (PSG) system for sleep monitoring at home. The system is designed to collect important physiological information for sleep study, including Electroencephalogram (EEC), Electrooculography (EOG), Electromyography (EMG), ECG, and airflow. The system aims to reduce the cost of attending in-laboratory PSG experiments for OSA diagnosis and offer the comfort and convenience of sleeping in the patient's own bed. It is based on an IoT framework.

- **Systems Using Novel Technologies**: Rachakonda et al. [28] proposed a system called SaYoPillow, which is a non-wearable device that monitors physiological parameters during sleep including sleep duration, snoring range, respiratory rate range, heart rate range, oxygen in blood range, eye movement rate or duration of time spent in REM, limb movement rate, and change in body temperature. The objective of the system is to detect and predict stress levels, classified into five levels: high stress, low stress, medium low stress, medium high stress, and medium stress. The system uses an edge processor to analyze and provide stress control mechanisms for the user. The data is securely stored in the cloud and can be accessed through a user interface. Additionally, the system incorporates blockchain technology for secure data transfer and storage. The proposed system achieved an accuracy of 96%. Yacchirema et al. [29] proposed a smart system to detect and support the treatment of OSA (Obstructive Sleep Apnea) in elderly people by monitoring multiple factors, including sleep environment, sleep status, physical activities, and physiological parameters. The system consists of three functional layers: the IoT layer, the Fog layer, and the cloud layer. The IoT layer receives and aggregates data from multiple heterogeneous sources and transmits it into the fog layer. The Fog layer for the pre-processing of the sensor data. The cloud layer for storing, processing, and analyzing the data collected from the IoT and fog layers. The system includes a web-based graphical user interface (GUI) for converting the analyzed information into rich content and displaying it to the users. The proposed methodology demonstrated an accuracy of 93.3%. Lazazzera et al. [29] suggested sleep monitoring platform that consists of a smart-glove called UpNEA, a mobile application, and a secure remote server called AeneA for cloud computing. The smart-glove collects PPG, SpO_2, and 3-axis accelerometer signals, which are transmitted to a remote server for cloud computing to estimate the status of sleep, breathing, and heart rate during the night. The analysis results are then stored in a SQL database and accessed by a smartphone or web application to show the overnight report. The accuracy of the algorithm used for detecting apnea and hypopnea was 75.1%. The classification task revealed 92.6% Acc in separating central from obstructive apnea, 83.7% in separating central apnea from central hypopnea, and 82.7% in separating obstructive apnea from obstructive hypopnea.

The works described above are summarized in Table 1, which provides a comprehensive overview of the research conducted in this field.

Thus, several sleep studies and analysis systems have been developed using various technologies, such as IoT, which collects necessary data such as blood oxygen levels, heart rate, body temperature, and rapid eye movement, etc. This information is stored and analyzed by a healthcare professional or by relying on artificial intelligence algorithms such as deep learning and machine learning, enabling the prediction and detection of sleep disorders. Recently, researchers have been focusing on developing AI and IoT-based systems due to their ease of use, low cost, and ability to provide accurate results.

Table 1. Sleep study systems.

Title/Publication year	Authors	Methodology and key features	Dataset	Signals	Results
Automatic Detection of Obstructive Sleep Apnea Events Using a Deep CNN-LSTM-2023	Zhang et al. [30]	-CNN-LSTM network -Use of a single-channel ECG for sleep monitoring	Apnea-ECG [31]	Lead ECG	%
A Sleep Apnea Detection System Based on a One-Dimensional Deep Convolution Neural Network Model Using Single-Lead Electrocardiogram-2020	Chang et al. [32]	-CNN -Use of a single-lead ECG for sleep apnea identification	Apnea-ECG [31]	Lead ECG	87.9%
Automatic sleep staging based on SVD, VMD, HHT and morphological features of single-lead ECG signal-2018	Yücelbaş et al. [6]	-Random Forest - Use of morphological features on ECG for sleep staging - Identification of 3 sleep stage classes	MITBPD [31]	ECG	77.02% for healthy people 76.79% for the patient
A Novel Multi-Class EEG-Based Sleep Stage Classification System-2018	Memar and Faradji [7]	-Random Forest -Use of EEG signals for sleep and wake classification -Decomposition of EEG into 8 sub-bands	UCDDB [31]	EEG	95.31% for 5-fold cross-validation 86.64% for subject cross-validation
IoT Healthcare: Design of Smart and Cost-Effective Sleep Quality Monitoring System-2020	Saleem et al.[8]	-Random Forest -Use of three sensors (accelerometer, microphone, and pulse oximeter) for sleep monitoring -Sleep quality prediction	sleep patterns dataset	SpO_2, ECG, Snoring, EMG	95.67%
An oximetry based wireless device for sleep apnea detection-2020	Mendonça et al. [9]	-Logistic regression -Use of Pulse oximeter with automatic scoring algorithm -Non-invasive home monitoring device for OSA diagnosis	70 subjects	SpO_2	88% for detection 95% for the diagnosis
IoT-Based Wireless Polysomnography Intelligent System for Sleep Monitoring-2017	Lin et al. [28]	-Polysomnography (PSG) system with IoT framework -PSG system for sleep monitoring at home -Collection of important physiological information for sleep analysis	/	EEG, EOG EMG, ECG, and airflow	/

(*continued*)

Table 1. *(continued)*

Title/Publication year	Authors	Methodology and key features	Dataset	Signals	Results
SaYoPillow: Blockchain-Integrated Privacy-Assured IoMT Framework for Stress Management Considering Sleeping Habits-2021	Rachakonda et al. [33]	-FCNN -Non-wearable device (SaYoPillow) -Edge processor for analysis and stress control mechanisms -Use of blockchain technology for secure data transfer and storage	-dataset from NSRR [34]	Snoring, SpO_2, ECG, EOG, EMG, Body temperature, Respiratory rate	96%
A smart system for sleep monitoring by integrating IoT with big data analytics-2018	Yacchirema et al. [35]	- ANN - consists of three functional layers: the IoT layer, the Fog layer, and the cloud layer - includes a web-based graphical user interface (GUI) - detect and support of treatment of OSA of elderly people	-the Valencia City Council portal -the Spanish Meteorological Agency AEMET	Snoring, heart rate, Temperature, humidity, air pollution, motion, Weight	93.3%
Proposal for a Home Sleep Monitoring Platform Employing a Smart Glove-2021	Lazazzera et al. [29]	- Gaussian Support Vector Machines (SVM) - wearable device (UpNEA) - acquires a 3-axis accelerometer signal, a photoplethysmography (PPG), and (SpO2) - sleep monitoring platform - The analysis results are stored in a SQL database	- UZ Leuven hospital dataset [36]	SpO2, photoplethysmography (PPG), 3-axis accelerometer signals	75.1% for detection 92.6% for classification in separating central from obstructive apnea 83.7% for classification in separating central apnea from central hypopnea 82.7% for classification in separating obstructive apnea from obstructive hypopnea

4 Problematic and Objective

The use of the Internet of Everything (IoE) and artificial intelligence in sleep recording and analysis has gained great attention in recent years, however, despite the promising potential of this technology, there are still many challenges that need to be addressed, and these challenges are the collection of physiological data in real time, decision-making and accurate diagnosis, as well as ensuring the security and confidentiality of personal data and infrastructure. This literature review aims to analyze current research on intelligent systems based on the Internet of Everything to record and analyze sleep. In addition, the review aims to identify challenges, gaps and potential future trends in the field of intelligent sleep systems that support the Internet of Things.

5 Methodology and Ongoing Work

In this literature review titled 'Internet of Everything-Based Intelligent System for Sleep Recording and Analysis,' a comprehensive search of academic databases such as Google Scholar and IEEE Xplore was conducted. Relevant keywords, including 'sleep study and analysis,' 'Artificial Intelligence,' 'Internet of Everything,' 'Big Data,' and 'cloud computing,' were utilized. The inclusion criteria for article selection were as follows: studies published between 2018 and 2023, written in English, with full-text availability, and focused on the development and application of intelligent systems and the Internet of Everything for sleep recording and analysis.

The initial search yielded a substantial number of articles, which were then filtered based on their titles, abstracts, and conclusions. Selected articles were thoroughly reviewed, and key information related to the components, datasets used, algorithms employed, and model accuracy was extracted and analyzed. The findings were then represented in a table for better presentation and understanding. Regarding the ongoing work, we are currently studying the systems that have been developed for studying sleep and collecting various references that will help us create a system for recording and analyzing individuals' sleep. A detailed study of the main features for sleep recording and analysis is currently underway.

6 Conclusion

In this paper, we have studied the latest advancements in sleep study systems, which involve the detection of sleep apnea, classification of different sleep stages, and evaluation of individual sleep quality. To achieve this, various technologies such as the Internet of Things (IoT) are utilized to collect physiological data in real-time, while artificial intelligence techniques including deep learning (CNN) and machine learning (RF, LR, ANN, SVM) are employed to analyze this data. Through the examination of these systems, we have gained a deeper understanding of the most effective techniques, physiological data, and dataset used in sleep study. However, challenges persist in ensuring data security, confidentiality, and accurate diagnosis. Addressing these challenges is essential for the development of an advanced and precise sleep recording and analysis system.

References

1. Why We Sleep_Unlocking the Power of Sleep and Dreams (PDFDrive)
2. Bangladesh University of Engineering and Technology, Department of Electrical and Electronic Engineering, IEEE Region 10, Institute of Electrical and Electronics Engineers. Bangladesh Section, and Institute of Electrical and Electronics Engineers, 5th IEEE Region 10 Humanitarian Technology Conference 2017: Smart Technologies for Humanity, BUET, Dhaka, Bangladesh, 21–23 December 2017
3. Lobato, F.M.F., de Resende, D.C.O., do Nascimento, R.P., Siqueira, A.L.C., Jacob, A.F.L., de Santana, Á.L.: Multimodal low-invasive system for sleep quality monitoring and improvement. In: Batalla, J.M., Mastorakis, G., Mavromoustakis, C.X., Pallis, E. (eds.) Beyond the Internet of Things. IT, pp. 223–242. Springer, Cham (2017). https://doi.org/10.1007/978-3-319-50758-3_9

4. Shustak, S., et al.: Home monitoring of sleep with a temporary-tattoo EEG, EOG and EMG electrode array: a feasibility study. J. Neural Eng. **16**(2), 026024 (2019). https://doi.org/10.1088/1741-2552/aafa05
5. Abad, V.C., Guilleminault, C.: Diagnosis and treatment of sleep disorders: a brief review for clinicians. Dialogues Clin. Neurosci. **5**(4), 371–388 (2003). https://doi.org/10.31887/dcns.2003.5.4/vabad
6. Yücelbaş, Ş, Yücelbaş, C., Tezel, G., Özşen, S., Yosunkaya, Ş: Automatic sleep staging based on SVD, VMD, HHT and morphological features of single-lead ECG signal. Expert Syst. Appl. **102**, 193–206 (2018). https://doi.org/10.1016/J.ESWA.2018.02.034
7. Memar, P., Faradji, F.: A novel multi-class EEG-based sleep stage classification system. IEEE Trans. Neural Syst. Rehabil. Eng. **26**(1), 84–95 (2018). https://doi.org/10.1109/TNSRE.2017.2776149
8. Saleem, K., Bajwa, I.S., Sarwar, N., Anwar, W., Ashraf, A.: IoT healthcare: design of smart and cost-effective sleep quality monitoring system. J. Sens. **2020**, 1–17 (2020). https://doi.org/10.1155/2020/8882378
9. Mendonça, F., Mostafa, S.S., Morgado-Dias, F., Ravelo-García, A.G.: An oximetry based wireless device for sleep apnea detection. Sensors (Switz.) **20**(3), 888 (2020). https://doi.org/10.3390/s20030888
10. Knutson, K.L.: Sleep duration and cardiometabolic risk: a review of the epidemiologic evidence. Best Pract. Res. Clin. Endocrinol. Metab. **24**(5), 731–743 (2010). https://doi.org/10.1016/j.beem.2010.07.001
11. Colten, H.R., Altevogt, B.M.: Sleep disorders and sleep deprivation: an unmet public health problem. National Academies Press (2006). doi: https://doi.org/10.17226/11617
12. João, K.A.D.R., de Jesus, S.N., Carmo, C., Pinto, P.: The impact of sleep quality on the mental health of a non-clinical population. Sleep Med. **46**, 69–73 (2018). https://doi.org/10.1016/j.sleep.2018.02.010
13. Grandner, M.A.: Sleep, Health, and Society. Sleep Med. Clin. **12**(1), 1–22 (2017). https://doi.org/10.1016/j.jsmc.2016.10.012
14. Ohayon, M.M.: Epidemiology of insomnia: What we know and what we still need to learn. Sleep Med. Rev. **6**(2), 97–111 (2002). https://doi.org/10.1053/smrv.2002.0186
15. Banno, K., Kryger, M.H.: Sleep apnea: Clinical investigations in humans. Sleep Med. **8**(4), 400–426 (2007). https://doi.org/10.1016/j.sleep.2007.03.003
16. Kerr, S., McKinon, W., Bentley, A.: Descriptors of restless legs syndrome sensations. Sleep Med. **13**(4), 409–413 (2012). https://doi.org/10.1016/j.sleep.2011.11.020
17. Fleetham, J.A., Fleming, J.A.E.: Parasomnias. Can. Med. Assoc. J. (CMAJ) **186**(8), E273–E280 (2014). https://doi.org/10.1503/cmaj.120808
18. Sofia, K., Ilham, K.: Multi-layer agent based architecture for internet of things systems. J. Inf. Technol. Res. **11**(4), 32–52 (2018). https://doi.org/10.4018/JITR.2018100103
19. Miraz, M.H., Ali, M., Excell, P.S., Picking, R.: A Review on Internet of Things (loT), Internet of Everything (IoE) and Internet of Nano Things (IoNT)
20. paper8
21. Di Martino, B., Li, K.-C., Yang, L.T., Esposito, A.: Trends and strategic researches in internet of everything. In: Di Martino, B., Li, K.-C., Yang, L.T., Esposito, A. (eds.) internet of everything. IT, pp. 1–12. Springer, Singapore (2018). https://doi.org/10.1007/978-981-10-5861-5_1
22. McCarthy, J., Minsky, M.L., Rochester, N., Shannon, C.E.: A proposal for the dartmouth summer research project on artificial intelligence (2006)
23. Mccarthy, J.: What is artificial intelligence? (2007). http://www-formal.stanford.edu/jmc/
24. Park, C., Took, C.C., Seong, J.K.: Machine learning in biomedical engineering. Biomed. Eng. Lett. **8**(1), 1–3 (2018). https://doi.org/10.1007/s13534-018-0058-3

25. Ellahham, S., Ellahham, N., Simsekler, M.C.E.: Application of artificial intelligence in the health care safety context: opportunities and challenges. Am. J. Med. Qual. **35**(4), 341–348 (2020). https://doi.org/10.1177/1062860619878515
26. Asha, R.B., Suresh Kumar, K.R.: Credit card fraud detection using artificial neural network. Glob. Transitions Proc. **2**(1), 35–41 (2021). https://doi.org/10.1016/j.gltp.2021.01.006
27. Zhang, J., et al.: Automatic detection of obstructive sleep apnea events using a deep CNN-LSTM model. Comput. Intell. Neurosci. **2021**, 5594733 (2021). https://doi.org/10.1155/2021/5594733
28. Lin, C.T., et al.: IoT-based wireless polysomnography intelligent system for sleep monitoring. IEEE Access **6**, 405–414 (2017). https://doi.org/10.1109/ACCESS.2017.2765702
29. Lazazzera, R., Laguna, P., Gil, E., Carrault, G.: Proposal for a home sleep monitoring platform employing a smart glove. Sensors **21**(23), 7976 (2021). https://doi.org/10.3390/s21237976
30. Automatic detection of obstructive sleep apnea events using a deep CNN-LSTM model. https://www.hindawi.com/journals/cin/2021/5594733/. Accessed 30 Mar 2023
31. Goldberger, A.L., et al.: PhysioBank, PhysioToolkit, and PhysioNet components of a new research resource for complex physiologic signals (2000). http://www.physionet.org
32. Chang, H.Y., Yeh, C.Y., Lee, C.T., Lin, C.C.: A sleep apnea detection system based on a one-dimensional deep convolution neural network model using single-lead electrocardiogram. Sensors **20**(15), 4157 (2020). https://doi.org/10.3390/S20154157
33. Rachakonda, L., Bapatla, A.K., Mohanty, S.P., Kougianos, E.: SaYoPillow: blockchain-integrated privacy-assured IoMT framework for stress management considering sleeping habits. IEEE Trans. Consum. Electron. **67**(1), 20–29 (2021). https://doi.org/10.1109/TCE.2020.3043683
34. Zhang, G.Q., et al.: The national sleep research resource: towards a sleep data commons. J. Am. Med. Inform. Assoc. **25**(10), 1351–1358 (2018). https://doi.org/10.1093/jamia/ocy064
35. Yacchirema, D.C., Sarabia-Jacome, D., Palau, C.E., Esteve, M.: A smart system for sleep monitoring by integrating IoT with big data analytics. IEEE Access **6**, 35988–36001 (2018). https://doi.org/10.1109/ACCESS.2018.2849822
36. Varon, C., Caicedo, A., Testelmans, D., Buyse, B., Van Huffel, S.: A novel algorithm for the automatic detection of sleep apnea from single-lead ECG. IEEE Trans. Biomed. Eng. **62**(9), 2269–2278 (2015). https://doi.org/10.1109/TBME.2015.2422378

Using Web Technologies to Implement a Modular Integrated System for Augmented Tourism Destinations

Stergios Palamas[1(✉)], Yorghos Voutos[1], Paraskevi Zigoura[2], Phivos Mylonas[3], and Vasileios Chasanis[1]

[1] Department of Informatics, Ionian University, Corfu, Greece
spalamas@ionio.gr
[2] iED - Institute of Entrepreneurship Development, Larissa, Greece
[3] Department of Informatics and Computer Engineering, University of West Attica, Athens, Greece

Abstract. Identifying and promoting attractions and points of interest to visitors, especially through multimedia and – lately - AR/VR technologies, has been shown to enhance the pre-traveling and on-site overall experience. Throughout this paper, the development of a modular, web-based, generally applicable platform, consisting of interconnected modules and based on open-source technologies is presented. Not long ago, the deployment of engaging, rich-media user experiences required special equipment and native development tools. The present work demonstrates, how recent advancements in web technologies and modern web browsers (e.g., the WebGL), enable the deployment of such cutting-edge services, on mainstream mobile devices, using exclusively well-established web technologies like HTML, CSS and JavaScript.

Keywords: Multimedia · Augmented Reality · Digital Tourism · Web Technologies · Mobile Applications

1 Introduction

Considering the wealth of information available on the types, components, and advantages of gamified applications, there is currently no established technique for creating applications suited to a given sector (such as educational or touristic). Future Immersive Media appears to be based on computer games, although there are both technological and practical challenges with this idea. Therefore, based on existing models of persuasion technologies, gaming methodologies, and cognitive processes, we decided to create an application running on contemporary computing devices and targeting mostly tourists' visitors, without disregarding the broader public. Immersive VR and AR technology can offer extremely high degrees of authenticity and a real-world experience [1]. Performance, as well as the perceived sensation of presence, engagement, and general usability, are all greatly impacted by both visual and interaction fidelity.

K. Kabassi et al. (Eds.): NiDS 2023, LNNS 783, pp. 338–346, 2023.
https://doi.org/10.1007/978-3-031-44097-7_36

Our strategy involves navigating an unfamiliar virtual environment in an intelligent way in order to improve both the physical and virtual visitor experiences. This research is a part of a larger context in which a system or platform can improve alternative/rural tourism destinations through the creation of interactive applications like self-guided tours & Augmented Reality models (AR) [2]. The target platform for the development of modern software applications to advertise are mobile computing devices such smart phones, tablets, and smart watches (for on-site visitors), as well as a web browser (for remote visitors) [3].

Most of the existing implementations are based on multimedia, augmented, and virtual reality apps, which, despite having grown mature and even being able to be deemed "mainstream" today, are still heavily dependent on proprietary, custom-made solutions. These implementations demand significant resources (such as funding) from the implementing entities as well as technical know-how and specialized capabilities. As a result, we suggest an open, standards-based platform that is flexible and relevant to all users. This platform enables end users to submit their own content and to give feedback by observing their behavior.

Beginning with the associated state-of-the-art, current work, based on the most recent literature, is how this study is structured. The system's design and architecture are then presented, together with a description of the goals, specifications, and fundamental components of the suggested platform. Finally, the system's deployment and difficulties will be covered.

2 State of the Art

In recent years, the concept of "smart tourism destinations" that make use of ICT infrastructure has grown in popularity. Such destinations are anticipated to boost the destination's competitiveness by spreading information and expertise among stakeholders and giving tourists with personalised experiences. Virtual reality and augmented reality are new technologies that are predicted to transform the tourism sector by providing multimodal experiences without the use of direct contact. The COVID-19 epidemic has also spurred the use of augmented reality and virtual reality technology in the tourism industry. Several research projects are under underway to explore new technologies to improve the tourism experience, such as context-based marketing strategies and game-based mixed museum learning services. It is critical to boost user confidence in the usage of new technology in order to achieve high levels of user satisfaction with the deployment of AR technologies in tourism.

To begin, Chiu et al.'s [4] study outlines the design and implementation of an augmented reality system for tourism that uses image-based recognition. The technology provides users with information about tourist locations via a smartphone app that recognises items or landmarks using image recognition. The software then shows augmented reality information about the object or landmark, such as historical and cultural value, as well as local attractions and events. The system was evaluated in a real-world environment and proved to be beneficial in improving the tourist experience and giving users' useful information. The authors conclude that image-based recognition technology has significant potential for improving tourism and advocate for additional research in this field.

Also, Shih et al.'s research [5] describes ARTS, an AR tourism system that combines 3D scanning and smartphone AR for cultural heritage tourism and pedagogy. Users can scan physical artefacts and receive augmented reality information on their smartphones, boosting their experience and comprehension of cultural heritage locations. The paper describes the system's technical aspects and examines its possible uses in tourism and education.

On top of that, Wei's study [6] analyses the integrated growth of countryside ecotourism in the context of AI applications with the wireless Internet of Things (IoT). The author contends that combining AI and IoT can improve the visitor experience as well as the efficiency and sustainability of rural tourism development. The study also presents a framework for the integrated development of rural ecotourism utilising AI and IoT technology, which encompasses smart tourism services, environmental monitoring, and smart management. The author believes that this framework will be a good guide for future rural tourist development.

Safari Bazargani et al.'s article [7] analyses the advantages and disadvantages of integrating GIS with AR and investigates different uses such as urban planning, disaster mitigation, and tourism. The authors also present an overview of the existing literature on GIS and AR integration and indicate relevant future research areas. Overall, the article emphasises the need of integrating GIS and AR for sustainable development and provides a roadmap for future research in this field. In particualar, several tourism-related uses of GIS and AR integration are mentioned by the authors, including producing interactive guides for cultural and historical places, upgrading museum exhibits, and facilitating virtual tourism experiences. They also point out that such technologies can deliver personalised and immersive experiences to tourists, which may lead to improved engagement and satisfaction. Furthermore, the authors argue that integrating GIS and AR can support sustainable tourism by informing tourists about ecologically beneficial activities and boosting conservation initiatives.

As a result, rather than relying on exclusive and costly implementations, we suggest a broadly applicable, modular, and open standards-based architecture. Potential beneficiaries and stakeholders can simply and affordably adopt VR and AR technologies in tourism sites, focusing on the content rather than the underlying technologies' complexity. Visitors to the site and remote users in general will have a richer, enhanced experience, while stakeholders will receive valuable feedback and metrics on visitor behavior and preferences.

3 System Design and Architecture

The system aims to provide stakeholders, an open and cost-effective solution to augment and enhance cultural, activities, tourism etc. destinations with meaningful information and multimedia content, attracting potential visitors and enhancing the experience of on-site travelers. Additionally, it aims to provide travelers with a useful tool assisting their pre-traveling selection and planning process.

3.1 Objectives

The specific objectives of the system can be summarized into providing:

- a back-end infrastructure/application, based on open source, established and proven web technologies, following the Content Management System model, content creators will be able to publish, link, and update multimedia content related to their marketed locations.
- a general applicable, mobile application, based on web, open-source technologies and featuring modern presentation techniques, such as Augmented Reality and 360 multimedia content). The application will be used by travelers, either on-site or remotely, to access the system content and provide useful feedback.

3.2 Requirements

The main system requirements are:

- Compatibility with modern mobile devices and popular web browsers.
- Support for modern multimedia content formats e.g., streaming video, 360 photos and video, 3D objects.
- Featuring of modern content presentation methods such as Augmented Reality
- Modular system design.
- Personalization
- Users feedback mechanisms

3.3 System Architecture and Components

The system follows a modular design, consisting of a set of interoperable components with discreet functionality, building upon the 3-tier architecture, most of the web applications are based on:

- Client Mobile Application: responsible for the content presentation and user feedback (presentation tier)
- Multimedia Content Repository/Database: responsible for the maintenance of the information (data tier)
- Content Management Application: providing the back – end system infrastructure/functionality: content administration web application, user profiling and content customization mechanisms, the API (application tier)

The system modules and functionality as well as the main user scenarios are presented in Fig. 1. System modules, functionality and user scenarios

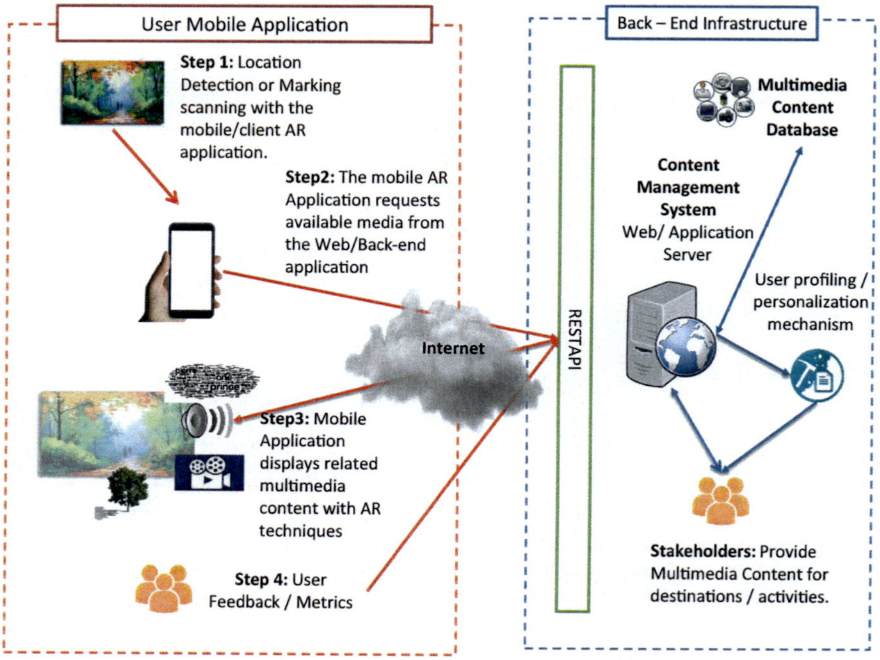

Fig. 1. System modules, functionality and user scenarios

4 Implementation

4.1 Implementation of the Back-End Infrastructure

The back-end infrastructure was developed using the MySQL RDBMS and the PHP, a programming language targeted, mainly, on web application development. A RESTful web API was developed as an interface for the Client Mobile Application to exchange information with the system's back-end services.

The data is transferred to the mobile client application using the JavaScript Object Notation (JSON) lightweight data-interchange format, a choice which facilitates their handling via JavaScript code on the Client mobile app.

The featured REST API calls, their input parameters and the resulting data are briefly presented on Table 1.

Table 1. Featured REST API Calls.

REST API Call	Input Parameters	Result
activities	UserID (integer)	JSON formatted data about the activities featured by the system
pois	UserID (integer)	JSON formatted data about the Points of Interest (POIs) featured by the system for each activity
media		JSON formatted data about the media sources featured by the system for each POI
postRating	UserID (integer) poiID (integer) poiRate (0–5)	Registration of a user's rating (1–5) for a POI
postVisit	UserID (integer) poiID (integer)	Registration of a user's physical visiting of POI (tracked by geolocation services)

4.2 User Mobile Application Implementation

The user mobile application acts as the system's client, utilizing the REST API to obtain information from the back-end services and provide user feedback/metrics to the system. It implements the presentation layer of the system's 3-tier architecture.

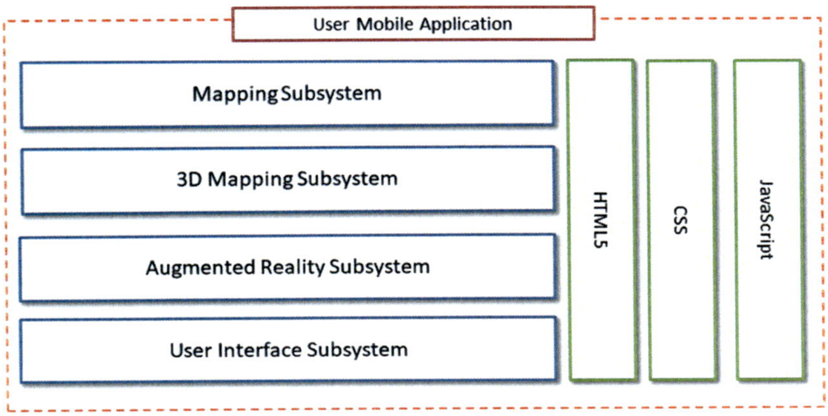

Fig. 2. Mobile Application - Subsystems

The mobile application was implemented as a mobile-friendly web application, utilizing exclusively web technologies and more specifically, HTML5, CSS and JavaScript, along with a collection of JavaScript frameworks which provide the basis and core functionality of the various subsystems which are presented on Fig. 2.

The JavaScript libraries/frameworks that were used for the implementation of each subsystem are listed on Table 2. The JavaScript API of each framework was used to implement functionality specific to each subsystem, while additional custom JavaScript code handles the integration of the subsystems and core functionality of the mobile application such as geolocation and data exchange with the back-end infrastructure.

Table 2. JavaScript libraries/frameworks used for the implementation of the mobile app. subsystems.

Subsystem	Library/Framework	Short Description
Mapping	Leafllet [8] & OpenStreetMap [9]	Leaflet provides a JavaScript programming interface for the development of interactive maps, while OpenStreetMap provides open-source GIS/Mapping services
3D Mapping	CesiumJS [10]	CesiumJS is an open-source JavaScript library for creating 3D globes and maps with performance, precision, and visual quality
Augmented Reality	AR.js [11]	AR.js is a lightweight library for Augmented Reality on the Web, featuring Image Tracking, Location based AR and Marker tracking
User Interface	OnsenUI [12]	OnsenUI provides a framework for the development of web and hybrid mobile applications, with the look-and-feel of native Android and iOS mobile apps

5 Discussion

The present study presents an approach for the implementation of a web based, modular system, multimedia material and augmented/virtual reality deployment are used to improve the experience of tourism places and attractions both before and during travel.

Rather than depending on proprietary technologies, the system solely relies on relevant, modular, and open-source solutions. The primary objective is to prioritize content over complex technological matters, aiming to facilitate the simple and cost-effective deployment of VR and AR technology in tourism destinations.

The system implements its functionality based on the classic 3-tier architecture and distributing the overall functionality on a set of sub-systems/modules, residing on the back end or front-end of the system. This approach enhances the upgradability, expandability, and serviceability of the system.

The implementation of the back-end services of the system is based on well-established and proven technologies for the development of web applications and services. A REST API has been developed as an interface between the back-end functionality and the client-mobile application, while JSON has been proven an effective means of data inter-exchange between the mobile application and the system's back-end services.

The approach followed for the implementation of the client/mobile application, has proven the feasibility of deploying complex, location/map-based, Multimedia and

Augmented Reality applications, using exclusively web technologies (HTML, CSS, JavaScript) and modern web browsers' features (e.g., WebGL). Sophisticated mobile applications that - not long ago - required development with custom SDKs and native development tools on the targeted mobile platforms, now can be deployed as cross-platform, web, or progressive web applications, while maintaining the required level of functionality and the user-interface of each platform (e.g., Android application).

A wide variety of modern JavaScript frameworks, provide the basis for developing demanding, modern mobile applications, implementing advanced technologies such as location-based services, interactive maps, location and tracker based augmented reality, 360° videos and 3D content. On the downside, programming – and especially JavaScript - frameworks' longevity and life cycle, has always been a concern and critical ingredient in project success. Metrics like community size, popularity, and big company support can justify the choice of framework to invest in and base application development on.

Moreover, the Virtual and Augmented Reality industry has consistently been characterized by rapid changes and fluidity. These traits are even more pronounced in web-based implementations, which are currently in a premature state. Therefore, it is essential to closely monitor emerging developments in the field to improve the provided services and capitalize on newly available resources such as updated or new Software Development Kits (SDKs), Application Programming Interfaces (APIs), Frameworks, and more.

Finally, some components of the system, such as multimedia material, may necessitate internet connectivity, which may result in user fees. However, as mobile internet becomes more accessible and affordable, and as free options such as public Wi-Fi become more widely available in urban areas, this issue may become less of a concern.

Acknowledgment. This research has been co-financed by the European Union and Greek national funds through the Competitiveness, Entrepreneurship and Innovation Operational Programme, under the Call "Research – Create – Innovate", project title: "Using Digital Tools and Applications for Outdoor Alternative Tourism Operators", project code: T2EDK-03168, MIS code: 5069920.

References

1. Oyelude, A.A.: Virtual reality & augmented reality in libraries and museums. Libr. Hi Tech News **35**(5), 1–4 (2018)
2. Dorcic, J., Komsic, J., Markovic, S.: Mobile technologies and applications towards smart tourism - state of the art. Tourism Rev. **74**(1), 82–103 (2019)
3. Beck, J., Rainoldi, M., Egger, R.: Virtual reality in tourism: a state-of-the-art review. Tourism Rev. **74**(3), 586–612 (2019)
4. Chiu, C.C., Wei, W.J., Lee, L.C., Lu, J.C.: Augmented reality system for tourism using image-based recognition. Microsyst. Technol. **27**(4), 1811–1826 (2021)
5. Shih, N.J., Diao, P.H., Chen, Y.: ARTS, an AR tourism system, for the integration of 3D scanning and smartphone AR in cultural heritage tourism and pedagogy. Sensors **19**(17), 3725 (2019)
6. Wei, H.: Integrated development of rural eco-tourism under the background of artificial intelligence applications and wireless Internet of Things. J. Ambient Intell. Humanized Comput. 1–13 (2021)

7. Safari Bazargani, J., Zafari, M., Sadeghi-Niaraki, A., Choi, S.M.: A survey of GIS and AR integration: applications. Sustainability **14**(16), 10134 (2022)
8. Leafletjs Homepage. https://leafletjs.com. Accessed 10 Apr 2023
9. OpenStreetMap Homepage. https://www.openstreetmap.org. Accessed 10 Apr 2023
10. CesiumJS Homepage. https://cesium.com. Accessed 10 Apr 2023
11. AR.js Homepage. https://ar-js-org.github.io/AR.js-Docs. Accessed 10 Apr 2023
12. OnsenUI Homepage. https://onsen.io. Accessed 10 Apr 2023

Using CAD to Preserve Local Traditional Weaving Craft in Teaching Mathematics

Crysali Therese R. Dayaganon[1(✉)], Judith M. Aleguen[1], Marichou L. Carreon[1], Virginia S. Albarracin[1], and Aurelio P. Vilbar[2]

[1] Cebu Technological University, Argao Campus, Philippines
crysalitherese.dayaganon@ctu.edu.ph
[2] University of the Philippines, Cebu, Philippines

Abstract. Addressing the dearth of research integrating Computer-Aided Design (CAD) software in teaching mathematics for cultural preservation, this study aims to determine the usefulness of using CAD in teaching the undergraduate course "Problem Solving, Mathematical Investigation and Modeling" with the output of creating traditional handloom weave drafts to be used as prototypes by the local handloom weaving community in Argao, Cebu, Philippines. It used the descriptive design, which utilized qualitative data-gathering procedures: online open-ended surveys and focus-group discussions to determine its impact on the students, designers, and the weavers. Findings show that the use of CAD developed the students' problem-solving skills, deeper appreciation of the traditional handloom weaving craft, and skills in creating weave designs. CAD promotes visual representation of the mathematical weave patterns that created authentic learning. The designers evaluated the student-CAD weave outputs to have aesthetic and commercial value, while the local weavers described the designs to exude sophistication, marketability, and uniqueness.

Keywords: handloom weaving · CAD in teaching · mathematics teaching · cultural preservation

1 Rationale

Many traditional forms of production have been left behind or forgotten due to the effects of industrialization, globalization, mass production, and digitalization on the weaving industry, resulting in new and faster technology and expanded access to markets [1]. There is a growing recognition of the value of traditional knowledge and the need to find ways to incorporate it into modern textile production to help preserve cultural heritage and promote sustainable economic development [2].

The effects of industrialization to weaving industry extended to the traditional handloom weaving craft in Argao, Cebu, Philippines. In Argao, handloom weaving, called 'hablon,' used to be a major source of income for the town during the Spanish colonial period [3]. However, the laborious task and complexity of the weaving process, insufficient financial and developmental support, and lack of efforts to preserve the industry caused its decline over the years [4].

K. Kabassi et al. (Eds.): NiDS 2023, LNNS 783, pp. 347–356, 2023.
https://doi.org/10.1007/978-3-031-44097-7_37

Through the efforts of Cebu Technological University-Argao Campus (CTU-AC), this ‘once-dying industry’ handloom weaving industry of Argao was revived [5] through extension projects participated by the students with the goal of developing a community of creatives who advocates on maintaining the relevance of the local handloom weaving in today’s generation.

Handloom weaving as a traditional craft can be a valuable teaching tool for mathematics classrooms, for it can develop an understanding of mathematical concepts like measurement, symmetry, and geometry [6]. Although there are some studies on ethnomathematics that explore the relationship between traditional crafts and mathematical concepts [7, 8], there is a lack of research that investigates the use of Computer-Aided Design (CAD) software in teaching mathematics with the aim of preserving non-tangible heritage. This gap highlights the need for further exploration of how technology can be utilized to enhance the preservation of traditional crafts through the teaching of mathematics.

This study aimed to determine the usefulness of using CAD software in teaching math competencies in the course “Mathematical Investigation, Problem Solving, and Modeling” with the output of creating handloom weave drafts to be used by the local weaving community. It answers the following questions: (1) How was the CAD WeavePoint integrated to develop mathematical investigation skills and in designing the weave drafts? (2) What is the feedback of the students on using CAD in learning mathematics? and (3) What is the feedback of the weavers and the designers of the weave drafts?

2 Review of Related Literature

2.1 Teaching Mathematics Using CAD and Weaving

Handloom weaving is a practical application of mathematical competence, for it involves intricate patterns and designs, and mathematical modeling and analysis can help students understand their underlying structures [6]. The use of weaving designs in teaching can enhance the understanding of symmetry, repetition, and geometric shapes [9], pattern analysis and fabric dimension calculation [10], mathematical modeling, and simulation [11].

For affective development, when students do hands-on weave designs in math classes, they develop their real-world application context [12], and a deeper appreciation and understanding of traditional craft, history, and culture [13]. For example, weaving is often used in traditional clothing, textiles [14], and art, and mathematical analysis can help students understand the cultural and historical significance of these works.

Computer-Aided Design (CAD) has the potential tool for teaching mathematics [15] because it gives students the chance to visually and interactively explore mathematical ideas by building simulations and 2D models. Using CAD can enhance students’ understanding of concepts and problem-solving abilities [16, 17], creativity [18] and develop a better understanding of geometric concepts with an engaging learning experience [19].

Weaving using CAD as an alternative assessment in mathematics is a relatively new approach. However, the use of authentic assessment and outcomes-based assessment has been around for decades. Authentic assessment is applying knowledge in real-life contexts and situations, which can involve real-world problems or presenting findings to audiences outside the classroom [20].

2.2 Traditional Weave Craft in Teaching Mathematics

Although there are few studies on cultural preservation and mathematics, there are no studies on the role of mathematics in traditional weave craft preservation. Culturally responsive mathematics teaching (CMRT) aims to provide students with the tools they need to use mathematics as a vehicle for self-expression and to view the world through a quantitative lens [21]. Weaving provides the students with opportunities for hands-on exploration to increase their chances of discovering mathematical concepts on their own, and it offers authentic, engaging activities that can aid children in understanding complex math concepts [22]. In addition, coconut weaving integrated into teaching mathematics can develop the mathematical ability and cultural preservation [23].

3 Method

This paper used the descriptive design, which used qualitative data-gathering procedures to determine the impact of using CAD in teaching mathematics and preserving the traditional handloom craft. First, the researchers designed the course syllabus, "Problem-Solving, Mathematical Investigation, and Modeling," which integrated CAD software as a tool for mathematical investigation to preserve the traditional weaving craft. The course was intended for Bachelor of Secondary Education-Mathematics. The syllabus underwent content validation from the University's Curriculum Review Committee. It used online open-ended surveys triangulated with focus-group discussions (FGD) to determine the impact of CAD integration and traditional craft preservation on the students. It also used FGD to determine the feedback of the weavers and the designers on the weave drafts created by the students.

The participants comprised one class of 35 third-year BSEd-Math students enrolled in the said course. The course was conducted in a blended learning modality. The teaching of the mathematical concepts was done online, while the use of CAD in designing the weave drafts was done face-to-face at the CTU-Argao Campus. The students were divided into five groups to create weave drafts for the Hablon Center. A weave draft is a diagram describing the mechanical variables of a woven fabric, illustrating the resultant design.

Two representatives from each group were trained in using the CAD software WeavePoint.8 Demo Version to create new handloom weave designs and edit weave patterns. The CAD software displays four quadrants of the drafting grid, composed of threading, treadling, tie-up, and drawdown. The software also features color mapping, fabric analysis, and design options to create weave drafts [24]. It allows the user to investigate and analyze the threading and treadling pattern where they can apply the concept of symmetry, as it can automatically display the drawdown while developing the weave plan.

4 Results and Discussion

Table 1 shows a portion of the mathematics course syllabus highlighting the integration of CAD in preserving the traditional weaving craft. Its learning outcomes included the demonstration of creative and critical thinking and the exhibit of problem-solving competence, as shown in Table 1. The syllabus underwent review and evaluation by the University's Curriculum Review Committee.

Table 1. Salient components of the course syllabus.

Lesson	Course Learning Outcome	Cultural Topics	CAD task	Outputs
1	Demonstrate creative and critical thinking in exploring, inventing problem-solving strategies, and posing new problems in collaboration with other students.	The History of Hablon The Need for Reviving Hablon Industry	Explore the CAD interface and functionalities.	Recognized the features and functionalities of CAD
2	Exhibit problem-solving competence by engaging in mathematical investigations anchored on real-life and non-routine problems.	The Hablon Pattern: Weave Structures	Discover how to convert known weave structures into weave draft	Converted common weave structures using CAD
3			Investigate handloom weave structures.	Converted handloom weave structures into weave drafts
4			Create original designs using CAD.	Created original weave drafts

The syllabus shows the integration of the preservation of culture, for it can provide an opportunity for students to learn about the significance of the traditional weaving craft and develop their skills in using CAD software to create new designs. It supports the method of culturally-responsive mathematics teaching [21] and the concept of providing students ts opportunities to explore and discover mathematical concepts [25].

For Lesson 3, the mathematical competency of "formulating and refining conjectures based on data from investigative tasks involving searching for a pattern" was demonstrated through the exploration of the handloom weave structures, particularly in the context of Hablon Pattern: Weave Structures discussing the pattern and symmetry.

Figure 1 shows the weave draft output of Student A in the competency, "mathematical investigation involving the search for a pattern."

The weave produced with a mini handloom and acrylic yarns is illustrated in Fig. 2, providing a visual representation of the actual fabric created.

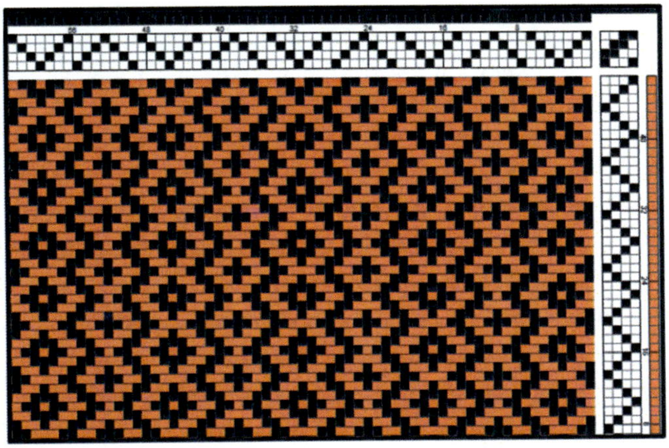

Fig. 1. The "Diamond Cross" weaving draft created by Student A

Fig. 2. The fabric view of the weave draft

As shown in the figures, Student A used WeavePoint to create a weave draft applying the concept of translation symmetry. The design shows diamonds and a cross-shaped symbol. The treadling sequence used a zigzag pattern, and the threading sequence used V and inverted V. The design shows a combination of black and orange colors.

Student A said, "My design represents my Roman Catholic faith using black to symbolize the ongoing conflict between day and night, good and evil, and right and wrong. Orange represents stability and emotional intelligence during difficult times, inspiring drive, optimism, and hope. This color reminds me that there's light at the end of a dark tunnel and promotes confidence, providing the courage to face fears." The statement of Student A proves that the teaching of creating patterns becomes more applied using CAD for the software can make visual representations of patterns and symmetry. This means that the CAD-integrated syllabus provides a powerful tool for teaching that allows students to visually explore mathematical concepts and construct simulations [15] and can improve problem-solving skills [16, 17] and creativity [18].

Impact of Using CAD on Students

Table 2 shows the summary of student feedback in using CAD. From the students' survey, the project developed their problem-solving skills, deeper appreciation for the local and traditional handloom weaving, and their skills in creating designs.

Table 2. Themes and Categories of Students' Responses

Themes	Categories	Sample Statements
Developed problem-solving skills	Enhanced exploration	"Yes, through using the software, my skills in exploration and discovery of what mathematics can really do in creating weave designs were enhanced."
	Developed critical thinking	"I noticed that there are no circle patterns, so I tried making designs with circle but it was not easy and I really studied hard on how I can make it."
	Enhanced creativity	"We need to think creatively on what design we want to create that is not only doable for the weavers but also showcase a very attractive products."
Fostered a deeper appreciation for traditional craft	Acquired an appreciation for traditional handloom weaving	"I learned to appreciate hablon when we did the prototyping of our designs. We encountered some difficulties, so we had to go back to fix them."
	Gained knowledge of the history of handloom weaving	"We also learned a lot about the history of hablon when we were talking to the weavers during the presentation of our new designs."

(*continued*)

Table 2. *(continued)*

Themes	Categories	Sample Statements
Developed design skills	Gained knowledge in using CAD software	I discovered a new application which could aid me in making new designs. It automatically translates my imaginations and it is time saving. I am amazed by how my designs turned into reality
	Enhanced visualization	Using the WeavePoint really develops my skills in mathematical investigation and designing weaves because it helps me to visualize my designs

Student A stated that because of the use of CAD software, she could easily create designs and experiment with many more designs. Student B also claimed that the project developed his critical thinking skills, as shown in his answers, "It developed my skills in mathematical investigation it led us to think deeper to develop such design while operating on the different aspects on this device".

The result validated the works of [16] and [17], stating that integrating CAD software in a mathematics course can improve problem-solving skills. It showed that the use of CAD software in mathematics education can enhance problem-solving skills by promoting creativity [18], innovation [26], critical thinking, and spatial visualization [17, 18]. This could be due to the fact that CAD software provides a hands-on, interactive learning experience that encourages students to explore and engage with mathematical concepts in a more meaningful way.

The use of CAD software in education has been found to have a positive impact on students' design skills, such as creating models and attention to detail. In addition, incorporating CAD modeling tools can increase students' design-thinking skills, which is an essential aspect of successful design [27].

Moreover, the students mentioned in the focus group discussion that they gained a deeper appreciation for the traditional craft of handloom weaving and the weavers. It is evident in Students C's statement stating, " Yes, I learn to appreciate handloom weaving so much, especially when we were doing the prototyping of our designs, and we encountered some difficulties, so we had to go back to fix them."

The students also expressed that they appreciated more about the history of handloom when they had to visit the hablon center during their Mathematical Investigation course. During the interview, some students also expressed that people around them got curious about their project using CAD, and they had to explain to them about handloom weaving. Thus, the CAD Hablon Project, where integrating weaving into the mathematics course, not only improves mathematical ability but also aids in the cultural preservation of the traditional craft [23].

Feedback from the Designer and the Weavers

The feedback from the weavers and the designers in the FGD and interview showed that they all agreed that the designs exuded intricate details, aesthetic appeal, and sophistication appropriate for the local and global market. Weaver A said, 'The designs are complex, but we will strive to translate them using the handloom". Weaver B added, "The designs were appealing. The CAD Weavepoint can surely create more creative designs from the students."

The statements made by the weavers demonstrated that students' designs are indicative of their visual understanding of mathematical concepts [17]. In using CAD in teaching, students could concretize math designs into actual handloom fabric. Moreover, the designer also commended that the designs had a strong marketability. She expressed, "The students really applied the CAD software in their mathematical modeling into wearable and aesthetically appealing designs. They have a very good eye for color. I appreciate the fact that they do not have constraints on anything and that is very important." The designer's appreciation validates that CAD can assist students in creating new handloom weave designs [28]. The use of CAD software deepens students' understanding of geometric concepts providing them with opportunities to explore and create more novel weave designs.

5 Conclusion

The use of CAD in teaching mathematics can develop students' mathematical abilities, problem-solving and design skills, and cultural heritage preservation. The ability of CAD to create visual representations of the weaving patterns developed the students' interconnection with the mathematical design of the handloom fabric. The abstraction of mathematics becomes concrete and practical through the hablon product, thus promoting mathematical investigation skills.

This research has limitations in terms of software utilization. It only used the Demo version, which featured threading, tie-up, treadling, and drawdown. These free features were sufficient to perform the mathematical investigation tasks like analysis of symmetry and pattern required in creating basic weave designs, which were the final outputs of the research. However, the software has free features like fabric analysis and project planning which were not utilized in teaching mathematical modeling due to the syllabus scope. These features can enrich teaching advanced concepts like mathematical relationships, optimization, and cost analysis [10, 11].

This research further recommends integrating CAD in teaching mathematics with the performance task of preserving traditional crafts. For example, in teaching "scaling," teachers can use CAD to let the students analyze the dimensions and cost of the fabric from the weave draft. Hence, this task can promote the two-pronged learning goals of developing mathematical investigation and promoting active students with local heritage preservation consciousness.

Acknowledgments. The authors would like to acknowledge the following: CTU administration, colleagues, panel of language and research experts, BSEd-Math students, handloom weavers of Shared Service Facility for Handloom Weaving Center, Holistic Coalition of the Willing (HoliCOW) designers and WeavePoint 8.0 Demo Version.

Research Ethics. All students, weavers, and designers gave their consent to be a part of the research and the publication. All participant information was kept confidential and identities were kept anonymous. Participants were treated with respect and their welfare was prioritized throughout the study.

References

1. Yang, Y., Shafi, M., Song, X., Yang, R.: Preservation of cultural heritage embodied in traditional crafts in the developing countries. a case study of Pakistani handicraft industry. Sustain. **10**(5) (2018). https://doi.org/10.3390/su10051336
2. SEAMEO SPAFA, Preserving Textiles : Indigenous Knowledge and Methods (2014). https://www.seameo-spafa.org/preserving-textiles-indigenous-knowledge-and-methods/
3. Sales, T.L.: Argao: 400 Years in Legend and History Sales, Todd L. The Municipality of Argao, Cebu (2008)
4. Mishra, S., Kumar, A., Mohapatra, D.: A study on the present condition of the weavers of handloom industry: a review. September 2019 (2020)
5. Libby, M.: Gresheen Gift, Hablon : History and culture woven together, 17 October 2019. https://cebudailynews.inquirer.net/264575/hablon-history-and-culture-woven-together
6. Eclarin, L.A., Ragual, C.T., Pagtaconan, W.C.R., Tagata, L.J.C., Rico, L.Y.: Developing new patterns for local weaving using a mathematical algorithm. In: Proceedings of 2nd International Conference Education Technology (ICETECH 2021), vol. 630, no. Icetech 2021, pp. 27–30 (2022). https://doi.org/10.2991/assehr.k.220103.005
7. Nur, A.S., Kartono, K., Zaenuri, Z., Waluya, S.B., Rochmad, R.: Ethnomathematics thought and its influence in mathematics learning. MaPan J. Mat. dan Pembelajaran **8**(2), 205–223 (2020)
8. Pradhan, J.: Mathematical Ideas in Mat Weaving: Connecting Ethnographic Field Study and Classroom Teaching (2020). https://www.researchgate.net/publication/346398768%0A
9. Baylas, N.A., Rapanut, T.A., Peñas, M.L.A.D.: Weaving Symmetry of the Philippine Northern Kankana-ey, pp. 267–274 (2012). https://archium.ateneo.edu/mathematics-faculty-pubs/39/
10. Satish, A., Rani, S.S., Saranya, B.: Fabric texture analysis and weave pattern recognition by intelligent processing. J. Telecommun. Electron. Comput. Eng. **10**(1–13), 121–127 (2018)
11. Wu, R., et al.: Weavecraft: an interactive design and simulation tool for 3D weaving, pp. 1–16 (2020). https://doi.org/10.1145/3414685.3417865
12. Cherinda, M.: Weaving exploration in the process of acquisition and development of mathematical knowledge, July 2012, pp. 926–944 (2012). https://www.researchgate.net/publication/274295226
13. Bellrichard, C.: Teaching math through cedar weaving, trade beads and oral history (2018). https://www.cbc.ca/news/indigenous/ts-msyen-indigenous-math-curriculum-cedar-weaving-1.4922447
14. Chen, Y.: Developments in leno-weave fabric, pp. 118–140 (2011). https://doi.org/10.1533/9780857093936.118
15. Ng, O.-L., Chan, T.: Learning as making: using 3D computer-aided design to enhance the learning of shape in STEM-integrated ways, p. 11 (2019). https://doi.org/10.1111/bjet.12643

16. Kohen, Z., Amram, M., Dagan, M., Miranda, T.: Self-efficacy and problem-solving skills in mathematics: the effect of instruction-based dynamic versus static visualization. Interact. Learn. Environ. 759–778 (2019). https://doi.org/10.1080/10494820.2019.1683588
17. Kösa, T., Karakuş, F.: The effects of computer-aided design software on engineering students' spatial visualisation skills. Eur. J. Eng. Educ. **43**(2), 296–308 (2018). https://doi.org/10.1080/03043797.2017.1370578
18. Robertson, B., Radcliffe, D.: The role of software tools in influencing creative problem solving in engineering design and education. In: Proceedings of ASME Design Engineering Technical Conference, vol. 42584, pp. 999–1007 (2006). https://doi.org/10.1115/detc2006-99343
19. Adelabu, F.M., Makgato, M., Ramaligela, M.S.: Enhancing learners' geometric thinking using dynamic geometry computer software. J. Tech. Educ. Train. **11**(1), 44–53 (2019). https://doi.org/10.30880/jtet.2019.11.01.006
20. Rule, A.C.: Editorial: The components of Authentic Learning. https://www.researchgate.net/publication/253893836_Editorial_The_Components_of_Authentic_Learning
21. Ellis, M.W.: Knowing and Valuing Culturally responsive mathematics teaching (2019). https://in.nau.edu/wp-content/uploads/sites/101/2022/05/Culturally-Responsive-Mathematics-Teaching.pdf
22. Peppler, K., Keune, A., Thompson, N.: Reclaiming traditionally feminine practices and materials for STEM learning through the modern maker movement. In: Designing Constructionist Futures, p. 1687288 (2020). https://doi.org/10.7551/mitpress/12091.003.0017
23. Rivera, M.N.: Using Coconut Weaving in Guam Classrooms to Improve the Mathematics Skills of Local Students (2010). https://macimise.prel.org/wp-content/uploads/2015/06/Rivera-Matilda-N-Using-Coconut-Weaving-in-Guam-Classrooms-to-Improve-the-Mathematics-Skills.pdf
24. WeavePoint (2023). http://www.weavepoint.com/. Accessed 26 Apr 2023
25. Connor, C.D.B., Mcdaniel, C., Carr, J.: Bringing math to life: provide students opportunities to connect their lives to math. Netw. Online J. Teach. Res. **21**(2), 3 (2019). 10.4148/2470-6353.1299
26. Pielsticker, F., Witzke, I., Vogler, A.: Edge models with the CAD software: creating a new context for mathematics in elementary school. Digit. Exp. Math. Educ. **7**(3), 339–360 (2021). https://doi.org/10.1007/s40751-021-00092-w
27. Taleyarkhan, M.R., Dasgupta, C., Mendoza-Garcia, J.A., Magana, A.J., Purzer, S.: Investigating the impact of an educational CAD modeling tool on student design thinking. In: ASEE Annual Conference Exposition Conference Proceedings (2016). https://doi.org/10.18260/p.25485
28. Kumar, M.: CAD and Weaving (2023). https://www.tutorialspoint.com/cad-and-weaving

Author Index

K. Kabassi et al. (Eds.): NiDS 2023, LNNS 783, pp. 357–358, 2023.
https://doi.org/10.1007/978-3-031-44097-7